Testing, Monitoring and Diagnostic of High Voltage Equipment

Testing, Monitoring and Diagnostic of High Voltage Equipment

Editors

Zhijin Zhang
Hualong Zheng

MDPI • Basel • Beijing • Wuhan • Barcelona • Belgrade • Manchester • Tokyo • Cluj • Tianjin

Editors

Zhijin Zhang
Xuefeng Mountain Energy
Equipment Safety National
Observation and Research Station
(Chongqing University)
China

Hualong Zheng
Chongqing University
China

Editorial Office
MDPI
St. Alban-Anlage 66
4052 Basel, Switzerland

This is a reprint of articles from the Special Issue published online in the open access journal *Energies* (ISSN 1996-1073) (available at: https://www.mdpi.com/journal/energies/special_issues/voltage_equipment).

For citation purposes, cite each article independently as indicated on the article page online and as indicated below:

LastName, A.A.; LastName, B.B.; LastName, C.C. Article Title. *Journal Name* **Year**, *Volume Number*, Page Range.

ISBN 978-3-0365-7274-1 (Hbk)
ISBN 978-3-0365-7275-8 (PDF)

Contents

About the Editors

Zhijin Zhang

Zhijin Zhang (Prof.) was born in Fujian, China, in 1976. He received his B.Sc., M.Sc., and Ph.D. degrees in electrical engineering from Chongqing University, China, in 1999, 2002, and 2007, respectively. He was appointed as a Professor of HV engineering and insulation in 2011, and the Head of the Department of High-Voltage Engineering, School of Electrical Engineering, Chongqing University in 2022. He has also worked as the vice station master of the Xuefeng Mountain Energy Equipment Safety National Observation and Research Station since 2021. His main research interests include high voltage, external insulation, icing of transmission lines, and the reliability of power equipment under extreme weather conditions.

Hualong Zheng

Hualong Zheng (Dr.) was born in Henan, China, in 1988. He received his B.Eng. degree in Electrical Engineering and Automation from the Changsha University of Science & Technology, China, in 2010, M.Sc. degree in Electrical Power Systems Engineering from the University of Manchester, UK, in 2011, and Ph.D. degree from the University of Leicester, UK, in 2015. He joined Chongqing University in 2019. His research interests lie in space charge phenomena in polymeric DC insulations and the corresponding measurement techniques, electrical treeing breakdown process in polymeric materials, PD measurements, and the reliability of power equipment under extreme weather conditions.

energies

MDPI

Review

A Review of Icing and Anti-Icing Technology for Transmission Lines

Zhijin Zhang [1,*], Hang Zhang [1], Song Yue [1,2] and Wenhui Zeng [3]

1 Xuefeng Mountain Energy Equipment Safety National Observation and Research Station of Chongqing University, Chongqing 400044, China
2 State Grid Tibet Electric Power Research Institute, Lhasa 850000, China
3 State Grid Sichuan Economic and Technology Research Institute, Chengdu 610000, China
* Correspondence: zhangzhijing@cqu.edu.cn; Tel.: +86-1388-3207-915

Abstract: This paper reviews the application of various advanced anti-icing and de-icing technologies in transmission lines. Introduces the influence of snowing and icing disasters on transmission lines, including a mechanical overload of steel towers, uneven icing or de-icing at different times, Ice-covered conductors galloping and icing flashover of insulators, as well as the icing disasters of transmission lines around the world in recent years. The formation of various icing categories on transmission lines, as well as the effect of meteorological factors, topography, altitude, line direction, suspension height, shape, and electric field on ice-covered transmission lines, are all discussed in this study. The application of various advanced anti/de-icing technologies and their advantages and disadvantages in power transmission lines are summarized. The anti/de-icing of traditional mechanical force, AC/DC short-circuit ice melting, and corona effect is introduced. Torque pendulum and diameter-expanded conductor (DEC) have remarkable anti-icing effects, and the early investment resources are less, the cost is low, and the later maintenance is not needed. In view of some deficiencies of AC and DC ice melting, the current transfer intelligent ice melting device (CTIIMD) can solve the problem well. The gadget has a good effect and high reliability for de-icing conductors in addition to being compact and inexpensive. The application of hydrophobic materials and heating coatings on insulators has a certain anti-icing effect, but the service life needs further research. Optimizing the shed's construction and arranging several string kinds on the insulators is advisable to prevent icing and the anti-icing flashover effect. In building an insulator, only a different shed layout uses non-consumption energy.

Keywords: icing; anti-icing; de-icing transmission lines; ice-melting; icing disaster; diameter-expanded conductor; current transferred

Citation: Zhang, Z.; Zhang, H.; Yue, S.; Zeng, W. A Review of Icing and Anti-Icing Technology for Transmission Lines. *Energies* **2023**, *16*, 601. https://doi.org/10.3390/en16020601

Academic Editor: Anna Richelli

Received: 14 December 2022
Revised: 27 December 2022
Accepted: 30 December 2022
Published: 4 January 2023

1. Introduction

Icing is a severe natural disaster of electrical power transmission lines. It will cause severe losses to the transmission systems in many aspects and bring serious threats to power system stability and even human energy security. In 1932, the United States recorded the first accident caused by icing on overhead lines in human history [1]. This was followed by ice damage to transmission lines in Britain in 1935 and 1962 and in Italy in 1964. Since then, icing disasters have occurred on transmission lines between 1980 and 2000. The U.S. cities of Ohio, Chicago and Idaho, Quebec, and Ontario of Canada, Russia, Norway, Yugoslavia, Japan, the United Kingdom, Sweden, Finland, and Iceland suffered power transmission line failures caused by snowing and icing [2–10]. Since the beginning of the 21st century, the Czech Republic, Alberta, and Canada, all had also occurred serious icing disasters on power transmission lines. In 2005 and 2008, China experienced two severe snow and icing disasters on transmission lines, which caused significant economic losses [11]. In February 2021, an icing disaster occurred on the power grid in Texas, USA, resulting in millions of people without power in the cold weather of snow and icing disaster. Figure 1 shows the scene of the accident [10].

Figure 1. The collapse of the power grid tower was caused by the snow and icing disaster [10].

To this end, scientists have conducted a significant amount of research on icing and frost resistance, made observations, engaged in theoretical analysis, conducted experiments, and studied the mechanisms underlying the icing of transmission lines and their flashover characteristics [12,13]. In addition, this study provides many effective methods to prevent and deal with the snow and icing disasters of transmission lines [14–17]. Conductor anti/de-icing technology includes mechanical de-icing [18–26], short-circuit ice-melting [27–31], torque corona discharge effect [32–35], pendulum method [36–38], DEC [39–43] and CTI-IMD [44–46], insulator anti/de-icing technology, including the application of hydrophobic materials [47–49], heating coating [50–52], optimization of shed structure [53–57], and string arrangement [58–61].

Snow and icing-related power interruptions have become less frequent thanks to people's efforts. However, in recent years, it is still happening somewhere. This paper will summarize and analyze the existing research to do a review.

2. The Failure Classification of Icing Lines

According to previous studies, transmission line icing disasters are mainly about the following four aspects [62]:

2.1. Mechanical Overloading

Mechanical overloading refers to the snow and ice attached to the steel towers, conductors, insulators, steel accessories, and other equipment, exceeding the maximum mechanical load designed for the line, causing deformation, damage, fracture, collapse, etc. For example, if the steel tower is coated in ice and the mechanical load exceeds the design, or if an unequal force causes conductors to discharge to the ground or between phases, or if the sag of the conductors is too low and causes conductors to discharge to the ground or between phases. Due to heavy ice accretion, conductors and steel towers broke because of exceeding the maximum stable mechanical load. In addition, accidents of twisting, colliding, and even breaking of insulator strings caused by heavy icing [63,64]. Seriously, a large area of transmission lines broke down, towers fell, and power outages were caused by the overloading of mechanical load. In addition, it can easily cause chain reactions, such as the icing disaster on the transmission lines in southern China in 2008.

2.2. Uneven Icing or De-Icing at Different Times

Due to the varied positions and directions of the neighboring conductors, which may result in unequal icing or de-icing at different times, the conductors caused a tension differential that slips inside the clamp, causing wear and tear, broken strands, and even the complete breakdown of the strand. Or when the ice melts, the three-phase line does not fall off at the same time, and the conductors do not swing synchronously, there will be a collision, burn and fuse, etc. If the torque is too large, it will distort the steel accessories or the steel tower, even collapse. The airflow on the windward side and the leeward side of the transmission line is different, which will lead to uneven ice accretion of all components and unilateral ice-covered conductors and insulators. Overturning, twisting or unilateral load is too heavy and the whole is not uniform, so the steel accessories and tower were out of shape, resulting in broken lines, broken strings, and tower collapse. The difference between uneven icing and de-icing at different times is that the tension difference of the former is static load, while the latter is dynamic load [65].

2.3. Ice-Covered Conductors Galloping

Under the action of wind, the conductors will be galloping, its essence is mechanical shock, for the stability of the tower and the electrical gap distance of the conductor is a great threat, it will also be considered at the beginning of the design. However, after the conductors are covered with ice, the mechanical load is further increased, self-excited oscillation and low-frequency galloping are generated under the ice wind load, and friction is generated between various objects and steel accessories, resulting in damage to the steel accessories, broken strands of the conductor, the leaning of the tower and even the collapsing of the tower [66–68]. When ice covered the conductor, the wind area increases, especially when it is unevenly covered, it will be galloping more violently and the amplitude is larger than that without icing, which is more likely to cause the flashover to ground or between phases. In the bealock of wind pass, it is more severe [69,70].

2.4. Icing Flashover of Insulator

When the insulators were covered with ice or the shed is bridging, the original electric field distribution will be changed, the insulators' leakage distance will be shortened, the insulation strength will be reduced, and flashover will occur. In addition, most of the icing flashover accidents occurred in the transmission lines while the ice and snow melted. The pollution before icing melts with the ice and snow to form high conductivity liquid, which lasts a long time and has a high probability of flashover. Meanwhile, the heavier the pollution before icing, the higher the probability of flashover. According to the literature, icing flashover accidents of insulators account for the largest proportion of snowing and icing disasters on transmission lines. From 1981 to 1997, multiple icing flashover incidents occurred in Quebec and Ontario, Canada, and from 2004 to 2005 and 2011, multiple continuous icing flashover accidents occurred in power grids in East and Central China [71–73].

3. Formation and Influencing Factors of Icing

3.1. Formation of Icing

When the temperature drops to -5 to 0 °C, the wind speed is 3 to 15 m/s, if the relative humidity of the air is above 85%, there is fog or drizzle, the transmission line will begin to form a glaze, and if the conditions remain the same, a short time will form a very thick layer and strong adhesion of pure glaze. If the temperature continues to drop, there will be rain and snow weather, freezing rain and snow rapidly grow on the glaze surface, forming a density of 0.6 g/cm^3 or more icing. If the temperature continues to drop from -15 to -8 °C, the ice layer will be covered with frost. This process will lead to the formation of glaze-rime accrete on the transmission line surface. In this process, if there are many sunny and cold weather reciprocating alternates, and the rime just begins to melt and then refreezes, it will strengthen the ice density, and will develop into a mixture of glaze and rime alternating overlap, that is, mixed rime [62,74].

The icing first grows on the windward side, and if the wind direction remains unchanged, the icing accretion thickness keeps increasing in the same direction. When the icing accretion thickness on the windward side of the conductor reaches a certain thickness, it will be twisted under the action of gravity, and the leeward side without icing will turn to the windward side and continue to grow. Finally, a circle or oval icing accretion will be formed to wrap the conductor. However, due to the differences in structure, material, arrangement, spatial position and surrounding influence, the conductors and insulators have different capturing capacities of supercooled water drop under the same environmental parameters [75]. Although the growth trend of ice thickness and icicle length of conductors and insulators is roughly the same, the growth rate is different. The ice thickness and icicle growth rate of the insulator are both higher than that of the conductor [76].

3.2. *Classification of Icing*

The icing of transmission lines can be generally divided into five categories: hoar frost, rime, mixed rime, glaze and snow [62,77–80].

3.2.1. Hoar Frost

When the transmission line is below 0 °C and the wind speed is low, the moisture in the air will come into contact with it and form hoar frost on the surface, which is mainly in the form of "needle" or "leaf" crystal form. The adhesion of frost is very weak, light vibration or sudden high wind speed can make the frost break away from the surface of conductors and insulators, so it can only form a thin layer on the surface of the contact. The density is about 0.05–0.3 g/cm^3. Because of this characteristic, compared with other types of icing accretion, white hoar frost only has a certain effect on the electrical strength of insulators. It hardly poses a hazard to conductors and steel towers.

3.2.2. Rime

Under the drag force of the wind, the supercooled water droplets in the fog constantly collide with the conductors, insulators, and the surface of the steel tower and are trapped and condensed into rime. Its appearance is mainly "shrimp tail" or "pine needle", which is white. Under the microscope, it mainly presents a granular structure mainly consisting of hard rime and soft rime, and the density is usually about 0.3–0.6 g/cm^3. The adhesion of rime on conductors and insulators is very weak, generally in windward growth. Under certain conditions, its growth rate is speedy and often appears in the winter high-altitude mountain transmission lines.

3.2.3. Mixed Rime

It is also called hard rime. As freezing rain and fog appear alternately in the environment and condense on the transmission lines, it appears in the form of hard ice with layered and plate structure, with dispersed tiny bubbles inside, alternating transparent and opaque, strong adhesion and density of 0.6–0.8 g/cm^3. Mixed rime is a composite icing process, and the transmission line is exposed to wet fog for a long time, the rime and the glaze together alternate icing, the growth rate is fast, and most of the time, the formation of mixed rime, for the transmission line equipment is more dangerous.

3.2.4. Glaze

The glaze is formed by freezing rain or large particle-size supercooled water droplets caught by transmission lines. When glaze icing, freezing rain, or supercooled water droplets hit the transmission lines and has not been completely frozen, water droplets continuously collide and icing continues to grow. The glaze is clear and transparent, occasionally mixed with impurities and bubbles. Its adhesion is firm, causing the most significant harm to the transmission line equipment, the density is close to the pure ice density of 0.917 g/cm^3, and usually, a density more significant than 0.9 g/cm^3 of icing is classified to glaze. Because of the short duration of freezing rain, the situation of transmission lines with pure glaze icing is less.

3.2.5. Snow

The snow on the transmission lines means that when the temperature is around 0 °C, and the wind speed is weak (less than 3 m/s), the "wet snow" particles in the air interact with the fog to attach to the equipment. Density is usually below 0.05 g/cm^3. When the "wet snow" falls on the transmission lines, the snow occurs, while the dry snow and the ice crystals are too weak and will be taken away when there is a strong wind.

3.3. Influence Factor of Icing

As icing is a natural meteorological phenomenon with numerous natural conditions, many complicated and changeable factors affect the icing, so people cannot fully grasp its law and influence. The main factors affecting transmission lines icing are: meteorological conditions, geographical terrain, altitude, transmission line direction and suspension height, shape size and electric field. The differences in icing on transmission lines mainly reflect in the thickness, density, shape and form [81].

3.3.1. Meteorological Conditions

Including the ambient temperature, wind speed and direction, the diameter of water droplets in the air, etc. When transmission lines are covered with ice, the environment is generally between −15 and 0 °C, and the ambient relative humidity is above 85% [82]. At −5 to 0 °C, the diameter of the droplet is large, generally 10 to 40 um and the glaze forms easily; the average temperature of the formation of the glaze is −2 °C, and the temperature is relatively high; at −9 to −13 °C, the diameter of the droplet is usually between 5 and 35 um, easy to form mixed rime, the average temperature at −7 °C. When the temperature is −15 to −10 °C, the diameter of water droplets is generally between 1 and 20 um, which is easy to form rime. When the ambient wind speed is between 0 and 4 m/s, the ice layer thickness of conductors and insulators increases with the increase in wind speed, and the icicle growth rate of insulators before bridging is greater than that of the conductors. When the ambient temperature is between −4 and −10 °C, the icing degree of conductors and insulators increases with the decrease in temperature [83].

3.3.2. Geographical and Topography

Due to the influence of the cold current at the poles of the earth, the ice accretion of the east–west mountains are more serious than that of the north–south mountains. Windward slopes are more serious than that of leeward slopes, and bealock is more serious than that of other terrains. Micro-topography and micro-climate characteristics are easily generated in these places [84]. The water environment around the transmission lines also greatly impacts the icing accretion. When rivers and lakes are around and water vapor is sufficient, the icing situation is more serious.

3.3.3. Altitude

The height at which icing begins to form in each area is called condensation height. Due to the different environmental parameters in each region, the condensation height is also different. Above the condensation height, the higher the altitude, the ice thickness will increase. The icing near the condensation height is mainly glazed. As the height increases, the icing form transitions to mixed rime and then to rime. Based on the analysis of the observed ice accretion data of conductors in the Donbass region of the former Soviet Union, the empirical formula of the average icing accretion thickness D (cm) and altitude h (m) in this region was obtained [84]:

$$D = Ae^{Bh} \tag{1}$$

e is the base of the natural logarithm, A and B are the coefficients of variation in ice accretion type and region. Rime: $A = 7.76$, $B = 0.032$, glaze: $A = 4.49$, $B = 0.0039$.

3.3.4. Transmission Lines Direction and Suspension Height

The more perpendicular the angle between the transmission line direction and the wind direction in the winter icing period, the more serious the icing. When the direction of the transmission line is parallel to the wind direction, the conductors have a feeble ability to capture water droplets and are relatively difficult to be icing; however, the effect on the insulator is not apparent. The higher the height above the ground, the higher the liquid water content in the air, the greater the wind speed, the higher the suspension height of the conductors, and the more serious the icing. Reference [85] gives the equation of the ice accretion ratio of conductors of two heights:

$$\xi = \frac{b_z}{b_0} = \left(\frac{z}{z_0}\right)^{\gamma_v+\gamma_w+\gamma_\alpha} \tag{2}$$

ξ is the ratio of two-height icing, z and z_0 represent the suspension height of the conductors, m. b_z and b_0 are the ice thickness of the conductors at the height z and z_0, respectively, cm. γ_v, γ_w, and γ_α, respectively, represent the corresponding values of wind speed, m/s, air humidity, and capture coefficient with height.

3.3.5. Shape and Size

The shape of transmission line equipment influences the degree of icing, including the diameter of the conductor, the diameter of a rod of the composite insulator, the size of the shed, the shed space and the angle of the shed, but the influence of the various parameters is different. Based on the data analysis of the observation in Xuefeng Mountain, the literature [42] found that the thickness of the icing accretion by the conductor decreases with the increase in the diameter within the range of 5 to 25 mm, no matter it is rime, mixed rime or glaze. In the rime condition, both the rod diameter of the composite insulator and the diameter ratio of the adjacent shed greatly influence the ice weight. The ice weight increases with the rod diameter increases. The inclination angle of the shed and the shed space has little effect on ice weight. The ice weight of rime and mixed rime decreases with the decrease in shed space [86,87].

3.3.6. Influence of Electric Field

When the electric field strength is small, the weight, thickness, and density of ice accretion increase with the increase in electric field intensity. When the electric field continues to increase, the ice accretion of conductors with energized is lighter than that of non-energized [88,89]. The electric field effects the appearance, density, and weight of insulator icing. For insulators with different dielectric surfaces, the electric field has different degrees of influence on the weight of ice. Due to the polarization effect, the electric field can inhibit the formation and growth of ice branches, thus contributing to the different anti-icing characteristics of various dielectric surfaces in the electric field. However, the effect of the electric field on insulator surface icing process is mainly in the initial stage. The effect is not evident in the later stage of icing and the heavy icing area [90].

4. Anti/De-Icing Technology of Conductors

Anti/de-icing of conductors mainly refers to the conductors of overhead transmission lines. When a transmission line cannot avoid the ice-covered area, it should ensure that the conductors have sufficient anti-icing strength or timely de-icing after ice-covered. There are six main anti-icing/de-icing methods.

4.1. Mechanical De-Icing

Mechanical de-icing refers to removing the ice from the conductors by applying external force through hard objects. It generally includes artificial knock de-icing (ad hoc de-icing), pulley scraper de-icing, blasting de-icing, electromagnetic vibration de-icing, robot de-icing, etc. [91,92].

4.1.1. Manual Deicing by Striking

The earliest de-icing method is artificial knock deicing, or ad hoc de-icing, was proposed by Phlman and Landers in 1982 and required on-site operation by operators. When the overhead line is out of power, operators can directly de-icings the line manually by climbing the tower, or throw hard insulating objects from the ground to the conductors to knock off the ice, or beating the ice-covered part with long wooden sticks. When the conductors are energized, the icing is removed by striking the ice accretion with an insulating rod. A U-shaped trap can also be made on the conductors and dragged along to scrape off the icing [18]. This method is relatively primitive and requires on-site operators to perform it according to the actual situation. It only applies to transmission lines with low conductors above the ground, resulting in low de-icing efficiency and low safety.

4.1.2. Icing Rolling

Alternatively, pulley scraper de-icing, was first developed and made by Manitoba Water and Power Department, Canada. It is mainly composed of pulley, traction rope and steel blade, and are installed on the conductors. Operators on the ground use the traction rope to drag the pulley to move along the conductors. On the one hand, the pulley drives the steel blade to scrape the icing on the moving path. On the other hand, the pulley generates stress to bend the conductors, so that the icing falls off [19]. Its advantages are simple device, simple operation, effect obviously, and practicability. The disadvantage is that it requires manual operation, time-consuming and laborious, limited by mountainous terrain, steel knife damage to the conductors.

4.1.3. Blasting De-Icing

Blasting de-icing is first to select appropriate detonating cables, connect explosives in sections according to a certain distance, and lay them on the transmission line conductors that need de-icing in advance. After ice covers the conductors, they will detonate. The impact force generated by the explosion will make the ice near the laying area of the detonating cables fall off. Its advantages are low cost, quick effect, and high efficiency, but the whole line needs to lay, the installation of detonating cable and explosives is arduous. The vibration amplitude of the line after explosion is relatively large, and the blasting location of the conductor is easy to receive irreversible damage. Xie Dongsheng et al. [20] obtained a good de-icing effect through the shrinkage blasting de-icing test. Cao Yonghong et al. [21] analyzed the dynamic response of the ice-covered tower system of the transmission line after blasting de-icing using numerical simulation and proposed the de-icing scheme of single conductor and double bundled conductors.

4.1.4. Electromagnetic Vibration De-Icing

Electromagnetic vibration de-icing mainly includes electromagnetic pulse deicing and electromagnetic impact de-icing. The electromagnetic pulse method first applies to de-icing aircraft wings [22]. Its principle is that the silicon-controlled rectifier controls the discharge of the energy storage capacitor to the pulse coil, and the coil generates a strong alternating magnetic field, which induces eddy current on the plate near the coil. The plate connects to the de-icing target. Under the combined action of eddy current and the coil magnetic field, a repulsive force with high amplitude and low duration is generated. The target is slightly expanded and then contracted to crush the icing and fall off [23], the schematic diagram is shown in Figure 2, but this method can only remove the local ice accretion of the conductors and has no obvious effect on the long line. Electromagnetic force impinging de-icing is a scheme proposed by the IREQ Laboratory in Canada for de-icing double bundled conductors of UHV lines, that is, short-circuit the transmission lines at rated voltage, and the electromagnetic force generated by the short-circuit current makes the conductors collide with each other and the ice falls off [24].

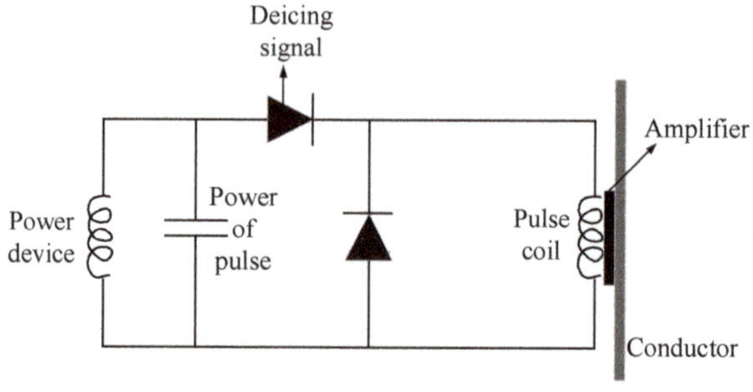

Figure 2. Schematic diagram of electromagnetic pulse de-icing.

4.1.5. Robot De-Icing

Robot de-icing method is a remote controllable de-icing mechanical device developed based on pulley scraper de-icing, with the conductor as the path and the pulley as the walking device. In 2003, Serge Montambault and Nicolas Pouliot of IREQ of Canada first proposed HQ LineROVer vehicle as a de-icing device for transmission lines [25], as shown in Figure 3. In 2010, China Shandong Electric Power Research Institute and Hydro-Quebec Research Institute cooperated to strengthen and improve HQ LineROVer de-icing vehicle, so it has infrared detection, waterproofing and long-distance communication functions [26]. The robot de-icing methods include impact type, milling type, scrape type and percussion type.

Figure 3. De-icing robot.

4.2. Short Circuit Ice Melting

4.2.1. Basic Parameters of Short Circuit Ice Melting

Transmission lines icing and melting is a complex physical process, which is difficult to describe by exact mathematical expressions because it occurs in the wild natural environment with numerous influencing factors and substantial variability. At present, scholars from all circles have yet to reach a unified consensus on the calculation method of basic parameters of melting ice. The melting process of the conductors includes two heat exchange processes: first is the conductors generate Joule heat through the current and transfer it to the ice layer; the second is the heat exchange between the ice surface and the air through heat radiation and convection.

1. The critical load current of icing

Under certain meteorological conditions, when the transmission line conductors are not covered with ice, the surface temperature is equal to 0 °C, and the conductors are in a critical state between being icing and not icing. At this time, the passing current is called the critical load current of icing [27]. The equation for calculating the critical current of AC line proposed in the literature [28] is as follows:

$$I_c^2 = \frac{2r}{R_0}\left\{W_s - 2\alpha wvc_w T_a - \left[\frac{h_p r_c v^2}{2c_a} + \frac{\alpha w v^3}{2}\right]\right\} \quad (3)$$

where r is the radius of the conductor, mm; R_0 is the resistivity per unit length of the conductor, Ω/m; c_w and c_a are the specific heat of water and air, respectively, $\mathrm{J}/(\mathrm{kg{\cdot}K})$; h_p is the forced convection coefficient, $\mathrm{W}/(\mathrm{m}^2{\cdot}\mathrm{K})$; r_c is the local recovery coefficient; I_c is critical load current, A; wind speed v, m/s; ambient temperature T_a, K; w is liquid water content in the air, $\mathrm{kg/m^3}$; α is collision coefficient of water droplets on the conductors; and W_S is the coefficient related to convective heat transfer and evaporation and blackbody radiation.

2. Critical ice melting current

Many factors affect the melting ice of transmission lines, including wind speed, ambient temperature, current density and ice thickness. In the past process of melting ice, there was a situation in which melting ice failed for several hours or even more than 20 h. The reason is that the current was too small, smaller than the critical melting current [29]. The critical melting-ice current is the minimum current that can melt the icing of conductors under certain meteorological conditions. Reference [30] proposed the equation of the critical ice-melting current of DC short-circuit ice-melting:

$$I_c = 3.735\sqrt{\frac{-t_i}{R_0 ln(1+2d/D)}} \quad (4)$$

t_i is ice surface temperature, K; d is the thickness of ice, mm; D is conductor diameter, mm; R_0 is the resistance per unit length of conductor, Ω/m.

Further, the relation of t_i was given in reference [31]:

$$t_i = \frac{r_i hln(r_i/r_c)T_a}{r_i hln(r_i/r_c) + \lambda_i} \quad (5)$$

r_i is the radius of the ice-covered conductor, mm; r_c is the radius of the conductor, mm; λ_i is the heat conductivity of the ice, s/m; and h is the heat exchange coefficient between the outer surface of the ice and the air, $\mathrm{W}/(\mathrm{m}^2{\cdot}\mathrm{K})$.

4.2.2. AC Short-Circuit Ice Melting

Since the 1950s, the former Soviet Union began AC short-circuit ice melting technology research and application to the transmission line to prevent icing. According to the number of short-circuit phases, AC short-circuit ice melting includes three phases, two phases, and single-phase short-circuit ice melting. According to the short-circuit object, it includes the phase–phase type and phase–earth type ice melting [93].

1. Three phases short-circuit ice melting

Three phases of short-circuit ice melting is a particular phase–phase type, which interrupts the first and last ends of the target line from the power grid. One end of the line is a three-phase short-circuit, and the other connects to the ice-melting power supply. The high energy Joule heat generated by the three-phase short-circuit current is used to melt the icing. The diagram shows in Figure 4.

2. Two phases short-circuit ice melting

Phase–phase type short-circuit ice melting is that two phase conductors are connected in series, and the first and last ends are, respectively, connected with the ice-melting power supply to form a loop.

3. Single phases short-circuit ice melting

One end of the single-phase conductor connects to the ice-melting power supply, the other end connects to the dedicated earth conductor, and the other end of the power supply connects to the earth electrode.

AC short-circuit ice melting needs to remove the line from the power grid, which requires high reactive power and power capacity and takes a long time to melt ice. It is not suitable for the de-icing of long transmission lines, and the ice-melting work of long lines with voltage above 500 kV is not feasible.

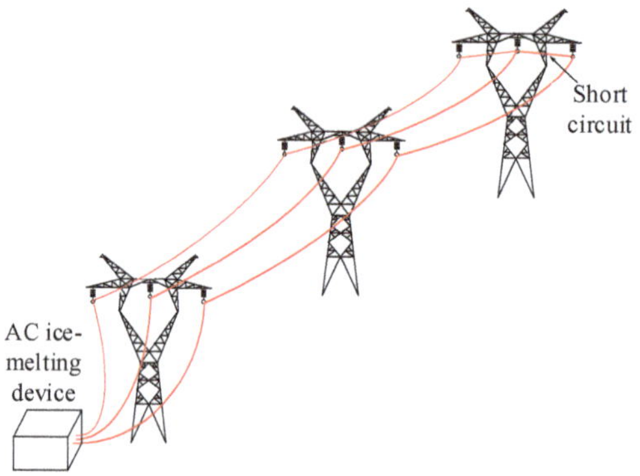

Figure 4. Three phases short-circuit ice melting diagram.

4.2.3. DC Short-Circuit Ice Melting

The research on DC short-circuit ice melting technology can be traced back to 1993 in Manitoba Water and Power Department, Canada. Since the AC impedance of the conductors is much greater than the DC resistance, the power supply capacity required by AC de-icing is 5 to 20 times that of DC. The DC de-icing only needs to consider the heating of the line resistance, without considering the influence of the insufficient reactive power, and the requirement on the power supply capacity is much smaller than that of AC. With the progress of controllable electrical rectifier components and HVDC technology, the technical obstacles of DC ice-melting devices have broken through, and development has been promoted [94,95]. The diagram shown in Figure 5A–C is the phase order.

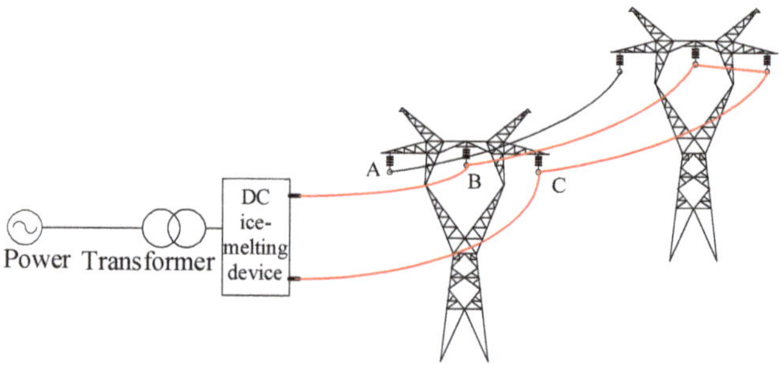

Figure 5. DC short-circuit ice melting diagram.

1. Melting ice for DC line

DC transmission lines generally use DC short-circuit melting ice. On the one hand the cross-sectional area of DC transmission line conductors is large, and AC ice-melting method cannot solve the problem effectively. On the other hand, DC ice-melting technology has many advantages over AC, and it is convenient to obtain DC ice-melting power supply.

When the DC line needs to melt ice, only use the existing equipment in the substation to change the wiring mode. So that the line current increases to above the critical melting current and wait for the completion of melting ice.

2. Melting ice for AC line

Due to the outstanding advantages of DC short-circuit ice melting method, it is also applied to AC transmission lines. In the process of DC ice melting, only resistance consumes active power and emits Joule heat, and only the rectifier itself consumes a small amount of reactive power. Therefore, DC short-circuit ice melting method can target AC lines of various voltage levels.

In this regard, the research and development of the DC ice melting device is the key. It is necessary to consider the first is the capacity of the DC ice melting device and the maximum DC ice melting current it can provide. Second is the harmonic impact generated during the converter's operation on the power grid. Third is matching the system design and power supply of the ice melting device when the substation is connected. Moreover, the cost of DC ice melting device is high, which needs to be further reduced. The length of various types of ice-melting lines determined by the capacity of DC devices is shown in Table 1 [96]:

Table 1. Length of various types of de-icing transmission lines determined by the capacity of DC devices [96].

Line Type	Line Parameters		DC Resistance (Ω/km)	Minimum Melting Current (A)	Effective Distance (km)			
	Voltage Class (kV)	Conductor Type			200 MW	100 MW	50 MW	10 MW
DC	800	LGJ-6×630/45	0.0077	7075	258.8	129.4	64.7	12.9
	500	LGJ-4×720/50	0.0100	5254	363.7	181.9	90.9	18.2
AC	500	LGJ-4×500	0.0148	3979	427.4	213.7	106.8	21.4
	200	LGJ-2×500	0.0296	1989	854.8	427.4	213.7	42.7
	110	LGJ-240	0.1198	609	2250.7	1125.3	562.7	112.5
	35	LGJ-150	0.1962	441	2620.7	1310.4	655.2	131.0

4.3. Corona Discharge Effect

Icing generation and growth can be inhibited by increasing the electric field intensity and corona discharge on the conductor surface. The corona discharge becomes more and more intense with the increase in electric field intensity, and the ion concentration in the surrounding space also increases. At the same time, more and more small water droplets are repulsed after being charged, and the electrical suction effect is cancelled out. The motion speed of water droplets decreases or reverses, and the collision with the conductors reduce, resulting in the reduction in the icing accretion [97].

The corona cage was used in the artificial climate room to conduct experimental research on the conductor icing and corona characteristics under AC electric field, and the surface electric field intensity of the conductor under the rime and glaze icing environment to measure the weight of icing accretion and icicle distance [32,33]. The relationship between the surface field intensity of conductor E_s(kV/m) and applied voltage U_a (kV) was obtained as follows [32]:

$$E_s = \frac{U_a}{r_c ln(R/r_c)} \tag{6}$$

where R and r_c are radius of corona cage and conductor, respectively, m.

When rime and glaze icing, the surface ice accretion trend is the same. When the electric field intensity of the conductor surface is less than 20 kV/cm, the ice accretion of the conductor surface increases with the increase in the electric field intensity. When the field intensity increases further, the ice accretion decreases. When covered with glaze, the percentage of icicle mass decreases with the increase in field intensity. The icicle spacing is independent of field intensity. With the increased rainfall, the icicle spacing decreases with the increase in field intensity. When the electric field is considered, the electric field intensity of the conductor is 15 kV/cm, the icing accretion of the conductor is less than that of the non-energized. The icing accretion is 0 kV/cm > 15 kV/cm > DC + 15 kV/cm > DC − 15 kV/cm. The icing condition under different electric field types shows in Figure 6, and the relationship between corona current, icing amount, icing density and electric field intensity shows in Figure 7.

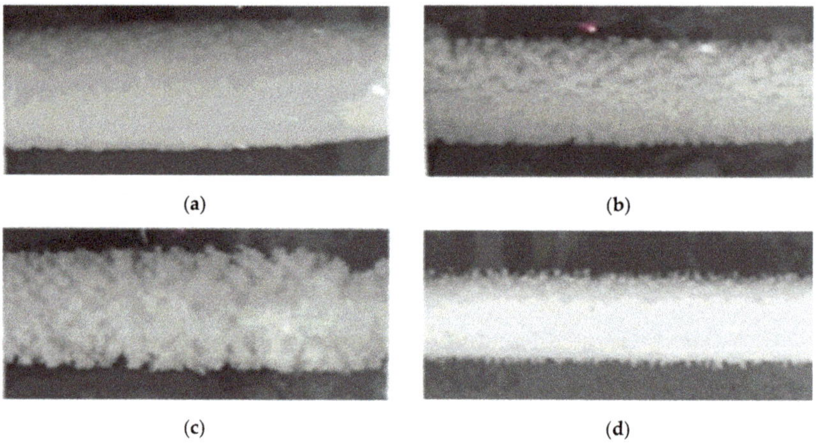

Figure 6. Accreted icing at different electric fields of conductors. (**a**) 0 kV/cm. (**b**) AC 15 kV/cm. (**c**) DC + 15 kV/cm. (**d**) DC − 15 kV/cm.

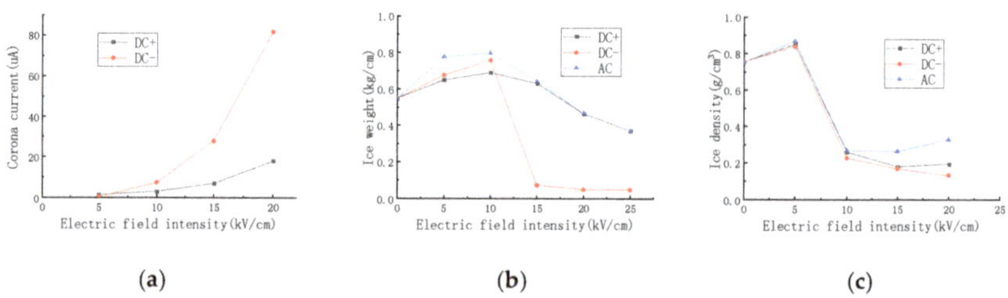

Figure 7. Relation between corona current, ice weight, ice density and electric field intensity. (**a**) Relation between corona current and electric field intensity. (**b**) Relationship between icing accretion and the electric field intensity. (**c**) Relation between ice density and electric field intensity.

The reference [34] conducted an experimental study on the rime icing characteristics of a conductor under AC electric field in an artificial climate chamber. It analyzed the influence of different AC electric field intensity on the shape, weight and density of ice. The conclusion is similar to that in the literature [32], which shows that the weight and ice density of ice accretion both increase first and then decrease with the increase in AC field intensity.

Within the permissible range of electromagnetic environment, the corona effect can be achieved by increasing the surface electric field intensity of ice-covered conductors with a smaller diameter, which can use to prevent icing disasters in local areas. However, at the same time, the loss caused by corona will increase, and strong corona discharge will always occur during the non-ice-covered period [35].

4.4. Torque Pendulum

When the air-flow blows the overcooled water droplets in the air onto the surface of the conductor and collides with it, then ice accretion increases along the windward side and the center of gravity shifts to generate torque. When the torque exceeds the critical value, the ice-covered conductors will be twisted and form a cylindrical or oval ice-covered after a certain period of time. This process repeats over and over again, so that all sides of the conductor continue to be icing. The ice-covered torsion process and the schematic diagram of the torque pendulum show in Figures 8 and 9:

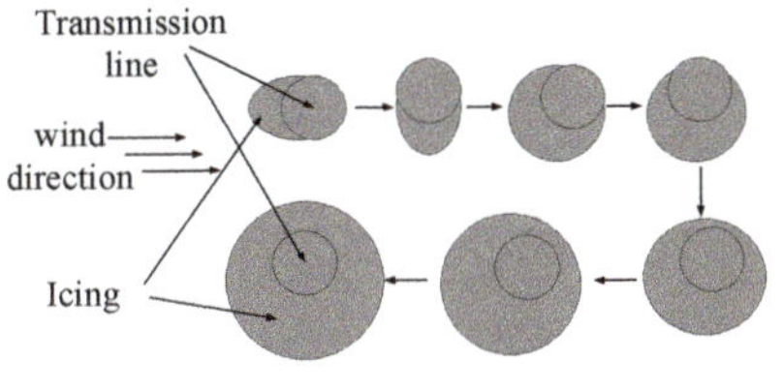

Figure 8. Icing torsion process of conductor.

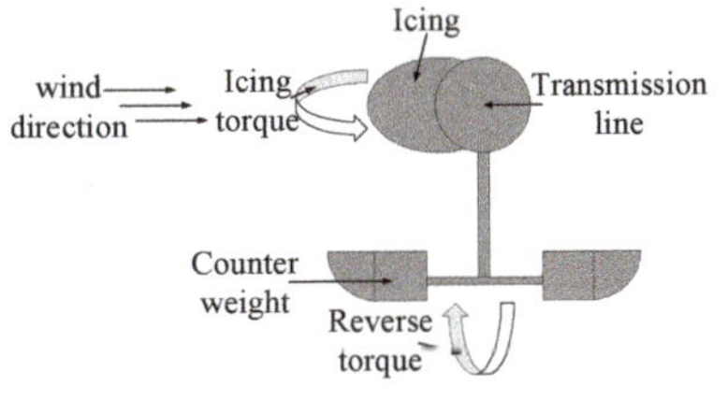

Figure 9. Schematic diagram of the torque pendulum.

The installation of a torque pendulum provides a torque of reverse force for the icing conductor to counter the twisting tendency during the icing process on the windward side, so that the icing will continue to occur on the side of the original wind direction. However, with the increase in the weight and thickness of the ice accretion, the phenomenon that the ice bonding moment cannot resist the gravity moment of the ice itself will appear, which is a macroscopic display as the icing falls off layer by layer. Especially under the action of natural wind, the falling off will be particularly intense, so as to reduce the degree of the ice accretion and the increase in the arc stress and the strength of the mechanical stress of conductor. Then improve the safety and stability of transmission lines, the effect is obvious.

In reference [36], according to the torsional vibration excitation galloping mechanism proposed by Nigol, the weight equation of the torque pendulum was constructed, and the arrangement coefficient was calculated. Furthermore, the arrangement distance of the torque pendulum on the conductor was determined, and a distributed arrangement method was developed to inhibit the torsion and galloping of the conductors. One torque pendulum was installed at an interval of 30 m. Torsional vibration is a crucial factor for the galloping of icing conductors. Torque pendulum can inhibit self-excitation torsional and change the torsional vibration characteristics of the system, which can effectively inhibit such galloping [37,38]. Application of torque pendulum and result shown in Figures 10 and 11.

4.5. Expanded the Diameter Conductor to Replace Bundled Conductors

The DEC has the same circular section and larger outer diameter than the bundled conductors. It is based on the standard conductor specifications that meet the technical conditions, and the technology of expanding the outer diameter is adopted. Under the premise of ensuring the outer diameter required by the corona, the weight of the structure, the ice load of the transmission line, and the total weight is reduced, then reduced the rate of the icing damage of the transmission line [40]. Cross sections diagram of DEC is shown in Figure 12. In a transmission line, when the cross sections area and the wave impedance of the bundled conductors and DEC are the same, the DEC can reduce the number of sub-conductors and the icing load of the transmission line [41]. However, the cost of the DEC is higher, its diameter is larger, the installation process is complicated, and the production process also required to be higher [42].

Figure 10. Application of torque pendulum.

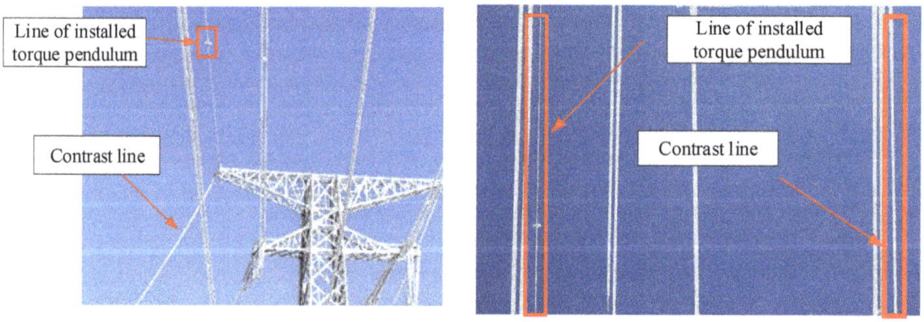

Figure 11. Natural icing contrast of overhead ground line.

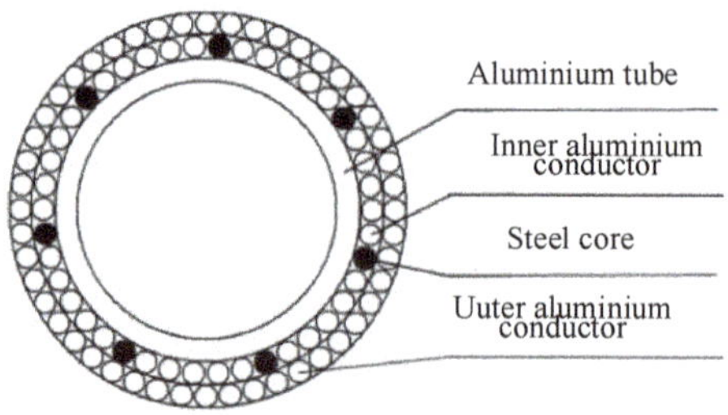

Figure 12. Cross sections diagram of DEC.

The DEC has mainly included three types: stranded type, high-density polyethylene supported type, and aluminum tube supported type. The advantages and disadvantages are summarized in Table 2.

Table 2. Comparison of advantages and disadvantages of different types of DECs.

Types	Advantages	Disadvantages
Stranded type	Production process is maturity, low cost	Poor structural stability, small diameter expansion
High-density polyethylene supported type	Large diameter expansion	Support core processing is complex, poor recovery after loose strands, high cost
Aluminum tube supported type	Metallic aluminum tubes made of conductive materials, need not add conductive material	The bending radius should be at least 20 times the diameter, difficult to construct, poor recovery after bending, high cost

Based on the concept of possible collision when the actual amounts of droplets colliding with the trajectory of droplets do not deflect, that is, the local collision coefficient (collision rate), the icing degree of DEC with different diameters under specific meteorological environment can be studied [39]:

$$\alpha = \frac{dy}{ds} \tag{7}$$

where, *ds* is the surface arc length of the upper and lower two trajectories of droplets between the collision points and the micro-element plane on the cylinder. *dy* is the distance between the collision points and the micro-element plane when the trajectories of droplets do not deflect. By analyzing the differential equation of water droplets in air movement trajectory in the atmosphere covered with ice, the local collision coefficient of conductor surface with different diameters is calculated, and the results are shown as follows [39].

It can be seen from Figure 13 [39] that the local collision coefficient reaches the maximum at the center point, and it decreases with the increase in the outer diameter of the conductor. Therefore, under the same icing condition, the larger the outer diameter, the smaller the collision coefficient of water droplets in the gas–liquid two-phase flow, the slower the icing growth rate, and the smaller the ice thickness at the same period.

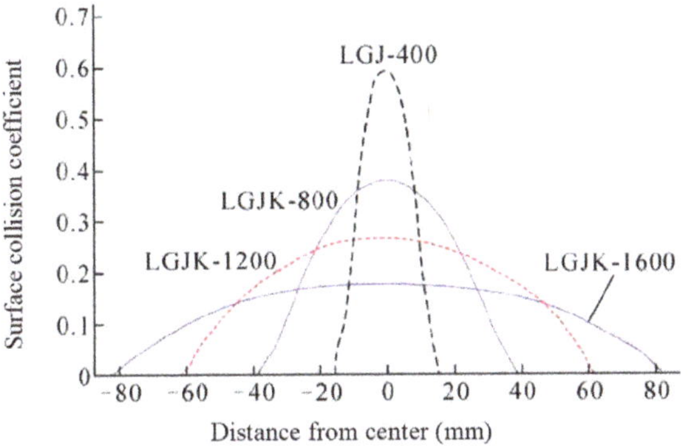

Figure 13. Local collision coefficient of different diameters conductor surface [39].

The icing accretion of the conductor M (kg) depends on the collision coefficient, capture coefficient, freezing coefficient, wind speed, liquid water content in the air and icing time. The expression is as follows [62]:

$$M = \alpha_1 \alpha_2 \alpha_3 w d v L \tau \tag{8}$$

where, α_1 is the collision coefficient, α_2 is the capture coefficient; α_3 freezing coefficient; w is the liquid water content, kg/m^3; d is conductor diameter, mm; v is wind speed, m/s; L is the length of the conductor, m; τ is icing time, s. The capture coefficient α_2 and freezing coefficient α_3 are basically equal to 1. Then the total weight of ice accretion n bundled conductors is $M_f = nM$. From the perspective of conductor diameter, the relationship between ice thickness and diameter shows in Figure 14:

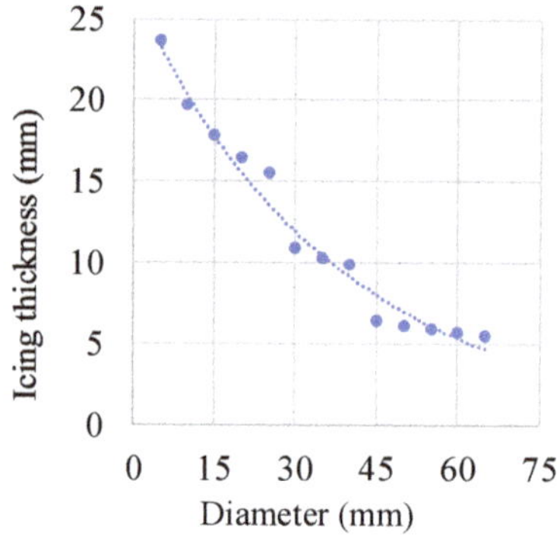

Figure 14. Relationship between ice thickness and conductor diameter.

The icing degree is studied and compared in the scheme of the DEC replace 500 kV quad bundled conductors [43]. It found that under the same environment, the bundled conductors have a more significant collision coefficient and a larger calculated diameter, and the ice accretion of the bundled conductors is higher than that of the expanded one. The ice accretion reduction in the DEC is 30~80%. The larger the diameter, the more ice accretion reduction, as shown in Figure 15 [43]. There are aluminum tubes to replace DECs to experiment.

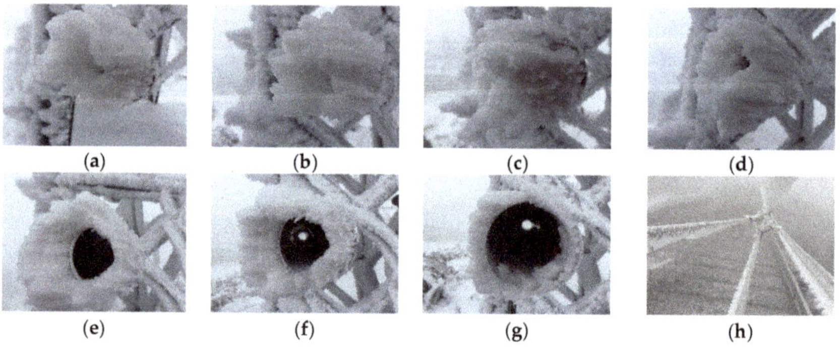

Figure 15. Icing picture of different diameter aluminum tubes and quad bundled conductors: (**a**) 40 mm, (**b**) 50 mm, (**c**) 60 mm, (**d**) 70 mm, (**e**) 80 mm, (**f**) 90 mm, (**g**) 100 mm, (**h**) quad bundled conductors [43].

Figure 16 is the simulation diagram of gas–liquid two-phase flow characteristics of the icing by the DEC and double bundled conductors. There have been application examples of replacing bundled conductors in engineering, as shown in Figure 17, and excellent results have been achieved.

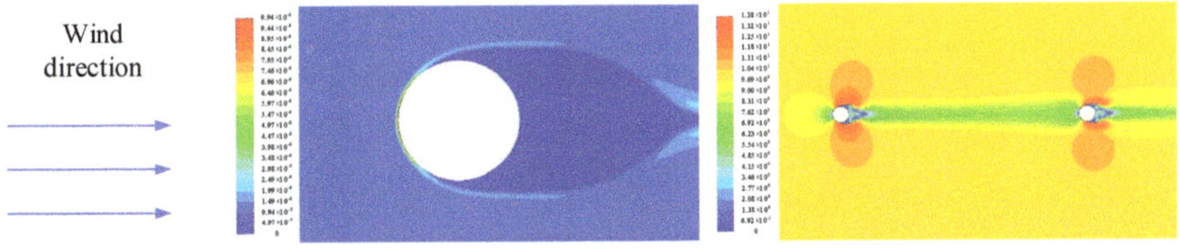

Figure 16. Gas–liquid two-phase flow characteristics of ice-covered DEC and double-bundled conductors. (**a**) DEC. (**b**) Double bundled conductors.

Figure 17. Application of DEC to replace double-bundled conductors.

4.6. CTIIMD

Due to many deficiencies of ice melting devices, some research institutes have developed load current transferred ice melting devices for bundled conductors of transmission lines [44]. When the m bundled conductors run normally, each bundled conductor undertakes the transmitted current. If the load current transmitted in all is I, the average current by each one is I/m. However, the current transferred ice melting method of bundled conductors is to concentrate the total load current on a group of conductors through a switch, which includes n ($m > n \geq 1$ and $n \in N$) conductors, to increase the transmitted current (I/n) on each conductor of this group. Use the overload current to generate more heat to heat and promote the ice to melt and fall off to achieve the purpose of ice melting on the conductors. After the ice melting of a group of bundled conductors are finished, the total load current is passed into the other groups successively. According to the actual line current and the required minimum ice melting current in the environment, one, two and other amount conductors can be set as a group. Finally, ice melting of the entire transmission line was realized [45,46]. The working mode of the current transferred ice melting device shows in Figure 18. The red conductors mean that there is current passing through, while the blue one means that there is no current:

In order to make the current transferred ice melting device more intelligent, the ice weight sensor is installed on the line. When the ice accretion reaches the set value, the ice melting device automatically starts the switch to work. CTIIMD has been proven to have a remarkable effect in the laboratory and transmission lines many times. Compared with other ice melting devices currently in use, the device is small, the manufacturing cost is low, and the DC ice melting device which costs tens of millions. The current flowing through the line is used as the heat source without consuming additional electric energy.

The operation is effortless and only requires the mobile phone APP to remotely control the switch inside the device and select the conductors that need to melt ice in batches. There is no need for a power outage and no impact on the stable operation of the power grid.

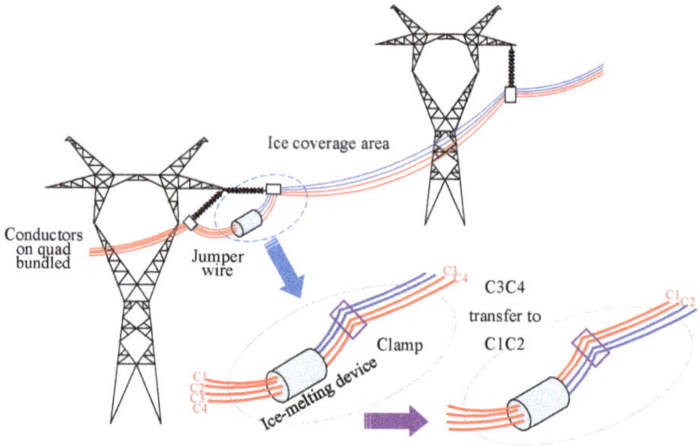

Figure 18. Working mode diagram of current transferred ice melting device.

The CTIIMD was opened to check the internal condition after three years of installation tests at the test site, and no changes were found. It indicates that the tightness, durability and reliability have been verified and can withstand various tests in complex environments. Although the premise of ice melting is to reach the minimum ice melting current when the total current of the line is transferred to one group, they are the transmission lines of high-power electric energy, so the ice melting current can fully meet. If the grid is widely operational, it will save vast sums of money. The CTIIMD has been successfully put into operation on the transmission lines, and outstanding results have been achieved. Figure 19 shows the application of CTIIMD on the transmission lines, and it in the squared red line.

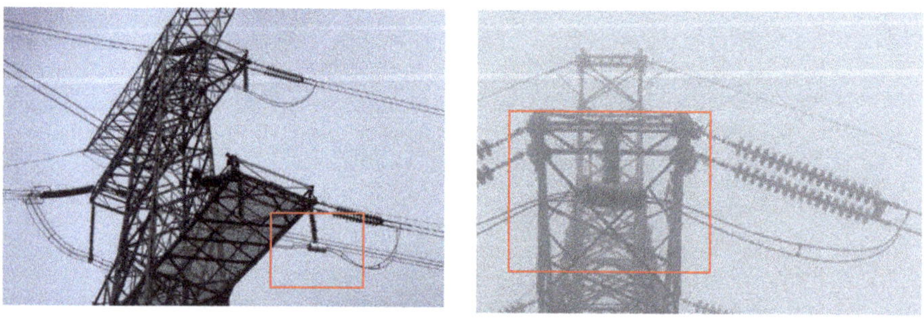

Figure 19. The CTIIMD operation diagram on site.

5. Anti/De-Icing Technology of Insulator

The purpose of insulator anti-icing/de-icing is not only to reduce the mechanical load but also to prevent the loss of insulation resulting in flashover and power failure. At present, the main method for insulators is anti-icing, mainly from the following four aspects:

5.1. Application of Hydrophobic Materials

Superhydrophobic phenomena similar to lotus leaf greatly arouse the interest of researchers in various research fields. Hydrophobic materials have unique properties of waterproofing, anti-icing and self-cleaning, so they have essential application prospects in insulators. When the hydrophobic material is applied to the surface of the insulator, the lower surface energy of the coating will cause a weak adhesion between the ice and the coating. Moreover, prevents the continuous water film from be icing, which is more conducive to shedding the ice under the action of natural wind and gravity. Hydrophobic materials mainly include silicone oil, silicone grease, ozocerite, room temperature vulcanized silicone rubber (RTV), and durable in-place forming antifouling flashover composite coating (PRTV) [98]. Microscopic view of a material surface with droplets is shown in Figure 20.

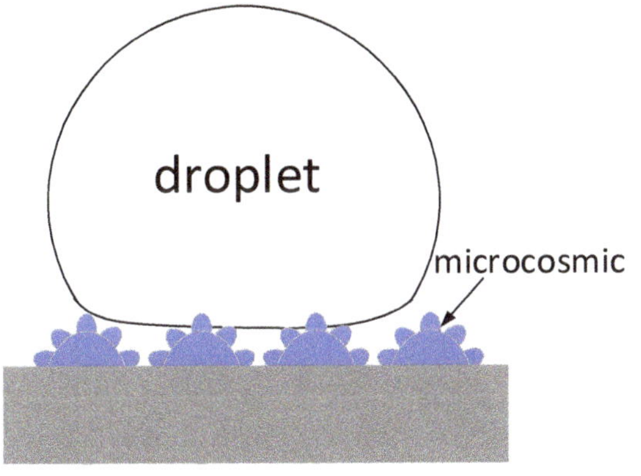

Figure 20. Droplets attached on the surface of superhydrophobic material in microcosmic.

Micro–nano rough structure and low surface energy are the fundamental reasons for the superhydrophobic properties of lotus leaves [99–101]. Simply put, it creates an extremely rough structure at the microcosmic to reduce the contact area between water droplets and the surface. The greater the contact angle θ between the overcooled water droplets and the surface, the stronger the hydrophobicity of the surface and the longer the freezing time on the surface [102]. θ is a sign that indicates interfacial tension and is one of the important factors in determining the hydrophobic property of a substance. The θ is the result of the surface tension balance between solid, gas, and liquid interface. Generally speaking, when $0^\circ < \theta < 90^\circ$, the solid surface has hydrophilic, $90^\circ < \theta < 180^\circ$, the solid surface has hydrophobicity, the fewer θ, the better hydrophilic, the greater θ, the better hydrophobicity. The relationship between the θ and the surface energy can be described by Young's equation:

$$COS\theta = \frac{(\gamma_{sv} - \gamma_{sl})}{\gamma_{lv}} \tag{9}$$

where, γ_{sv} is the tension of solid and gas surface, γ_{sl} is the tension of solid and liquid surface, γ_{lv} is the tension of gas and liquid surface, θ is the contact Angle of the material. The real object of water droplets on the surface is shown in Figure 21a,b [47], and the model analysis is shown in Figure 21c. The red triangle in (a) represents the angular relationship of water droplets on the surface of a material, it is also shown in (c).

Different hydrophobic materials show different contact angles. The LXP-70 insulator string coated with silicone acrylate resin (SAR) hydrophobic material with a contact angle of 110–150° was subjected to artificial icing and flashover tests [47]. After studying the ice morphology, density, weight growth and AC flashover voltage, it was found that the more hydrophobic the coating was, the more ice particles are formed, the ice density on the coated insulator surface is lower than that on the uncoated insulator surface. SAR coating with the θ greater than 130° can significantly reduce the ice weight of the glass insulator at the initial icing stage and increase the flashover voltage of the insulator.

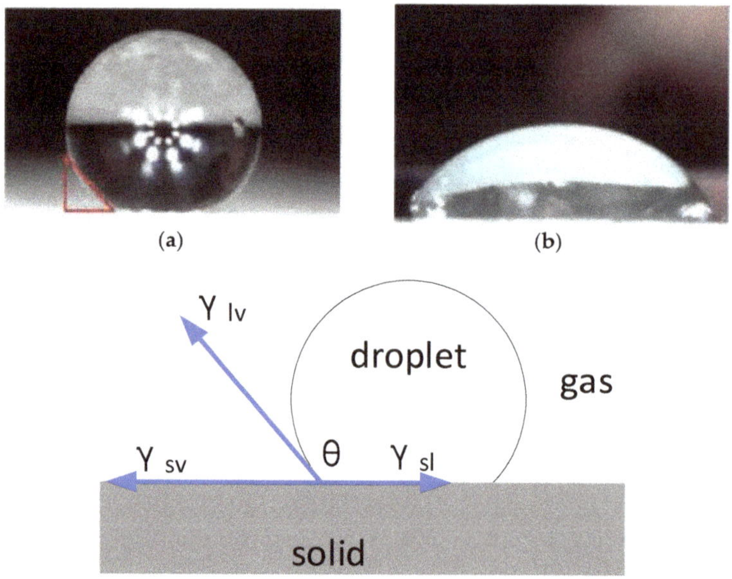

(a) (b)

(c)

Figure 21. Droplets on the surface. (**a**) Droplet on the hydrophobicity. (**b**) Droplet on the hydrophilic. (**c**) Model analysis of water droplets on solid surfaces [47].

In reference [48], a superhydrophobic material based on SiO_2 was prepared with a θ of up to 163.6°. As the contact area and contact time between droplets and the superhydrophobic surface decreased significantly, icing on the insulator surface was prevented. Under the experimental conditions of −8°, vapor flow rate 70 L/h and 4-h icing, the coating insulator has about 50% less ice accretion than the uncoated one, and no icicle on the insulator string at low temperature, as shown in Figure 22 [49].

(a) (b)

Figure 22. Results of 4-h icing with and without a coating of insulators. (**a**) With coating. (**b**) Without coating [49].

5.2. Heating Coating

Some scholars have proposed applying a heating coating to the surface of the insulator. When the insulator is icing, it will increase the leakage current on the surface and give off heat to prevent the formation icing, so as to convert energy from the power system into heat for anti-icing [103]. This method also does not require an additional heat sources or equipment, which reduces the input of resources. It does not require transmission line shutdown or short circuit for cooperation, which can ensure that the transmission of electric energy is not affected. The application of the electric thermal coating to insulators is one of the ideas for dealing with the disaster of rain and snow freezing [104,105].

Carbon black is added to RTV coating and coated on the surface of the insulator to make it slightly conductive. By studying the black carbon content of RTV coating, the resistivity and leakage current of insulator can be controlled, and the anti-icing performance of the coating with different leakage currents can be compared [50]. The result of 2-h icing shows in Figure 23 [50]. The black carbon content of the a–d insulator is 6–9%, and the coating is 0.4 mm thick.

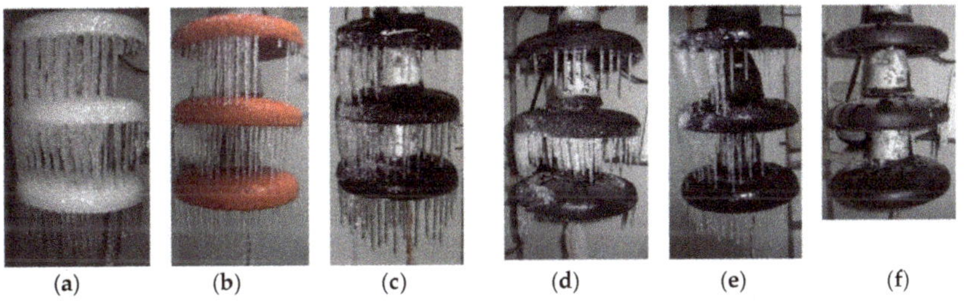

Figure 23. Comparison of insulator ice accretion with different carbon black content coatings. (**a**) No coating. (**b**) RTV coating. (**c**) Coating a. (**d**) Coating b. (**e**) Coating c. (**f**) Coating d [50].

However, if the whole insulator is fully coated, the insulator will always be in a state of large leakage current and heating. Therefore, the "switching effect" coating method is generally adopted, that is, the lower surface is coated completely, the upper surface is coated with the outer ring, and the connection between the inner ring and the steel cap is not coated, resulting in a "blank zone". When the icing starts, the ice conducts electricity to fill the "blank band". Connecting the steel cap with the electrothermal coating will generate large leakage current heating, inhibiting the icing generation and growth. The literature [51] quantified the influence of coating position on the anti-icing effect and studied the position of semiconductor coating on the surface of large-diameter insulators. It found that the size and position of the "blank zone" had a great influence on the leakage current on the surface of insulators and are the important factors affecting the anti-icing effect of the electrothermal coating. The "blank zone" should be set at the edge of the shed, and the width should be moderate, 5 to 7 cm can achieve a better anti-icing effect. "Switching effect" coating insulators are shown in Figure 24:

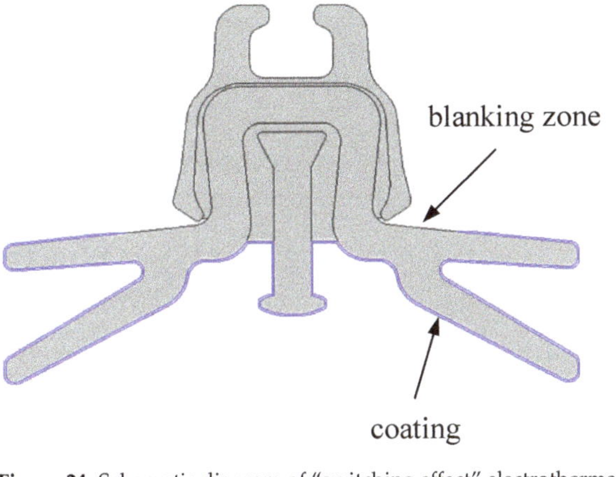

Figure 24. Schematic diagram of "switching effect" electrothermal coating.

There is also photoelectric mixed coating for insulator anti-icing [52]. The important parameter of photothermal anti-icing coating is solar radiation energy absorption rate (SREAR). The SREAR is the ratio of the solar energy absorbed by the surface of an object per unit area to the total solar energy reached by the sun. The higher the absorption rate of solar energy, the stronger the ability of the object to absorb solar thermal radiation. Common photothermal materials are: intrinsic absorption materials, ceramic–metal composite materials and spinel materials.

The heating coating is useful for preventing glaze on insulators, but it cannot reduce the accumulation of rime [106]. Meanwhile, the durability of both heating coating and hydrophobic coating under continuous high pressure and outdoor insulator stress needs further research.

5.3. Optimization of the Shed Structure

Many studies have shown that [107–112] insulator shed configuration has a great influence on the icing and the flashover voltage, but under heavy icing conditions, the influence of shed structure is not obvious [53,76,113]. When the natural icing occurs, the influence law of shed structure on insulator icing growth is consistent with the results of the artificial experiment. For porcelain and glass insulators, it can be considered to change the shed structure by positing and mixing large and regular-size insulators in proportion. For composite insulators, large (medium) and small sheds can be made in proportion to change the shed structure, which increases the difficulty of ice bridging and increases the icing flashover voltage [114]. However, determining the value and proportion of shed diameter under different external environments is the focus of researchers. The discharge path of composite insulators is shown in Figure 25 [114]:

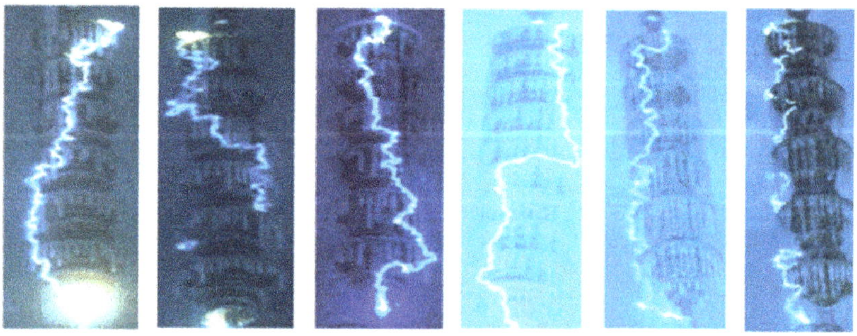

Figure 25. Discharge path of composite insulators of different types [114].

Adding an appropriate number of large sheds to insulators can improve their electrical performance under icing conditions. In the study of icing on 330 kV composite insulators, in the condition of the same icing severity, the ice accretion on composite insulators with large shed is higher than that on insulator strings of the standard ceramic insulators. When the ice thickness is 13 mm, the U50% flashover voltage of composite insulators with large sheds is about 10% higher than that of those without large sheds [53]. In the literature [54], DC ice flashover characteristics of two composite insulators were tested. The ice thickness in the test was 5–30 mm, and the two samples were different only in the arrangement of large shed spacing. The test results show that the average flashover voltage of insulator samples with large shed spacing is higher than that of small spacing as the ice thickness increasing. When the ice thickness is 5 mm, the average flashover voltage is 9.2% higher, but the percentage decreases with the ice thickness increases. When the ice thickness is 30 mm, the average flashover voltage is only 3.0% higher.

Through simulation and experiment, the influence of several parameters of UHV composite pillar insulator shed tilt angle, rod diameter, shed spacing, and the ratio between the large-shed diameter and small-shed diameter on icing were studied [55]. It was found that in order to minimize the weight of icing, the insulator up tilt angle, down tilt angle, rod diameter, and shed ratio of insulator should be smaller, and the shed distance is around 48 mm. The icicles between the sheds are not easy to bridge by using the extended shed structure, and the ice flashover voltage gradient increases along the creepage distance. The extended shed structure is adopted to replace part of the small shed, which can effectively prevent ice bridging and improve the ice flashover voltage.

In the literature [56], 12 typical 220 kV composite insulators were tested of energized and non-energized icing, shown in Figure 26 [56]. It was found that when the shed spacing is larger, the length of icicles and the ice thickness become larger, while the icicle diameter is relatively smaller. The greater ratio of shed spacing and shed spread is beneficial to delay the icicle bridging and promote the growth of the ice layer on the shed surface. Although this will lead to an increase in the conductivity of the water film on the shed surface, the larger shed spacing hinders the flashover more. The differences in the ratio of icicle air gap arc during the flashover are one of the main reasons causing differences in icing flashover voltage for composite insulators with different shed configurations [57].

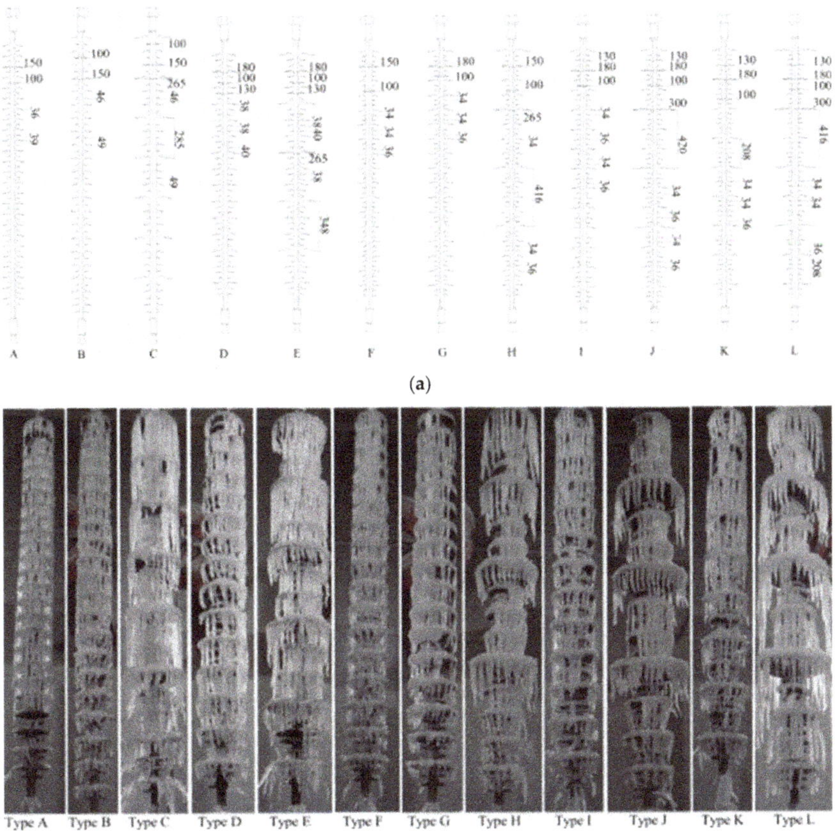

Figure 26. The contrast of parameters and icing situation of composite insulators of different types. (**a**) Insulator types. (**b**) Icing situation of different types of composite insulators [57].

5.4. String Arrangement

The literature has shown that insulator string arrangement has a great influence on ice flashover voltage, including standard arrangement, connected with alternately large and small diameter sheds (CALSDS) arrangement, V and Λ shaped arrangement, horizontal arrangement and inverted T-shape (⊥ shape) arrangement [115,116].

Due to the effect of gravity, all the rainwater falls straight down from the surface of the insulator. When the insulators are icing seriously, icicles bridge the insulator shed through the gap to greatly shorten the creepage distance. In this case, the icicle bridging can be effectively blocked by redesigning the arrangement of insulator strings or inserting large-diameter insulators, which will also affect the electric field and voltage distribution. The general relation between icing flashover voltage $U_{50\%}(kV)$ and string length of icing insulator string can be expressed as [117]:

$$U_{50\%} = AN^d \quad (10)$$

where: A is a constant related to the degree of icing, the structure of insulators, the degree of pollution before icing, etc. N is the number of insulators. d is the linearity coefficient. When d = 1, it is linear; when d < 1, it is nonlinear. A and d can be obtained by fitting test data.

The equation of $U_{50\%}$ and N and the surface salt density ρSDD (mg/cm^2) of the insulator:

$$U_{50\%} = KN^{d'}(\rho_{SDD})^{-b'} \quad (11)$$

K is the coefficient. d' is the characteristic index of the effect of N on $U_{50\%}$. b' is the characteristic index of the effect of ρSDD on $U_{50\%}$.

Connected with alternately large and small diameter sheds (CALSDS) ("2 + 1 ", "3 + 1" and "4 + 1", etc.) have a great influence on the icing flashover voltage of insulator string. Under the polluted icing condition, the flashover voltage of insulator strings arranged by "3 + 1" CALSDS is about 14.6–18.1% higher than that of standard strings, and arranged by "4 + 1" CALSDS is about 10.1–13.1% higher than that of standard, respectively. The "2 + 1" is lower than that of "3 + 1" and "4 + 1", Figure 27 [58]. For the V type and Λ type, "2 + 1" and "3 + 1" arrangement, the icing flashover voltage is influenced by the ice thickness. The thicker the ice, the less improvement. Both type V and Λ arrangement can significantly improve the icing flashover voltage, mainly because of the following reasons: the insulator shed is not easy to be bridged by icing, and due to thermal buoyancy, part of the arc can easily float upward from the insulator, so the arc can dissipate heat more easily [59].

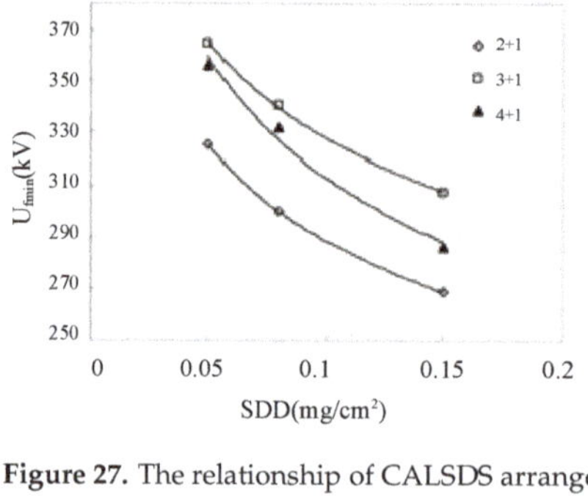

Figure 27. The relationship of CALSDS arrangement icing flashover voltage and SDD [59].

For II arrangement, increasing the distance between the two insulators can improve flashover voltage performance. When the distance increases from 450 to 650 mm, the flashover voltage of the standard insulator string and with large shed increases by 3.2% and 7.1%, respectively [60]. When the string connection angle widely used in the transmission line is 60°, the flashover situation of the insulator string of V and Λ shape arrangement under wet snow conditions is shown in Figure 28 [60].

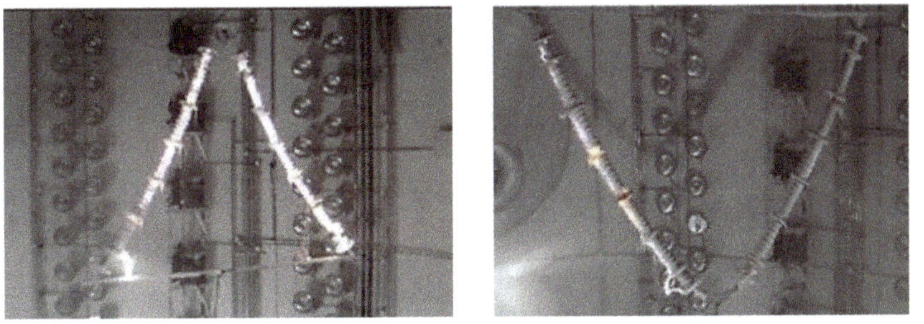

(a) (b)

Figure 28. Flashover situation of insulator string of 60° Λ and V arrangement. (**a**) Λ shape arrangement. (**b**) V shape arrangement [60].

For the inverted T-shape arrangement of the insulator string, the icicle growing down has less impact on the gap between the insulator shed, the arc development along the insulator surface in the process of flashover, there is no arc bridge shed phenomenon. The authors of [61] studied the icing and flashover of glass insulator strings arranged in inverted T shape, and compared the inverted T-shape arrangement of "8 + 1", "7 + 2", "6 + 3" and "5 + 4" with type I. When the strings were arranged in inverted T shape, the U50% increased by 6.66~15.37% compared with type I. The leakage current of the inverted T string is reduced by 3.84~18.50% compared with that of type I. Meanwhile, the higher the proportion of horizontal insulators, the higher the icing flashover voltage. The icing situation is shown in Figure 29 [61], and the *D* is icing thickness. Figure 30 is the application of inverted T-shape insulator strings in transmission lines.

(a) (b) (c) (d)

Figure 29. Inverted T-shaped arrangement icing situation of insulator string with different degrees. (**a**) D = 5 mm. (**b**) D = 10 mm. (**c**) D = 15 mm. (**d**) D = 20 mm [61].

Figure 30. Application of inverted T-shaped insulator strings in transmission lines.

6. Conclusions

Based on the literature, this review comprehensively expounds on the current situation and technical application of transmission lines icing, including the aspects of snowing and icing disasters of the power grid, icing characteristics and anti/de-icing methods of conductors and insulators. The conditions or influences of the formation of icing mainly include temperature, wind speed, humidity and water droplets bonded on the transmission line. As long as one of the links is hindered, the icing can be effectively prevented. When the icing has formed, it can be melted and make it falls off from the surface of the transmission line. For conductors, the most widely used method is DC ice melting, the State Grid Corporation and Southern Power Grid Corporation of China have used it on a large scale in recent years, effectively solving multiple icing disaster crises in the power grid. However, the cost of the DC ice melting device is very high, and the power needs to be cut off. AC ice melting is not applicable to long-distance transmission lines with high voltage levels. The current transfer intelligent ice melting device can effectively solve various problems of AC/DC ice melting and has a wide application prospect. It has been running in the transmission lines of China Southern Power Grid Corporation and achieved good effect. Traditional mechanical de-icing has developed to the present, which mainly relies on a robot with a blade on its conductors to remove the ice. The de-icing process causes great damage to the conductors. The corona discharge effect has a certain effect on reducing ice accretion, but corona loss is too large and limited by electromagnetic environment, so the application is limited. DEC and torque pendulum has obvious anti-icing effects, simple structure and great potential. As far as insulators are concerned, there are very few devices targeted to de-icing. Insulators coated with hydrophobic coatings will weaken the adhesion of water droplets in the air on the surface, thus hindering the development of icing, but the hydrophobic long-term performance and service life need to be further tested. When the ice is covered on the blank zone of the heating coating insulator, the leakage current increases and the heat is generated, which hinders the occurrence of icing, but it greatly reduces the pollution flashover voltage of the transmission line. The shed structure is applicable to composite insulators, and the CALSDS arrangement applies to porcelain and glass insulators. Changing the size shed ratio and disk diameter can hinder the ice bridging and improve the flashover voltage during light icing, but the effect is not obvious when the icing is heavy. Icing flashover voltage can be improved by changing the string arrangement, in which the inverted T, Λ and V arrangements have obvious effects. However, the string arrangement may be restricted by the environment in actual transmission lines. Each of these methods can play a good role in anti-icing insulators, widely used, but there are also disadvantages. In the future, the development of anti/de-icing technology of transmission lines, first is to explore fast, efficient and intelligent device, but also need it to have the function of non-manual intervention and non-power failure; the second is that the anti/de-icing research on such new energy power systems, such as wind turbine blade icing.

Author Contributions: Conceptualization, formal analysis, review, Z.Z.; references collecting and original draft preparation, editing, H.Z.; editing, review, S.Y.; editing, review, W.Z. All authors have read and agreed to the published version of the manuscript.

Funding: This research is supported by the Science and Technology Project of State Grid Corporation (Grant No. 5108-202218280A-2-322-XG).

Institutional Review Board Statement: Not applicable.

Informed Consent Statement: Not applicable.

Data Availability Statement: Not applicable.

Conflicts of Interest: The authors declare no conflict of interest. The funders had no role in the design of the study; in the collection, analyses, or interpretation of data; in the writing of the manuscript, or in the decision to publish the results.

References

1. Imai, I. Studies on Ice Accretion. *Res. Snow Ice* **1953**, *1*, 35–44.
2. Farzaneh, M.; Chisholm, W.A. *Insulators for Icing and Polluted Environments*; John Wiley & Sons: Hoboken, NJ, USA, 2009.
3. Barat, A.; Guo, Q.; Fan, F.; Guo, X.; Zhang, Q. Analysis and Design of Transmission Line Icing Monitoring System Based on Optical Fiber Sensing. In Proceedings of the 2022 IEEE 10th Joint International Information Technology and Artificial Intelligence Conference (ITAIC), Chongqing, China, 17–19 June 2022; pp. 1846–1849. [CrossRef]
4. Richardson, A. Dynamic analysis of lightly iced conductor galloping in two degrees of freedom. Generation, Transmission and Distribution. *IEE Proc. C* **1981**, *128*, 211–218.
5. Morgan, V.; Swift, D.A. Effect of ice loads on overhead-line conductors. *Electron. Power* **1965**, *11*, 22–23. [CrossRef]
6. Makkonen, L. Estimation of wet snow accretion on structures. *Cold Reg. Sci. Technol.* **1989**, *17*, 83–88. [CrossRef]
7. Ihara, S.; Yamabe, C.; Ushio, S. Breaking of ice using pulsed power. In Proceedings of the Conference Record of the 2006 Twenty-Seventh International Power Modulator Symposium, Arlington, VA, USA, 14–18 May 2006; p. 342.
8. Baliberdin, L.; Kozlova, M.; Shershnev, Y. Model group of controlled installation for melting ice on transmission line conductors. In Proceedings of the 2005 IEEE Russia Power Tech, St. Petersburg, Russia, 27–30 June 2005; pp. 1–5.
9. Kannus, K.; Lahti, K. Laboratory Investigations of the Electrical Performance of Ice-covered Insulators and a Metal Oxide Surge Arrester. *IEEE Trans. Dielectr. Electr. Insul.* **2007**, *14*, 1357–1372. [CrossRef]
10. Yang, J.; Li, Z.; Yang, F.; Huang, T. Analysis on the characteristics of ice-covered power grid disaster and tower toppling in 2008. *Adv. Power Grids Hydropower* **2008**, *4*, 4–8. (In Chinese)
11. Dong, B.; Jiang, X.; Yin, F. Development and prospect of monitoring and prevention methods of icing disaster in China power grid. *IET Gener. Transm. Distrib.* **2022**, *16*, 4480–4493. [CrossRef]
12. Zhou, X.; Zhu, Y.; Zhang, Y.; Li, H. A Review of Anti-icing and De-icing Technology of Overhead Ground Wire. In Proceedings of the 2022 International Symposium on Electrical. Electronics and Information Engineering (ISEEIE), Chiang Mai, Thailand, 25–27 February 2022; pp. 282–286. [CrossRef]
13. Peng, S.; Hao, W.; Zhai, Y. Review of the research on icing mechanism of transmission lines and ice-melting technologies. In Proceedings of the 2015 5th International Conference on Electric Utility Deregulation and Restructuring and Power Technologies (DRPT), Changsha, China, 26–29 November 2015; pp. 1648–1652. [CrossRef]
14. Yan, X.; Yuebin, Z.; Guobang, B.; Wanyu, C.; Liuqing, Y.; Wenyong, Y.; Qiansu, L.; Ronghua, Z.; Shukai, X.; Chuang, F. Research on MMC de-icing system capable of power grid interconnection. In Proceedings of the 2021 Annual Meeting of CSEE Study Committee of HVDC and Power Electronics (HVDC 2021), Beijing, China, 28–30 December 2021; pp. 29–38. [CrossRef]
15. Liu, Y.; Farzaneh, M.; Du, B. Investigation on shed icicle characteristics and induced surface discharges along a suspension insulator string during ice accretion. *IET Gener. Transm. Distrib.* **2017**, *11*, 1265–1269. [CrossRef]
16. Guo, J.; Li, J.; Rong, C.; Dong, Z.; Guan, W.; Zheng, Y.; Tan, L. Effectiveness Evaluation System and Evaluation Method for Anti-ice Damage Technical Transformation Project of Overhead Lines. In Proceedings of the 2019 IEEE 3rd International Electrical and Energy Conference (CIEEC), Beijing, China, 7–9 September 2019; pp. 314–319. [CrossRef]
17. Yan, X.; Li, J.; Li, L.; Huang, Z.; Hu, J.; Lu, M. An OH-PDMS-modified nano-silica/carbon hybrid coating for anti-icing of insulators part II: Anti-icing performance. *IEEE Trans. Dielectr. Electr. Insul.* **2016**, *23*, 2165–2173. [CrossRef]
18. Zhu, Y.; Tan, Y.; Huang, Q.; Huang, F.; Zhu, S.; Mao, X. Research on Melting and De-icing Methods of Lines in Distribution Network. In Proceedings of the 2019 IEEE 3rd Conference on Energy Internet and Energy System Integration (EI2), Changsha, China, 8–10 November 2019; pp. 2370–2373. [CrossRef]
19. Liu, C.; Qin, J.; Hao, Y.; Li, G.; Chen, Y. Research on Verification Test of Tension Stringing for UHV Transmission Line Project. In Proceedings of the 2020 IEEE Sustainable Power and Energy Conference (iSPEC), Chengdu, China, 23–25 November 2020; pp. 1766–1771. [CrossRef]
20. Xie, D.; Song, W.; Wang, W.; Sun, T.; Huang, T.; Cai, X. Research on ice removal blasting parameters of HV Transmission Lines. *J. North Univ. China* **2018**, *39*, 746–751. (In Chinese)
21. Cao, Y.; Xue, K.; Miao, L.; Li, H.; Guan, X.; Zhang, J.; Li, G. Dynamic response analysis of ice tower line system after blasting deicing. *J. North Univ. China: Nat. Sci. Ed.* **2019**, *040*, 198–205. (In Chinese)
22. Zumw, A.; Muwller, A. Flight and Wind Tunnel Tests of an Electro-Ipulse De-icing System. In Proceedings of the AIAA /NASA General Aviation Technology Conference, Hampton, VA, USA, 10–12 July 1984; pp. 10–12.
23. Jiang, X.; Chen, Y.; Huang, T.; Wang, M.; Huang, H. Electro-impulse de-icing (EIDI) test of aircraft wing leading edge. In Proceedings of the 22nd International Symposium on High Voltage Engineering (ISH 2021), Xi'an, China, 21–26 November 2021; pp. 66–71. [CrossRef]
24. Landry, M.; Beauchemin, R.; Venne, A. De-icing EHV overhead transmission lines using electromagnetic forcess generated by moderate short-circuit current. C. In Proceedings of IEEE 9th International Conference on Transmission and Distribution Construction, Operation and Live-line Maintenance, Montreal, QC, Canada, 8–12 October 2000; pp. 94–100.
25. Montambault, S.; Pouliot, N. The HQ LineROVer: Contributing to innovation in transmission line maintenance. In Proceedings of the 2003 IEEE 10th International Conference on Transmission and Distribution Construction, Operation and Live-Line Maintenance, Orlando, Orlando, FL, USA, 6–10 April 2003; pp. 33–40.
26. Zhao, J.; Guo, R.; Cao, L.; Zhang, F. Improvement of LineROVer: A mobile robot for de-icing of transmission lines. In Proceedings of the 2010 1st International Conference on Applied Robotics for the Power Industry, Montreal, QC, Canada, 5–7 October 2010; pp. 1–4.

27. Li, K.; Du, L.; Han, W. Icing Analysis of Transmission Lines Considering the Current Heat. In Proceedings of the 2011 Asia-Pacific Power and Energy Engineering Conference, Wuhan, China, 25–28 March 2011; pp. 1–4. [CrossRef]
28. Sun, C.; Jiang, X.; Xiong, Q.; Yi, H. Analysis of Critical Conditions for Ice Coating and Dry-Wet Growth of Conductors. *Proc. CSEE* **2003**, *3*, 141–145. (In Chinese)
29. Jiang, X.; Fan, S.; Zhang, Z.; Sun, C.; Shu, L. Simulation and Experimental Investigation of DC Ice-Melting Process on an Iced Conductor. *IEEE Trans. Power Deliv.* **2010**, *25*, 919–929. [CrossRef]
30. Jiang, X.; Fan, S.; Hu, J.; Zhang, Z.; Sun, C. Critical Current Analysis of DC Short-circuit Ice Melting in Transmission Lines. *Proc. CSEE* **2010**, *30*, 111–116. (In Chinese)
31. Shu, L.; Luo, B.; Jiang, X.; Hu, Q.; Li, T.; Lan, Q. Intelligent loop current critical melting ice thawing method and its current research. *J. Electrotech.* **2012**, *27*, 26–34. (In Chinese)
32. Yin, F.; Farzaneh, M.; Jiang, X. Corona investigation of an energized conductor under various weather conditions. *IEEE Trans. Dielectr. Electr. Insul.* **2017**, *24*, 462–470. [CrossRef]
33. Yin, F.; Farzaneh, M.; Jiang, X. Influence of AC electric field on conductor icing. *IEEE Trans. Dielectr. Electr. Insul.* **2016**, *23*, 2134–2144. [CrossRef]
34. Huang, Y.; Jiang, X.; Hou, L.; Zhu, M.; Han, X. Experimental Study on Reducing Icing on Conductor Using Self-heating Ring. In Proceedings of the 2018 IEEE International Conference on High Voltage Engineering and Application (ICHVE), Athens, Greece, 10–13 September 2018; pp. 1–4. [CrossRef]
35. Yin, F.; Farzaneh, M.; Jiang, X. Laboratory investigation of AC corona loss and corona onset voltage on a conductor under icing conditions. *IEEE Trans. Dielectr. Electr. Insul.* **2016**, *23*, 1862–1871. [CrossRef]
36. Zhang, Z.; Jiang, X.; Liu, J.; Lei, Y.; Li, R.; Fu, H. Distributed Hanging Weight Arrangement Method for Restraining Torsion and Dancing of Conductor. Patent CN112332353B, 21 December 2021. (In Chinese).
37. Desai, Y.; Yu, P.; Popplewell, N. Perturbation-based finite element analysis of transmission line galloping. *Sound Vibr.* **1996**, *191*, 469–489. [CrossRef]
38. Florea, G.; Florea, M.; Tibuliac, S.; Vaju, M.; Oltean, M.; Mateescu, E. Upgrading the Romanian 400 kV lines with 2 and 3 subconductors per phase to reduce the risk of galloping occurance and the galloping amplitudes by the installation of torsional dampers and detuners, live-line procedures. In Proceedings of the 2016 IEEE PES 13th International Conference on Transmission & Distribution Construction, Operation & Live-Line Maintenance (ESMO), Columbus, OH, USA, 12–15 September 2016; pp. 1–5. [CrossRef]
39. Wang, X.; Yu, J.; Liu, Z.; Qin, Z.; Jiang, X.; Hu, Q. Comparison of Icing on Equivalent DEC and Bundled Conductors. *High VOLTAGE Technol.* **2022**, *48*, 2698–2705. (In Chinese) [CrossRef]
40. He, J.; Li, X. Application Analysis of Expanded Conductor in 1000kV UHV Double-Circuit Line. *AMM* **2014**, *664*, 318–325. [CrossRef]
41. Si, J.; Zhu, K.J.; Wan, J.C.; Yang, J.L.; Liu, L.; Ding, Y. Parametric Study on Simulation Modeling of DEC. In *Advanced Materials Research*; Trans Tech Publications, Ltd.: Bäch, Switzerland, 2013; Volume 815, pp. 69–72. [CrossRef]
42. Jiang, X.; Fan, C.; Xie, Y. New method of preventing ice disaster in power grid using expanded conductors in heavy icing area. *IET Gener. Transm. Distrib.* **2019**, *13*, 536–542. [CrossRef]
43. Bi, C.; Jiang, X.; Han, X.; Yang, Z.; Ren, X. Anti-icing Method Using Expanded Wire instead of Split Wire. *Trans. China Electrotech. Soc.* **2020**, *35*, 2469–2477. (In Chinese) [CrossRef]
44. Huneault, M.; Langheit, C.; St-Arnaud, R.; Benny, J.; Audet, J.; Richard, J.-C. A dynamic programming methodology to develop de-icing strategies during ice storms by channeling load currents in transmission networks. *IEEE Trans. Power Deli.* **2005**, *20*, 1604–1610. [CrossRef]
45. Jiang, X.; Wang, Y.; Shu, L.; Zhang, Z.; Hu, Q.; Wang, Q. Control scheme of the de-icing method by the transferred current of bundled conductors and its key parameters. *IET Gener. Transm. Distrib.* **2015**, *9*, 2198–2205. [CrossRef]
46. Meng, Z.; Wang, Y.; Jiang, X.; Fan, S. Asynchronism of ice shedding from the de-iced conductor based on heat transfer. *IET Sci. Meas. Technol.* **2016**, *10*, 389–395.
47. Hu, J.; Lan, B.; Xu, K.; Jiang, X.; Zhang, Z.; Shi, B.; Yang, H.; Wu, Y. Artificial icing and AC flashover tests on glass insulators with silicone acrylate resin hydrophobic coatings. *IEEE Trans. Dielectr. Electr. Insul.* **2016**, *23*, 1038–1047. [CrossRef]
48. Li, X.; Yang, B.; Zhang, Y.; Gu, G.; Li, M.; Mao, L. A study on superhydrophobic coating in anti-icing of glass/porcelain insulator. *Sol-Gel Sci. Technol.* **2014**, *69*, 441–447. [CrossRef]
49. Liao, W.; Jia, Z.; Guan, Z.; Wang, L.; Yang, J.; Fan, J.; Su, Z.; Zhou, J. Reducing Ice Accumulation on Insulators by Applying Semiconducting RTV Silicone Coating. *IEEE Trans. Dielectr. Electr. Insul.* **2007**, *14*, 1446–1454. [CrossRef]
50. Zhao, Z.; Chen, H. Development of high-efficient synthetic electric heating coating for anti-icing/de-icing. *Surf. Coat. Technol.* **2018**, *349*, 340–346. [CrossRef]
51. Sun, Z.; Jia, Z.; Wei, X.; Guan, Z. Optimization of Anti-icing Structure of Insulators Partially Coated with Semi-conductive Coating. *Proc. CSEE* **2012**, *32*, 132–138+201. (In Chinese) [CrossRef]
52. Yin, F.; Jiang, X.; Farzaneh, M.; Hu, J. Electrical performance of 330-kV composite insulators with different shed configurations under icing conditions. *IEEE Trans. Dielectr. Electr. Insul.* **2015**, *22*, 3395–3404. [CrossRef]
53. Lan, S. *Study on the Anti-Icing Effect of Composite Insulator Coated with Photoelectric Mixed Coating*; Chongqing University: Chongqing, China, 2019; (In Chinese). [CrossRef]

54. Farzaneh, M.; Zhang, J.; Volat, C. Effect of insulator diameter on AC flashover voltage of an ice-covered insulator string. *IEEE Trans. Dielectr. Electr. Insul.* **2006**, *13*, 264–271. [CrossRef]
55. Huang, Y.; Jiang, X.; Virk, M. Ice accretion study of FXBW4-220 transmission line composite insulators and anti-icing geometry optimization. *Electr. Power Syst. Res.* **2021**, *194*, 107089. [CrossRef]
56. Hu, Q.; Wang, S.; Shu, L.; Jiang, X.; Qiu, G.; Li, H. Influence of shed configuration on icing characteristics and flashover performance of 220 kV composite insulators. *IEEE Trans. Dielectr. Electr. Insul.* **2016**, *23*, 319–330. [CrossRef]
57. Hu, Q.; Wang, S.; Shu, L.; Jiang, X.; Liang, J.; Qiu, G. Comparison of AC icing flashover performances of 220 kV composite insulators with different shed configurations. *IEEE Trans. Dielectr. Electr. Insul.* **2016**, *23*, 995–1004. [CrossRef]
58. Zhang, Z.; Jiang, X.; Sun, C.; Hu, J.; Huang, H.; Gao, D. Influence of insulator string positioning on AC icing flashover performance. *IEEE Trans. Dielectr. Electr. Insul.* **2012**, *19*, 1335–1343. [CrossRef]
59. Jiang, X.; Zhao, S.; Hu, J.; Zhang, Z.; Shu, L. Study of DC flashover performance of ice-covered insulators at high altitude. *IEEE Trans. Dielectr. Electr. Insul.* **2013**, *20*, 391–400. [CrossRef]
60. Xu, J.; Yin, F.; Li, L.; Wen, Q.; Wang, H.; Liu, S.; Jia, Z.; Farzaneh, M. Wet Snow Flashover Characteristics of 500-kV AC Insulator Strings with Different Arrangements. *Appl. Sci.* **2019**, *9*, 930. [CrossRef]
61. Huang, Y.; Jiang, X.; Virk, M. Study of inverted T-shape insulator strings in icing conditions. *Cold Reg. Sci. Technol.* **2020**, *173*, 103021. [CrossRef]
62. Jiang, X.; Yi, H. *Transmission Line Icing and Protection*; China Electric Power Publisher: Beijing, China, 2002. (In Chinese)
63. Lu, J.; Guo, J.; Jian, Z.; Yang, Y.; Tang, W. Dynamic Assessment of Resilience of Power Transmission Systems in Ice Disasters. In Proceedings of the 2018 International Conference on Power System Technology (POWERCON), Guangzhou, China, 6–8 November 2018; pp. 7–13. [CrossRef]
64. Runhua, W. Analysis of the accident based on low temperature, snowstorm, ice, coagulation. In Proceedings of the 2008 China International Conference on Electricity Distribution, Guangzhou, China, 10–13 December 2008; pp. 1–7. [CrossRef]
65. Kunpeng, J.; Jingshan, H.; Jialun, Y.; Bin, Z.; Jian, W.; Bin, L.; Xueping, Z.; Lichun, Z. Dynamic analysis and suppression method of bundled conductors following ice shedding. In Proceedings of the 2021 IEEE Sustainable Power and Energy Conference (iSPEC), Nanjing, China, 23–25 December 2021; pp. 1596–1601. [CrossRef]
66. Hu, J.; Yan, B.; Zhou, S.; Zhang, H.; Li, H.; Miao, C. Parameter Study on Galloping of Iced Bundled Conductors. In Proceedings of the 2010 Asia-Pacific Power and Energy Engineering Conference, Chengdu, China, 28–31 March 2010; pp. 1–4. [CrossRef]
67. Lu, J.; Wang, Q.; Wang, L.; Yang, L.; Xu, X.; Li, L. Research on galloping and insulation characteristics of 500kV crescent-shaped iced quad bundle conductor. In Proceedings of the 2018 12th International Conference on the Properties and Applications of Dielectric Materials (ICPADM), Xi'an, China, 20–24 May 2018; pp. 941–947. [CrossRef]
68. Wu, D.; Cao, H.; Li, D.; Yang, S. Energy-Efficient Reconstruction Method for Transmission Lines Galloping with Conditional Generative Adversarial Network. *IEEE Access* **2020**, *8*, 17310–17319. [CrossRef]
69. Akagi, Y.; Koyama, S.; Ohta, H.; Nishizawa, H.; Nagata, Y.; Oka, T. Development of anti-galloping device for UHV transmission line. *IEEE/PES Transm. Distrib. Conf. Exhib.* **2002**, *3*, 2158–2161. [CrossRef]
70. Huang, Q.; Tan, Y.; Mao, X.; Zhu, S.; Zhou, X. Study on Melting Ice Current Characteristics under Low Temperature and High Wind Speed. In Proceedings of the 2021 IEEE 5th Conference on Energy Internet and Energy System Integration (EI2), Taiyuan, China, 22–24 October 2021; pp. 4382–4385. [CrossRef]
71. Cigre, T. Influence of ice and snow on the flashover performance of outdoor insulators, part I: Effects of snow. *Electra* **2000**, *188*, 55–69.
72. Lu, J.; Zeng, M.; Zeng, X.; Fang, Z.; Yuan, J. Analysis of Ice-Covering Characteristics of China Hunan Power Grid. *IEEE Trans. Ind. Appl.* **2015**, *51*, 1997–2002. [CrossRef]
73. Juan, W.; Chuang, F.; Yiping, C.; Hong, R.; Shukai, X.; Tao, Y.; Licheng, L. Research and Application of DC De-Icing Technology in China Southern Power Grid. *IEEE Trans. Power Deliv.* **2012**, *27*, 1234–1242. [CrossRef]
74. Guo-te, L.; Yan-peng, H.; Yan, C. A comparative study of icing mechanism between soft rime and hard rime for the operating insulators. In Proceedings of the IEEE Annual Report Conference on Electrical Insulation and Dielectric Phenomena (CEIDP), Chenzhen, China, 20–23 October 2013; pp. 374–377.
75. Jiang, X.; Xiao, D. Analysis of the Physical Process of Conductor Surface Icing. In Proceedings of the 2010 Academic Conference of Chongqing Society of Electrical Engineering, Chongqing, China, 29 October 2010; p. 89. (In Chinese).
76. Zhang, Z.; Zhang, D.; Huang, H.; Jiang, X.; Gao, D. Study of the icing growth characteristic and its influencing factors for different types of insulators. *Turk. J. Electr. Eng. Comput. Sci.* **2016**, *24*, 63. [CrossRef]
77. Zhuang, W.; Zhang, L.; Yang, Y. Research on Ice Accumulation of Transmission Lines in Xinjiang. In Proceedings of the 2022 Power System and Green Energy Conference (PSGEC), Shanghai, China, 25–27 August 2022; pp. 1012–1015. [CrossRef]
78. Yiwei, X.; Yanpeng, H. Research of morphology of soft rime on suspension insulator for transmission line. In Proceedings of the 2013 Annual Report Conference on Electrical Insulation and Dielectric Phenomena, Chenzhen, China, 20–23 October 2013; pp. 319–322. [CrossRef]
79. Bin, F.L.G.; Rong, Z.F.; Gang, L.; Hong, Y.; Chao, Q.G.; Liangchi, S.; En, Y.; Xiang, L. A Research of Drawing and Application of Distribution Diagram of Yunnan Ice Region Based on the Typical Ice Model of Low Latitude Plateau Area. In Proceedings of the 2018 International Conference on Power System Technology (POWERCON), Guangzhou, China, 6–8 November 2018; pp. 3440–3447. [CrossRef]

80. Huang, W.; Hu, B.; Shahidehpour, M.; Sun, Y.; Sun, Q.; Yan, M.; Shao, C.; Xie, K. Preventive Scheduling for Reducing the Impact of Glaze Icing on Transmission Lines. *IEEE Trans. Power Syst.* **2021**, *37*, 1297–1310. [CrossRef]
81. Bin, F.; Lixing, Z.; Zhe, T. The Study on Factors Influencing the Ice-Melting Performance of Transmission Line. In Proceedings of the 2013 Fourth International Conference on Intelligent Systems Design and Engineering Applications, Zhangjiajie, China, 6–7 November 2013; pp. 710–713. [CrossRef]
82. Shen, Q.; Li, X. Method to Calculate the Critical Non-Icing Current on Transmission Line Considering the Impact of Humid Air Parameters. In Proceedings of the 2010 Asia-Pacific Power and Energy Engineering Conference, Chengdu, China, 28–31 March 2010; pp. 1–4. [CrossRef]
83. Gu, X.; Wang, H.; Liu, H.; Zhao, L.; Li, H. Expert System of Ice Prevention on Overhead Transmission Lines. In Proceedings of the 2010 International Conference on Intelligent Computation Technology and Automation, Changsha, China, 11–12 May 2010; pp. 273–276. [CrossRef]
84. Jiang, X. *Research on Icing Mechanism of Transmission Lines and Icing Law and Influencing Factors in the Three Gorges Area*; Chongqing University: Chongqing, China, 1997. (In Chinese)
85. Yunnan Electric Power Design Institute; Yunnan Institute of Meteorological Science. *Study on Icing of Power Lines in High Altitude Area of Yunnan*; Yunnan Science and Technology Publisher: Yunnan, China, 1983. (In Chinese)
86. Xingliang, J.; Quanlin, W.; Zhijin, Z.; Lichun, S.; Jianlin, H.; Qin, H.; Yang, P.; Yi, C. Estimation of rime icing weight on composite insulator and analysis of shed configuration. *IET Gener. Transm. Distrib.* **2018**, *12*, 650–660. [CrossRef]
87. Ren, X.; Jiang, X.; Li, Z.; She, Q. Effect of Diameter and Height on Ice Thickness of Cylinder. In Proceedings of the 2019 IEEE 3rd International Electrical and Energy Conference (CIEEC), Beijing, China, 7–9 September 2019; pp. 1343–1347. [CrossRef]
88. Farzaneh, M.; Laforte, J. Ice Accretion on Conductors Energized by AC or DC-Laboratory Investigation of Icing treeing. *Int. J. Offshore Polar Eng.* **1994**, *4*, ISOPE-94-04-1-040.
89. Farzaneh, M.; Drapeau, J. AC Flashover Performance of Insulators Covered with Artificial Ice. *IEEE Transm. Power Deliv.* **1995**, *10*, 1038–1051. [CrossRef]
90. Yang, Z.; Jiang, X.; Huang, Y.; Hu, J.; Han, X. Influence of electric field on the ice-coating process of insulators with a different dielectric surface. *IET Sci. Meas. Technol.* **2020**, *14*, 585–592. [CrossRef]
91. Hong, Z.; He, K.; Xu, Y.; Fang, H.; Zuo, Q.; Li, Z. Design and Research on Impact Deicing Mechanism of Cable Climbing Robot. In Proceedings of the 2021 International Conference on Computer, Control and Robotics (ICCCR), Shanghai, China, 8–10 January 2021; pp. 70–74. [CrossRef]
92. Huang, X.; Wei, X. A new on-line monitoring technology of transmission line conductor icing. In Proceedings of the 2012 IEEE International Conference on Condition Monitoring and Diagnosis, Bali, Indonesia, 23–27 September 2012; pp. 581–585. [CrossRef]
93. Yunqing, B.; Kongjun, Z.; Ke, Z. The Research of DC Deicing Technology in Power Line. In Proceedings of the 2006 International Conference on Power System Technology, Chongqing, China, 22–26 October 2006; pp. 1–7. [CrossRef]
94. Li, R.; Zhao, W.; Bai, T.; Zhang, Q.; Zhou, F.; Pu, Y. Influence of harmonics of DC ice melting device on MOA leakage current. In Proceedings of the 2020 4th International Conference on HVDC (HVDC), Xi'an, China, 6–9 November2020; pp. 468–471. [CrossRef]
95. Jianming, W.; Chuandong, L.; Qi, W.; Xiaojie, C.; Jian, Z. Research on operation control measure of UHVDC line and its nearby power grid. In Proceedings of the 2016 IEEE International Conference on Power System Technology (POWERCON), Wollongong, NSW, Australia, 28 September–1 October 2016; pp. 1–4. [CrossRef]
96. China Southern Power Grid Corporation. *Technology and Application of Anti-icing and Melting in Power Grid*; China Electric Power Publisher: Beijing, China, 2010. (In Chinese)
97. Li, Y.; Zhang, X.; Gao, S.; Tang, X.; Li, P.; Li, Y.; Guo, Y.; Wu, G. Effect of Water Droplets on the Corona Discharge Characteristics of Composite Insulators in Arid Areas. In Proceedings of the 2019 2nd International Conference on Electrical Materials and Power Equipment (ICEMPE), Guangzhou, China, 7–10 April 2019; pp. 467–472. [CrossRef]
98. Arshad; Momen, G.; Farzaneh, M.; Nekahi, A. Properties and applications of superhydrophobic coatings in high voltage outdoor insulation: A review. *IEEE Trans. Dielectr. Electr. Insul.* **2017**, *24*, 3630–3646. [CrossRef]
99. Kako, T.; Nakajima, A.; Kato, Z.; Uematsu, K.; Watanabe, T.; Hashimoto, K. Adhesion and Sliding of Snow on Hydrophobic Solid Surface. *J. Ceram. Soc. Jan.* **2002**, *110*, 186–192. [CrossRef]
100. Feng, L.; Li, S.; Li, Y.; Li, H.; Zhang, L.; Zhai, J.; Song, Y.; Liu, B.; Jiang, L.; Zhu, D. Super-hydrophobic Surfaces: From Natural to Artificial. *Adv. Mater.* **2002**, *14*, 1857–1860. [CrossRef]
101. Arianpour, F.; Farzaneh, M.; Kulinich, S. Ice Adhesion and Hydrophobic Properties of Coatings Based on Doped RTV Silicone Rubber. In Proceedings of the IWAIS XIII, Andermatt, Switzerland, 8–11 June 2009. Section 6.
102. Tadanaga, K.; Katata, N.; Minami, T. Super-Water-Repellent A12O3 Coating Films with High Transparency. *J. Am. Soc.* **1997**, *80*, 1040–1042.
103. Wei, X.; Jia, Z.; Sun, Z.; Guan, Z.; Macalpine, M. Development of anti-icing coatings applied to insulators in China. *IEEE Electr. Insul. Mag.* **2014**, *30*, 42–50. [CrossRef]
104. Momen, G.; Farzaneh, M. Study of ice accumulation on nanocomposite semiconducting coatings. In Proceedings of the 2010 Annual Report Conference on Electrical Insulation and Dielectic Phenomena, West Lafayette, IN, USA, 17–20 October 2010; pp. 1–4. [CrossRef]
105. Xu, Z.; Jia, Z.; Li, Z.; Wei, X.; Guan, Z.; Macalpine, M.; Zhao, Y.; Li, Y. Anti-icing performance of RTV coatings on porcelain insulators by controlling the leakage current. *IEEE Trans. Dielectr. Electr. Insul.* **2011**, *18*, 760–766. [CrossRef]

106. Wei, X.; Jia, Z.; Sun, Z.; Liao, W.; Qin, Y.; Guan, Z.; Xu, Z.; Peng, X. Study of anti-icing performance of insulator strings bottom-coated with semiconductive silicone rubber coating. *IEEE Trans. Dielectr. Electr. Insul.* **2012**, *19*, 2063–2072. [CrossRef]
107. Cherney, E. Flashover performance of artificially contaminated and iced long-rod transmission. *IEEE Trans. Power Appar. Syst.* **1980**, *99*, 46–52. [CrossRef]
108. Yang, Q.; Sima, W.; Deng, J.; Sun, C.; Hu, J. Shed Configuration Optimization for Ice-Covered Extra High Voltage Composite Insulators. *J. Adhes. Sci. Technol.* **2012**, *26*, 575–591. [CrossRef]
109. Hu, Q.; Shu, L.; Jiang, X.; Sun, C.; Zhang, Z.; Hu, J. Effects of shed configuration on AC flashover performance of ice-covered composite long-rod insulators. *IEEE Trans. Dielectr. Electr. Insul.* **2012**, *19*, 200–208. [CrossRef]
110. Lu, J.; Hu, J.; Fang, Z.; Qiao, X.; Zhang, Z. Electric Field Distribution and AC Breakdown Characteristics of Polluted Novel Lightning Protection Insulator under Icing Conditions. *Energies* **2021**, *14*, 7493. [CrossRef]
111. Li, P.; Fan, J.; Li, W.; Su, Z.; Zhou, J. Flashover performance of HVDC iced insulator strings. *IEEE Trans. Dielectr. Electr. Insul.* **2007**, *14*, 1334–1338. [CrossRef]
112. Ale-Emran, S.; Farzaneh, M. Experimental Design of Booster-Shed Parameters for Post Insulators under Heavy Icing Conditions. *IEEE Trans. Power Deliv.* **2015**, *30*, 488–496. [CrossRef]
113. Hu, J.; Sun, C.; Jiang, X.; Zhang, Z.; Shu, L. Flashover performance of pre-contaminated and ice-covered composite insulators to be used in 1 000 kV UHV AC transmission lines. *IEEE Trans. Dielectr. Electr. Insul.* **2007**, *14*, 1347–1356. [CrossRef]
114. Jiang, X.; Dong, B.; Zhang, Z.; Yin, F.; Shu, L. Effect of shed configuration on DC flashover performance of ice-covered 110 kV composite insulators. *IEEE Trans. Dielectr. Electr. Insul.* **2013**, *20*, 699–705. [CrossRef]
115. Yan, S.; Wan, Q.; Xu, Z. Experimental study on the flashover characteristics of 500 kV AC ice-covered insulator string under different arrangement. *High Volt. Eng.* **2011**, *37*, 3129–3134.
116. Porkar, B.; Farzaneh, M. DC Flashover Performance of an Inclined Insulator String Under Heavy Icing Conditions. In Proceedings of the 2015 IEEE 11th International Conference on the Properties and Applications of Dielectric Materials (ICPADM), Sydney, Australia, 19–22 July 2015.
117. Jiang, X.; Lu, J.; Du, Y.; Yuan, J. AC Ice Flash Characteristics of Insulator String with Interplug Arrangement. *Proc. CSEE* **2009**, *29*, 12–16. (In Chinese)

energies

Article

Improvement of the Electrical Performance of Outdoor Porcelain Insulators by Utilization of a Novel Nano-TiO_2 Coating for Application in Railway Electrification Systems

Pichai Muangpratoom [1,*], Issaraporn Khonchaiyaphum [2] and Wanwilai Vittayakorn [3]

1 High-Voltage Insulation Technology and Innovation Research Unit (HVRU-RMUTI), Department of Electrical Engineering, Faculty of Engineering, Rajamangala University of Technology Isan, Khon Kaen Campus, Khon Kaen 40000, Thailand
2 Department of Mathematics, Faculty of Engineering, Rajamangala University of Technology Isan, Khon Kaen Campus, Khon Kaen 40000, Thailand
3 Electroceramics Research Laboratory (ECRL), College of Materials Innovation and Technology, King Mongkut's Institute of Technology Ladkrabang, Bangkok 10520, Thailand
* Correspondence: pichai.mu@rmuti.ac.th

Abstract: The present study aimed to develop the electrical performance of outdoor insulators using a nano-TiO_2 coating for railway electrification systems. The prototype design of porcelain insulators with normal coatings and using a nano-TiO_2 coating is based on IEC 60815-1. The first test was performed to measure the low-frequency flashover AC voltage under both dry and wet conditions. In addition, the other test was conducted to measure the lightning impulse critical-flashover voltage at positive and negative polarity under dry-normal and wet-contaminated conditions. X-ray diffraction (X-RD) and Scanning electron microscopy (SEM) were used to examine the micro surface and show that the nano-TiO_2 coating was adhered to the surface of the outdoor porcelain insulator and exists in an amorphous state. Additionally, it was observed and discovered that scattered nano-TiO_2 strengthens the glassy matrix and creates a sturdy barrier that causes flashover voltage to be reduced under conditions of high dielectric strength. Nanostructured ceramic formulations outperform ordinary porcelain in terms of breakdown voltage strength, particularly for the insulators' low-frequency flashover performances under dry and wet test conditions. However, a significant change in the lightning impulse critical-flashover voltage characteristics is observed and is not much better when adding the nano-TiO_2 coating to the porcelain insulators.

Keywords: railway electrification system; outdoor porcelain insulators; low-frequency flashover AC voltage; lightning impulse critical-flashover voltage

Citation: Muangpratoom, P.; Khonchaiyaphum, I.; Vittayakorn, W. Improvement of the Electrical Performance of Outdoor Porcelain Insulators by Utilization of a Novel Nano-TiO_2 Coating for Application in Railway Electrification Systems. *Energies* **2023**, *16*, 561. https://doi.org/10.3390/en16010561

Academic Editors: Zhijin Zhang and Hualong Zheng

Received: 8 December 2022
Revised: 23 December 2022
Accepted: 28 December 2022
Published: 3 January 2023

1. Introduction

An important component of any electrical system is insulation. This includes gas insulation, such as SF6, and liquid insulation, such as transformer oil or mineral oil, and solid insulation in transformers or cables, such as kraft paper or pressboard. In addition, there are important insulators that are used on the outside, such as the high-voltage porcelain insulators on the outside of transformer bushings, cutouts, arresters, line posts, etc. [1]. To last for years, they must meet strict electrical, mechanical, and chemical requirements, such as not breaking down when exposed to UV light or when in a dirty environment [2–4]. Although polymeric insulators have just lately been available, porcelain and glass have been used as outside insulators in electric power systems for over 160 years [5–8]. In the past, triaxial porcelain compositions of alumina, silica, clay, and feldspar were used to create ceramic insulators [1,9]. Insulators are one of the essential elements of railway electrification systems; however, the popular viewpoint does not give this essential component much attention. Economic considerations must be made in addition to the strict criteria for availability and dependability, which is why this topic is given a lot of attention [10]. The dependability of the performance of insulators, which are widely used in transmission power systems, is primarily influenced by environmental pollution. Composite insulators have been used for transmission lines for a long time because they work well even in dirty environments [11]. As compared to other ceramic materials, outdoor porcelain insulators are the subject of little research today that deals with nanotechnology applications. Nevertheless, it has been said that they have produced some very intriguing and encouraging outcomes. Porcelain is a policrystalline ceramic created by combining and burning clays, feldspathic minerals, and silica or alumina, and may be characterised as a conventional ceramic formed by heating raw materials based on clay [6,12]. In 2010, Zhuang et al. [13] investigated a novel use of nanoscale titanium dioxide (TiO_2) for improving the performance of the porcelain insulator in contaminated settings. The researchers found that applying n-TiO_2 films to the porcelain insulator's surface maintains and even enhances the electrical characteristics of the AC wet flashover voltage (6%). To enhance the anti-icing performance of ceramic insulators utilised in China's colder regions, Gou et al. created three novel experimental formulations of polymeric nanocoatings [14]. In order to enhance the mechanical and electrical performance of siliceous porcelain insulators, Contreras investigated a novel nanotechnology concept in 2014 [1].

At present, the rail system is of interest in the transportation system of Thailand. Therefore, it is important to research and develop the design of the components used in the rail system. In particular, the insulation in the railway electrification systems is one of the details of the overhead line electrification system for the railway, as can be seen in Figure 1. In addition, Thailand's rail transport system uses an overhead power supply system to operate. Therefore, it is necessary to instal insulating materials, including outdoor porcelain insulators.There is very little research about the design and build of the porcelain insulator used in the electrification system for the railway in Thailand. Furthermore, most of these studies were about testing the durability of insulators, which included electrical tests.

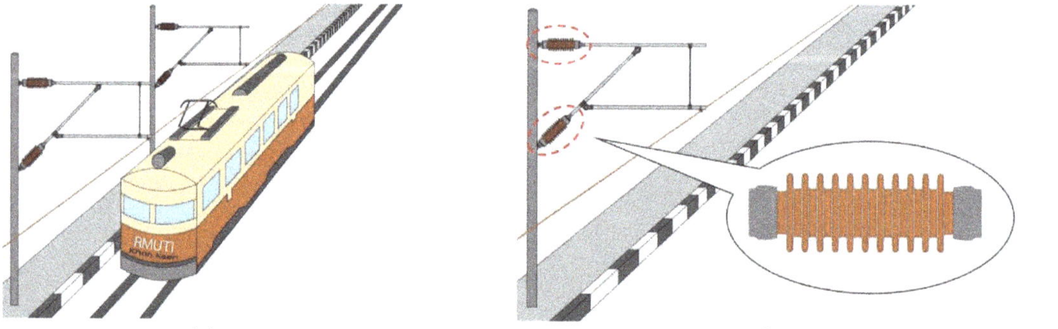

(a) (b)

Figure 1. Railway electrification system: (**a**) Overview of railway electrification system; (**b**) Design of overhead line insulation equipment as porcelain insulator.

This work was performed to improve the quality of the outdoor porcelain insulators used in the railway system and make it possible for electric trains to run on Thailand's railway system. The porcelain insulator used in this railway system is extremely important because there is no manufacturer in Thailand, as they are primarily interested in producing insulators for the alternating current distribution system. As a result, the goal of this research is to develop the performance of outdoor insulators used in the railway's overhead line electrification system. The performance-developed outdoor insulators of overhead line insulation equipment for railway electrification systems that use a nano-TiO_2 coating would be the quality insulator that passes the standard electrical properties test.

2. Materials and Methods

2.1. The Design for Outdoor Insulators on Railway Electrification Systems

The prototype design of porcelain insulators based on international standards and the outdoor insulators on railway electrification systems used in Japanese and Spanish railway systems was studied, which included the equations used in the design of the leakage distance of insulators used in rail systems. Furthermore, the reference to the design of porcelain insulators, the insulator profile parameters as defined in IEC 60815-1 [15], the design considers the level of pollution that is used. Moreover, this research will design insulators that can be used at pollution levels of 1 and 2 based on IEC 60815-1. To determine the leakage distance of insulators, it is necessary to calculate the profile factor (PF) and the creepage factor (CF), which are important in the design of the fin length (measured from the core) and the distance between the fins, which includes the creepage distance and arcing distance of the insulators. Meanwhile, the values calculated from Equations (1) and (2) are evaluated to determine the suitability of insulating insulators for the design or application area. Previously, the sizes were as shown in Figure 2.

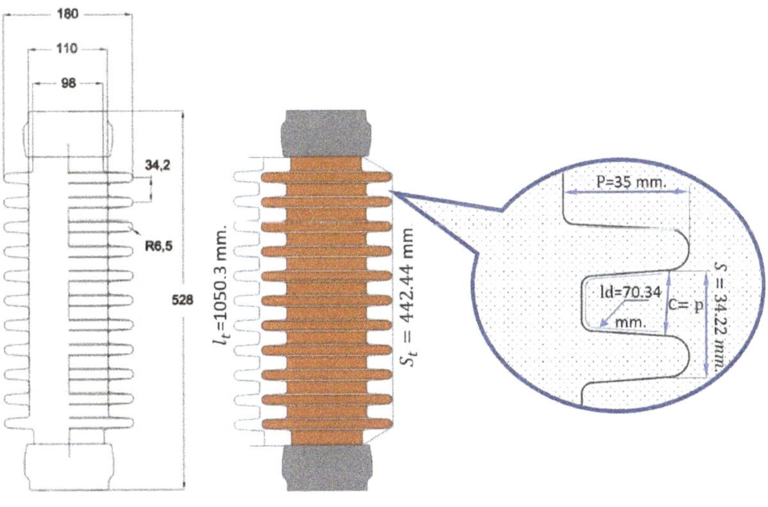

Figure 2. Schematic diagram of the main nanotechnology design concepts for outdoor insulators.

Profile Factor (P.F.)

$$P.F. = \frac{2P + S}{l_d} \tag{1}$$

when

P.F. > 0.8 for the pollution levels of 1 and 2 (light to medium pollution areas)
P.F. > 0.7 for the pollution levels of 3 and 4 (heavy to very heavy pollution areas)

Creepage Factor (C.F.)

$$C.F. = \frac{l_t}{S_t} \tag{2}$$

when

C.F. $\leq$ 3.5 for the level of pollution 1 and 2 (light to medium pollution areas)
C.F. $\leq$ 4.0 for the level of pollution 3 and 4 (heavy to very heavy pollution areas)

The outdoor insulator profile parameters in this work, as defined in IEC60815-1, are as follows:

P: shed projection—the maximum shed overhang (35 mm).

S: shed spacing—the vertical distance between two points that are similar on successive sheds. (34.22 mm).
l_d: Measured between the two places that constitute d is the creepage distance (70.34 mm).
l_t: the insulator's overall creepage distance (1050.30 mm).
S_t: the arcing distance of the insulator (442.44 mm).

The design variables were substituted into Equation (1) for P.F. and Equation (2) for C.F. So, the profile factor (P.F.) is equal to 1.482, which is more than 0.8 for the level of pollution in 1 and 2 (light to medium pollution areas). Moreover, the creepage factor (C.F.) is equal to 2.374, which is less than 3.5 for the level of pollution in 1 and 2 (light to medium pollution areas).

Therefore, the prototype outdoor insulators designed in this research were at pollution levels 1 and 2. It is standard practice on IEC60815.

2.2. Nanotechnology Concepts for Outdoor Insulators

This work presents new nanostructured porcelain composition coats for improving the electrical performance of outdoor porcelain insulators by using a novel application of nano-TiO_2 coatings, including evaluating the effect of nano-TiO_2 on the properties of conventional siliceous electrical porcelain for use in railway electrification systems. All the unprocessed nanomaterials employed in this investigation had an average particle size of ~40 nm. Additionally, they had a spherical form, and their analytical reagent quality was measured by TESCAN, Model: MIRA3, as seen in Figure 3.

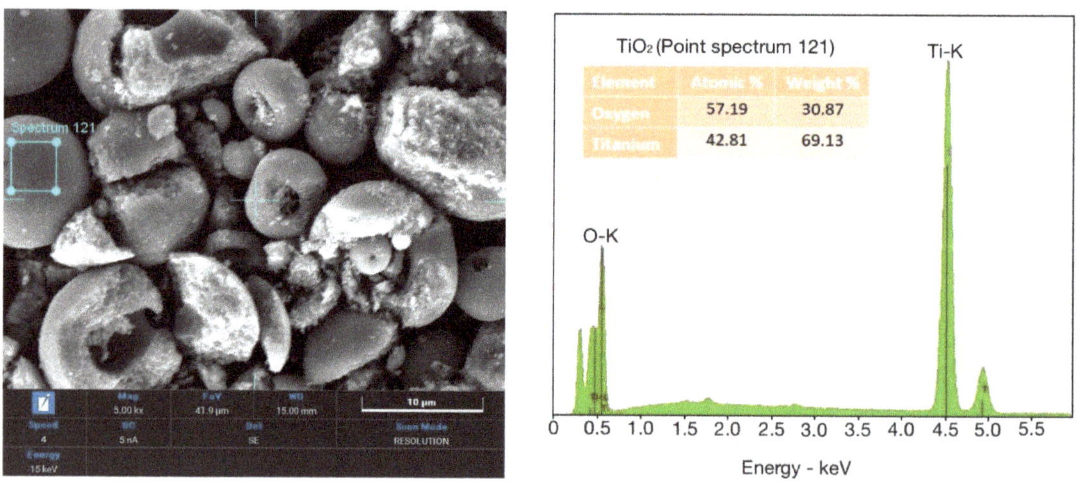

(a) (b)

Figure 3. SEM nano-TiO_2 are as follows: (**a**) The morphology of nano-TiO_2; (**b**) The electron dispersive X-ray spectroscopy (EDS) result of nano-TiO_2 used in this work.

The shape of nano-TiO_2 is also shown in Figure 3a, where a quasi-spherical form is often seen. There could be some nano-TiO_2 agglomerates present since no dispersant was utilised. Electron dispersive X-ray spectroscopy (EDS) was used to confirm the chemical composition of nano-TiO_2, which is shown in Figure 3b. The properties of the nanoparticles employed in this work are listed in Table 1 [16]. Detailed properties of nano-TiO_2 are specified by the manufacturer [16]. Titanium dioxide (titania) nanopowder, anatase. Specification: 99.9%; anatase; primary particle size ~40 nm (from SEM and BET surface area); BET multi-point specific surface area (SSA) >40 m^2/g; spray-dry agglomerated to ~50 microns for easy handling.

Table 1. The main properties of TiO_2 nanomaterial used to coat porcelain insulators are listed.

Characteristics	Specification
Parameters	Titanium dioxide (TiO_2)
Average particle size	~40 nm
Purity	99.9%
Specific surface area	40 m^2/g
Structure	anatase
Colour	white powder
Density	3.89 g/cm^3

The key ideas in this study for the outdoor porcelain insulators used in the railway system represent a novel approach to improving the final characteristics of outdoor high-voltage insulators. In addition, the porcelain raw materials utilised in this study were of industrial quality and were quartz, kaolinite clay, and sodium feldspar, which are all employed in the production of outdoor insulators according to the expertise of Asian Insulators Public Company Limited. Figure 4 shows that the main use of nanotechnology in this outdoor insulator by nano-TiO_2 coating is to improve efficiency, mostly by making the electrical performance better.

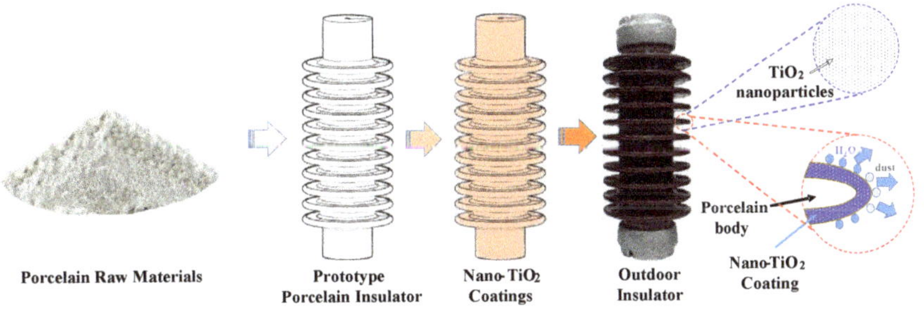

Figure 4. Schematic of the design concepts for nano-TiO_2 coatings of outdoor insulators in this study.

Figure 4 illustrates the design idea for the aforementioned procedure. The link between a material's resistivity, length, and the area is shown by the well-known Equation (3) [17].

$$R = \rho \frac{L}{A} \tag{3}$$

where R stands for resistance, ρ for resistivity, L for the distance that charges travel before leaving the material, and A for the cross-sectional area of the material that faces the direction of the current. The R will decrease if we shorten L, raise A, or decrease ρ, as shown by Equation (3).

For the purposes of this study, we will refer to ρ as the resistivity of the nano-TiO_2 coating, L as the thickness of the nano-TiO_2 film, and A as the surface area of the graphite felt used for the nano-TiO_2 film deposition. Thus, lowering the size and L of the nano-TiO_2 film is one of the development paths that might be explored in this study. In this work, a nano-TiO_2 coating was formed on the outside insulators to shorten L in order to decrease the resistance (R) of the coating. On the other hand, a two-step procedure was employed to create a nano-TiO_2 coating, which included first generating an amorphous film at a low temperature, and then increasing the temperature to crystallise the film. It is noted that a two-step technique may be used to create a nano-TiO_2 coating [18–20]. Since a large number of grain boundary defects in the nano-TiO_2 coating, the resistivity (ρ) of the nano-TiO_2 coating can be effectively reduced. Besides, the nano-TiO_2 coating structure can achieve low resistance.

2.3. Surface Modification of Outdoor Insulators

A thin nano-TiO_2 layer was applied to the surface of outdoor insulators via the development of an insulator nano-TiO_2 coating method for surface modification. The process includes six steps. The nanoparticles are mixed with deionized water. The nanoparticles must be weighed on a digital balance, and the particle content required is 0.1% wt of deionized water. TiO_2 nanoparticles (nano-TiO_2) combined with deionized water and a surfactant. The surfactants are organic compounds, which consist of two parts: the hydrophilic group and the hydrophobic group. Two different phases can come together. When a small amount of surfactant is added, two phases can be combined to form a single phase. In this research, a surfactant type, TERGITOL™ NP-6 Surfactant (Nonylphenol Ethoxylate) [21], which has properties as a non-ionic surfactant, was selected to be used as an ingredient in the preparation of nano-TiO_2 coatings for outdoor porcelain insulators. Then, it is placed in the stirrer for 30 min at 60 °C, or until it diffuses, and put in the high-frequency ultrasonicator at 60 °C for 1 h to ensure the homogeneity of the liquids, as shown in Figure 5. The step, as shown in Figure 5, begins with weighing the normal coatings and nano-TiO_2 coatings at the prescribed amount (a weight ratio of 2:1 between the normal coatings and nano-TiO_2 coatings). The next process involves mixing two substances, including measuring, and controlling the coating viscosity at 30–45 Cps. and waiting for the next process. After that, the nano-TiO_2 coatings are put into a chamber containing the coatings. The formed prototype insulator is then put on an automatic coating machine to process the coating. Lastly, the prototype insulator is put into a temperature-controlled kiln under control at 1300 °C for 60 h.

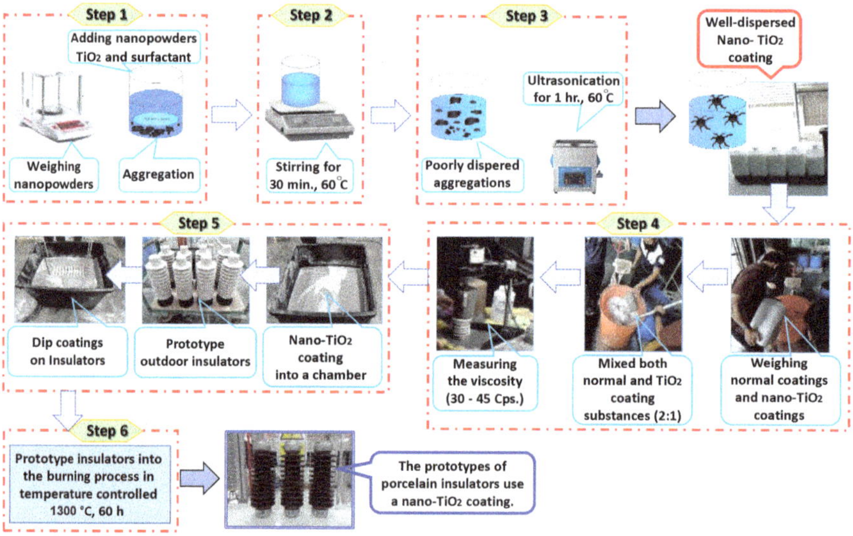

Figure 5. Preparation of nano-TiO_2 coating of outdoor insulators in this study.

3. Experimental Descriptions

3.1. Performances of Standard Porcelain Insulators in Low-Frequency Flashover in Dry and Wet Test Conditions

When referring to low frequency, as defined in these standards, anything between 15 and 100 cycles per second is considered to be low frequency. The low-frequency testing described in these standards may be performed at any commercial frequency that is currently in use within this range. In Figure 6, a clean, dry test specimen voltage application's root-mean-square (rms) voltage that results in a prolonged disruptive discharge through the air between electrodes is known as a low-frequency dry-flashover voltage. Approximately

75% of the anticipated average dry-flashover value may be rapidly reached by rapidly increasing the first applied voltage. The rate of voltage rise must remain constant after achieving 75% of the flashover value in order for there to be no time delay of more than 30 s or less than 5 s before the flashover occurs. An arithmetic mean, given in rms volts, of at least five different flashovers collected in a row constitutes the dry-flashover voltage value for a specimen. There must be between 15 s and 5 min of time between each flashover.

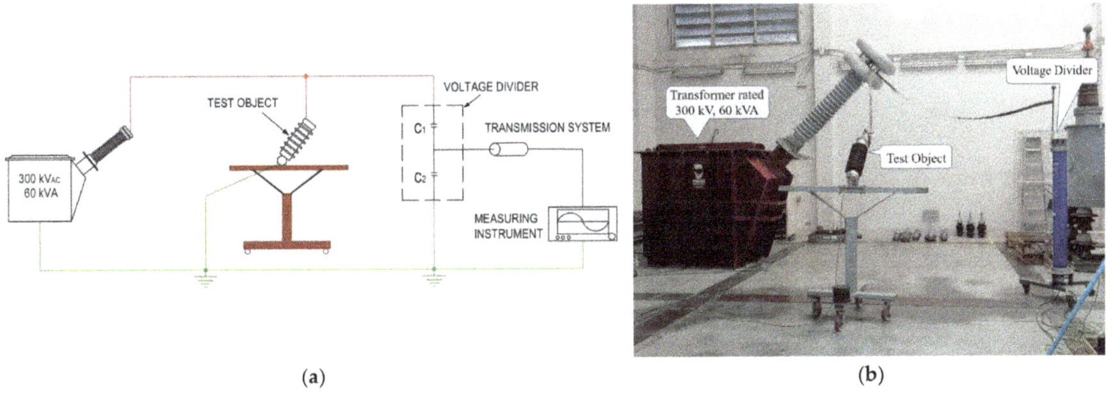

Figure 6. The experimental circuit of low-frequency flashover AC voltage under dry test. (**a**) The circuit diagram on low-frequency dry-flashover test; (**b**) The test circuit set-up on low-frequency dry-flashover test.

The low-frequency wet-flashover voltage is the root-mean-square (rms) voltage that causes a long, disruptive discharge between the electrodes of a clean, wet test specimen. The area where the insulator is being tested must have enough spray nozzles of the kind illustrated in Figure 7 or an equivalent supply of artificial precipitation that is fairly uniform. Within the limits of what is practically possible, the spray must be directed such that it strikes the insulator downward at a 45-degree angle from the vertical and parallel to the vertical plane across its axis. Large specimens, such as the outdoor porcelain insulators, must have spray that is 45 degrees off vertical in the middle and as near to this angle at the ends as feasible to provide adequate precipitation uniformity over the length of the specimen. At the specimen, the standard average rate of precipitation must be 0.2 inches (5.08 millimetres) per minute when measured. The low-frequency flashover AC voltage testing apparatus is shown in Figures 6 and 7 for both dry and wet experiments, respectively. Phenix Technologies' variants for the frequency output device are rated at 300 kV, 60 kVA, respectively, based on ANSI/NEMA C29.7-2015 [22], TIS 2623-2560 (2017) [23], and IEEE-4 (2013) [24].

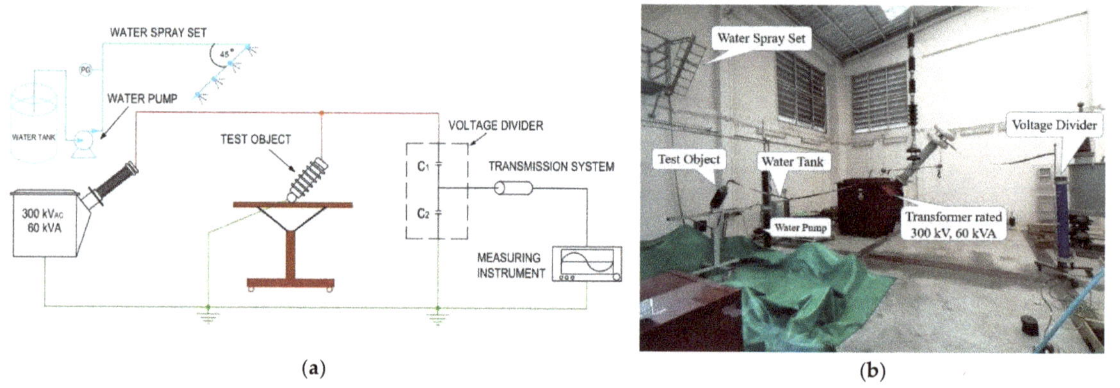

Figure 7. The experimental circuit of low-frequency flashover AC voltage under wet test. (**a**) The circuit diagram on low-frequency wet-flashover test; (**b**) The test circuit set-up on low-frequency wet-flashover test.

3.2. The Lightning Impulse Critical-Flashover Voltage Performances of Standard Porcelain Insulators under Normal and Contaminated Test Conditions

A test specimen's impulse critical-flashover voltage is equal to the wave's peak value in 50% of applications. A lightning impulse critical-flashover voltage test in polluted circumstances was used in this study to assess the performance of prototype outdoor porcelain insulators. Normally, impulse-flashover experiments are only carried out in dry environments. There must be 1.2/50 µs of time between each impulse voltage wave. Applying impulses to the specimen of varied intensities and monitoring the peak voltage of the impulse that, in about 50% of applications, will result in flashover allows for the determination of the critical impulse flashover voltage value. Impulses must be applied in stages of rising or decreasing voltage such that at the step with the greatest voltage, flashover occurs on every impulse, and at the step with the lowest voltage, there is no flashover. At least three impulses must be applied at each voltage step, and the difference in voltage between steps must not be more than 5% of the anticipated critical impulse flashover value. When the voltage range between the highest and lowest step is quite large, it is possible to calculate the crucial impulse flashover voltage by plotting an average curve of crest voltage vs flashover frequency.

The experimental setup for testing the critical-flashover voltage for a lightning impulse in both dry and polluted conditions is shown in Figures 8 and 9. The impulse voltage generator, whose rated voltage is 400 kV and 40 kJ under both normal and contaminated circumstances based on TIS 1077-2535 (1992) [25], TIS 2623-2560 (2017), and IEC 60507 (2013) [26], is used to acquire test power. The primary capacitor is charged to the desired voltage. The circuit is then automatically initiated by the spark gap to provide a positive polarity lightning discharge voltage of 1.2/50 µs. The experiment uses the up-and-down test methodology. A test was conducted in order to determine how effectively conventional porcelain insulators can tolerate impulses at both positive and negative polarities and in unclean circumstances. Furthermore, the surface of the specimens is polluted using test techniques on pollution flashover the suspension of sodium chloride and kieselguhr, which, respectively, imitate electric and inert materials. Based on the results of [27–29], the relationship between the critical flashover voltage representations of the electrical withstand value and the equivalent salt deposit density (ESDD) values is recommended as a contamination severity criterion for each case, as shown in Equation (4):

$$E_s = \lambda(ESDD)^{-\mu} \tag{4}$$

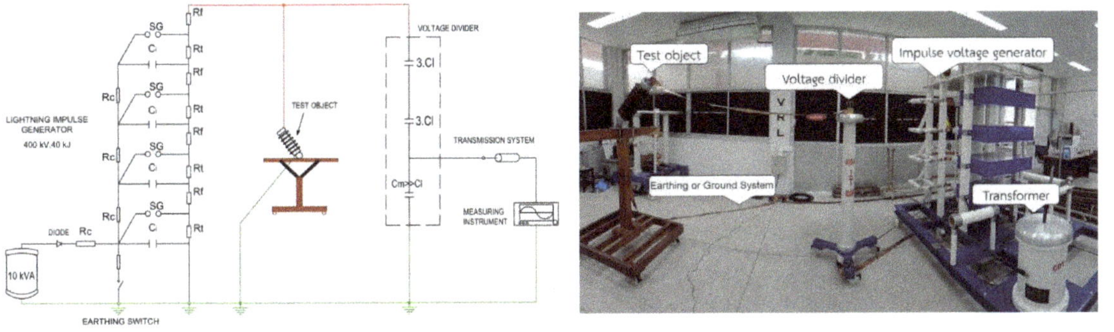

(**a**) (**b**)

Figure 8. The experimental circuit of the lightning impulse critical-flashover voltage under normal condition test.(**a**) The circuit diagram on the lightning impulse critical-flashover voltage test; (**b**) The test circuit set-up on the lightning impulse critical-flashover voltage test.

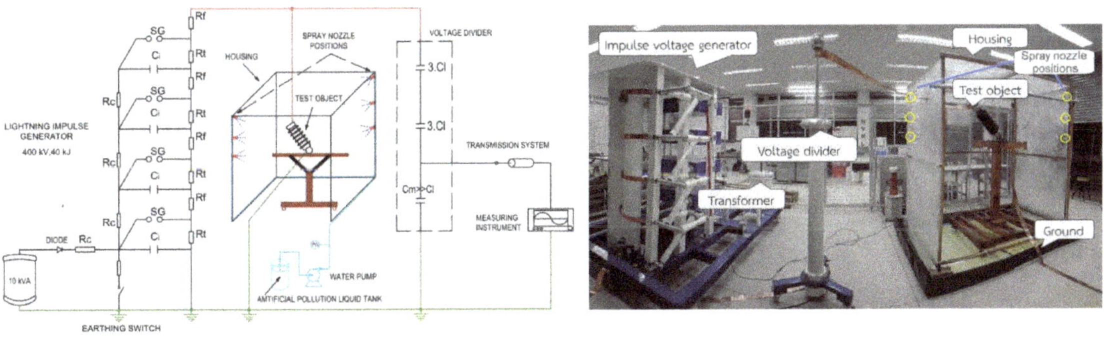

(**a**) (**b**)

Figure 9. The experimental circuit of the lightning impulse critical-flashover voltage under contaminated condition test. (**a**) The circuit diagram on the lightning impulse critical-flashover voltage test; (**b**) The test circuit set-up on the lightning impulse critical-flashover voltage test.

When the electrical voltage is withstood (Es), which is measured using experimental data from research and curve fitting, it is written as kV/cm and is related to the material and shape of the insulators. This research provides samples of contaminated liquid for testing under pollution conditions. By using the data for the contaminated liquid samples, values of ESDD can be calculated based on the procedure described in IEC 60507. The ESDD value obtained in this work is 0.4 mg/cm^2.

For all experiments, the salt deposit density (SDD)/equivalent salt deposit density ratio is 1:6. At the top of the artificial climate chamber, the test specimen is mounted on a solid base made of non-conductive wood. Spray nozzles are used to spray the test specimen. A spray pipe that is located not far below the surface of the body of the housing carries the spray, and the spray mist input rate is (0.05 ± 0.01) kg/(h.m^3) based on IEEE 4 (2013), and the fog chamber's humidity is almost saturated to a wet condition [30]. Through exhausting and misting water, the mist chamber is cooled to match the ambient temperature. Once the insulator's surface is suitably moist, the sample is left to stand for 15 to 20 min before the test is begun [26].

4. Results and Discussion

4.1. The Dielectric Strength

4.1.1. Low-Frequency Flashover Performance in Dry and Wet Testing Conditions

The influence of nano-TiO_2 on the low-frequency flashover AC voltage under dry and wet tests of the prototype outdoor porcelain insulator sample is shown in Table 2. It was discovered that adding nano-TiO_2 encouraged an improvement in the dielectric capacity of regular porcelain.

Table 2. Test results of low-frequency flashover performances under dry and wet test condition (kV).

Test Conditions	Insulator Types	1st	2nd	3rd	4th	5th	Average	SD
	Standard valve of condition test, kV (125 kV, based on ANSI/NEMA C29.7-2015)							
Dry	Normal coating	174.33	172.44	171.49	172.44	173.38	172.81	1.07
	Nano-TiO_2 coating	176.21	174.30	175.26	176.21	174.30	174.28	0.95
	Standard valve of condition test, kV (95 kV, based on ANSI/NEMA C29.7-2015)							
Wet	Normal coating	156.79	158.81	157.80	159.06	157.80	158.05	0.90
	Nano-TiO_2 coating	159.90	160.41	158.59	159.40	160.42	159.74	0.77

SD = Standard deviation of the data on low-frequency flashover performances tests.

Figure 10a depicts dry test results with the nano-TiO_2 concentration used in this study (0.1 wt.%), with the breakdown voltage reaching an average of 174.28 kV at 0.1 wt.% nano-TiO_2 addition and 172.81 kV at 0 wt.% nano-TiO_2 addition (uncoated nano-TiO_2). Using a new way to use nano-TiO_2 coatings on samples made the electrical performance of outdoor porcelain insulators 0.85% better than with regular porcelain.The effect of the nano-TiO_2 on the low-frequency flashover AC voltage under wet tests of the porcelain insulator sample is depicted in Figure 10b. It was found that the addition of nano-TiO_2 impacts the electrical performance of the porcelain insulator. When nanostructured porcelain with a nano-TiO_2 coating was compared to regular porcelain (without a nano-TiO_2-coating), it was better by 1.06%. Due to the inclusion of nanoparticles, the sintered microstructure has changed, which may be the cause of this behavior. According to [31,32], a larger concentration of the vitreous phase has a negative impact on the dielectric breakdown because the crystalline/glassy phase ratio affects the electric characteristics of porcelain systems. Due to a lower proportion of glassy phase than normal porcelain, nanostructured porcelain exhibits greater breakdown voltage values. This behaviour may be made possible by the optimization of the electrical design of insulators. Without a doubt, adding nano-TiO_2 to the prototype outdoor porcelain insulator makes its electrical properties better.

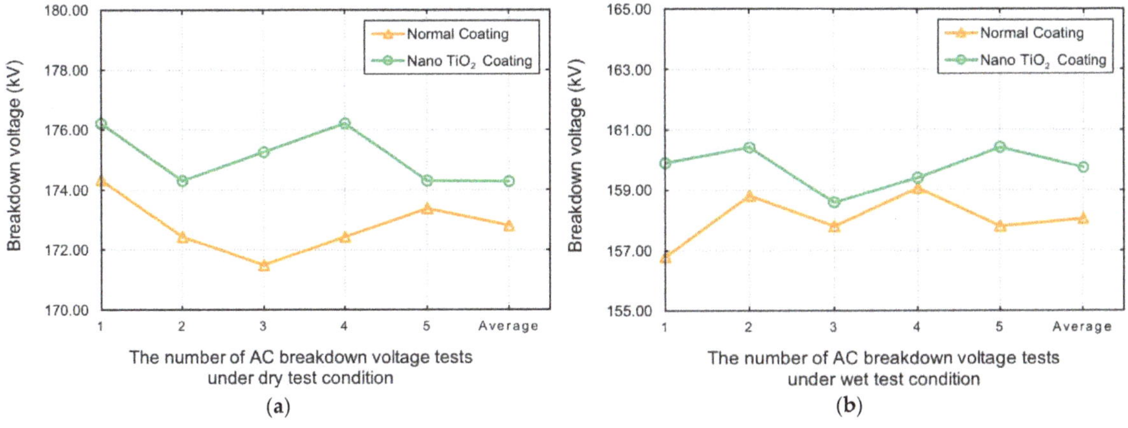

Figure 10. Test results of low-frequency flashover performances. (**a**) The low-frequency flashover performances under dry test condition; (**b**) The low-frequency flashover performances under wet test condition.

4.1.2. The Lightning Impulse Critical-Flashover Voltage Performance under Normal and Contaminated Test Conditions

A comparison of the critical impulse flashover voltage is illustrated in Table 3. In Figure 11, under normal test conditions, both positive and negative lightning impulses, the nano-TiO_2 coated insulators had a slight increase in dielectric strength compared to traditional coated insulators. In the same way, Figure 12 shows the critical impulse flashover voltage value under contaminated test conditions. When the nano-TiO_2 coating was added to the porcelain insulators, there was only a slight improvement in the changes under positive polarity testing, and in the negative polarity testing, there was no significant change in the characteristics of the lightning impulse critical-flashover voltage. The electrical performance test results are less than 2%, according to the results. It may be because the nano-TiO2 content is so low (only 0.1 wt.%) that it does not show a clear effect on the electrical properties of the coated insulators with improved coating as much as it should. In the future, we may add more nano-TiO2 to show higher electrical values. Additionally, it is well known that under both positive and negative lightning impulses, the distribution of the electric field on the insulator's surface is not uniform. The accumulation and growth of space charges would greatly distort the electric field across the air gap and reduce the flashover voltage [13,33]. Undoubtedly, one of the most effective ways to reduce the likelihood that charges will accumulate on the surface of an insulator is to decrease the surface resistivity [34]. Insulators are coated with nano-TiO2 to make the insulator surface diffuse the electric field and thus increase the lightning impulse voltage. Contrarily, the surface conductance of the TiO_2-coated insulator is higher than that of the uncoated one, which might lead to an increase in leakage current and a decrease in flashover voltage. During testing, the U50% stays pretty much the same because these two opposing effects often cancel each other out [13].

Table 3. Test results of critical impulse flashover voltage value under normal and contaminated test conditions.

Test Conditions		Insulator Types	1st	2nd	3rd	4th	5th	Average	SD
Normal (dry)	LI Positive	Test condition of critical impulse flashover voltage value, kV (210 kV, based on TIS. 1077-1992)							
		Normal coating	210.6	211.2	211.3	211.2	211.3	211.12	0.27
		Nano-TiO_2 coating	210.9	211.5	211.3	211.3	211.8	211.36	0.22
	LI Negative	Test condition of critical impulse flashover voltage value, kV (260 kV, based on TIS. 1077-1992)							
		Normal coating	260.6	260.7	261.3	260.7	260.6	260.78	0.29
		Nano-TiO_2 coating	262.5	262.4	262.5	261.9	262.6	262.30	0.27
Contaminated (wet)	LI Positive	Comparison of critical-flashover voltage performance, kV							
		Normal coating	147.1	148.2	147.4	149.2	148.4	148.06	0.83
		Nano-TiO_2 coating	149.4	148.3	148.4	149.4	148.5	148.80	0.55
	LI Negative	Comparison of critical-flashover voltage performance, kV							
		Normal coating	185.5	184.9	185.6	185.2	183.7	184.98	0.76
		Nano-TiO_2 coating	185.4	184.0	183.7	183.9	183.6	184.12	0.73

SD = Standard deviation of the data on lightning impulse (LI) tests.

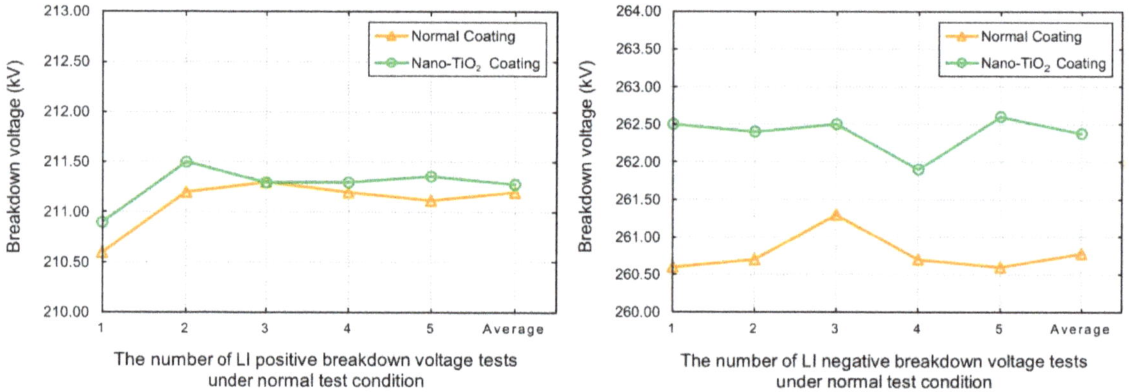

(**a**) (**b**)

Figure 11. Test results of critical impulse flashover voltage value under normal test conditions are as follows. (**a**) The lightning impulse critical-flashover voltage performance under positive polarity testing; (**b**) The lightning impulse critical-flashover voltage performance under negative polarity testing.

4.2. Phase Analysis

The prototype scale study's comparative XRD analysis of traditional porcelain systems and nanostructured porcelain systems in the sintered condition. Figure 13 shows the XRD patterns of prototype porcelain insulators with conventional coatings and nano-TiO_2 coatings measured by the Bruker D2 phaser at 30 kV, 10 mA. The results of this study revealed that the primary crystalline phases in the porcelain insulator prototype composition were quartz (SiO_2) and mullite ($3Al_2O_3 \bullet 2SiO_2$). The primary crystalline phases of the nanostructured porcelain's diffraction pattern remained the same, and corundum, an unaltered form of alumina, was also present. Figure 13a shows alumina peaks in formulations with a high proportion of traditional porcelain insulators. Furthermore, as illustrated in Figure 13b, adding a TiO_2 layer to the surface of outdoor insulators improves mullite peak intensity. Additionally, the fraction of crystalline phases rise with the inclusion of nanoparticles. It is important to note that the phase concentrations of all the samples match the published values for porcelain insulators [6,35,36]. The data for the prototype porcelain insulators with regular coatings and with a nano-TiO_2 coating were refined using the Rietveld method to produce crystallographic data, which was then used to further characterise the XRD findings. The lattice parameters, atomic coordinates, and atomic occupancies of manufactured materials may all be examined using the Rietveld refinement method, are displayed in Figure 14.

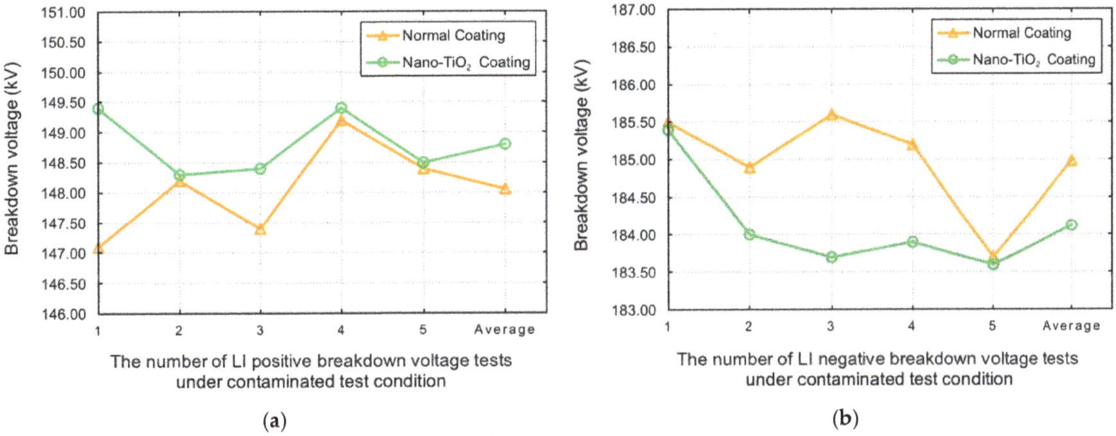

Figure 12. Test results of critical impulse flashover voltage value under contaminated test conditions are as follows. (**a**) The lightning impulse critical-flashover voltage performance under positive polarity testing; (**b**) The lightning impulse critical-flashover voltage performance under negative polarity testing.

The Rietveld refinement was carried out on porcelain insulators with normal coatings, and a nano-TiO_2 coating was used to analyse the crystal structure of the porcelain materials. Figure 14 shows the Rietveld refinement plot of the triaxial ceramic fired at 1300 °C. In this work, the observed XRD patterns and the structural model showed convergence with good weight profile R values (Rwp). The R values, Rietveld refinement, Residual values, or Rietveld discrepancy indices are all names for the numerical approach of evaluating the quality or goodness of fit, which is often expressed in terms of agreement indices [37]. The weighted profile R values are the easiest to understand because they come straight from the square root of the minimum amount, which is scaled using weighted intensities. Typical values of Rwp range from a few per cent for very accurate refinements to 20–30% for X-ray refinements, depending in part on the counting times used, the degree of preferred orientation, and the number of variable parameters [38,39]. The refinements were adequate. Indeed, the Rwp values were, in all cases, between 10 and 30%. Furthermore, in all samples, the estimated standard deviation of weight per cent derived from the estimated standard deviations of individual scale factors for the respective phases was below 1.5%. In particular, when considering the comparisons from Figures 14a and 14b, the graphs show the percentage of the substance content on all the samples of insulating coating. Moreover, it can be seen that when nano-TiO_2 coatings are applied to insulators, they result in higher percentages of mullite, and sillimanite compared to insulators using traditional coatings. The results obtained from Rietveld refinement are consistent with the XRD results shown in Figure 13.

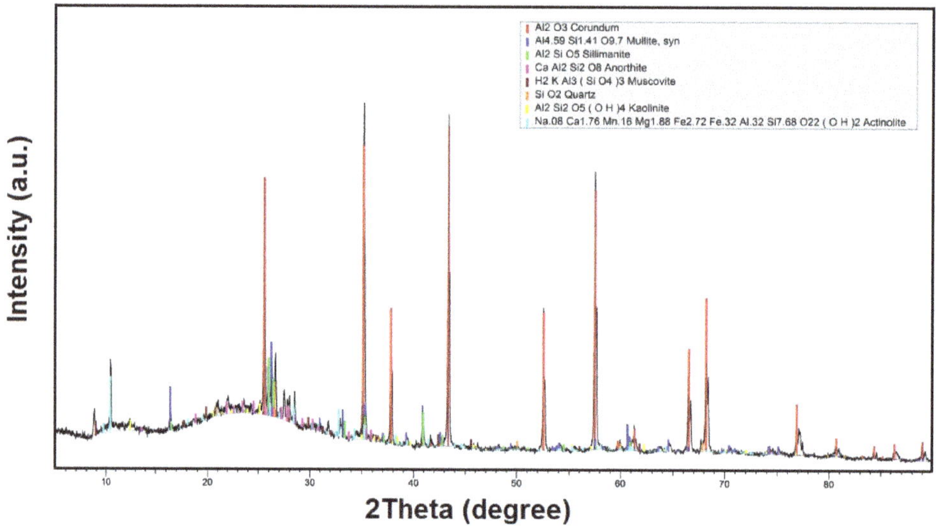

(**a**)

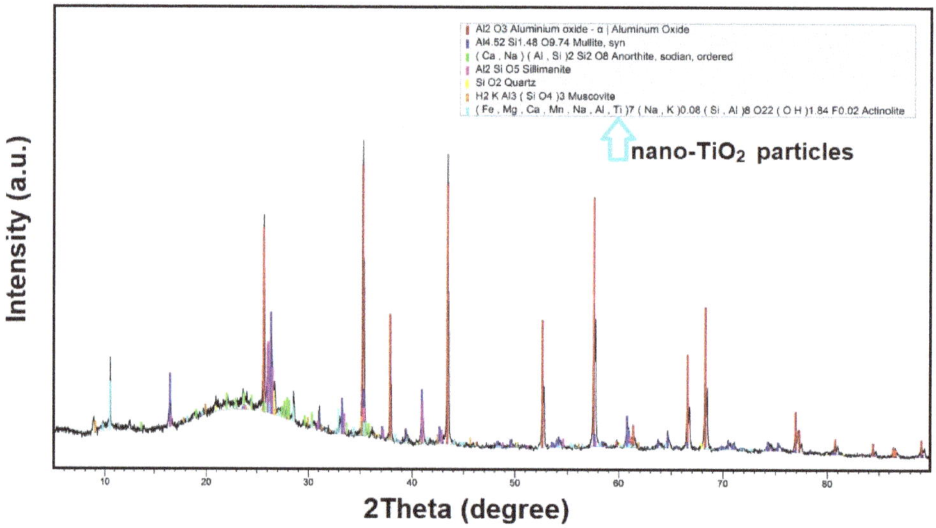

(**b**)

Figure 13. X-ray diffraction (X-RD) phase analysis. (**a**) X-RD pattern of a normal porcelain insulator prototype; (**b**) X-RD pattern of a nano-TiO_2 coated porcelain insulator prototype.

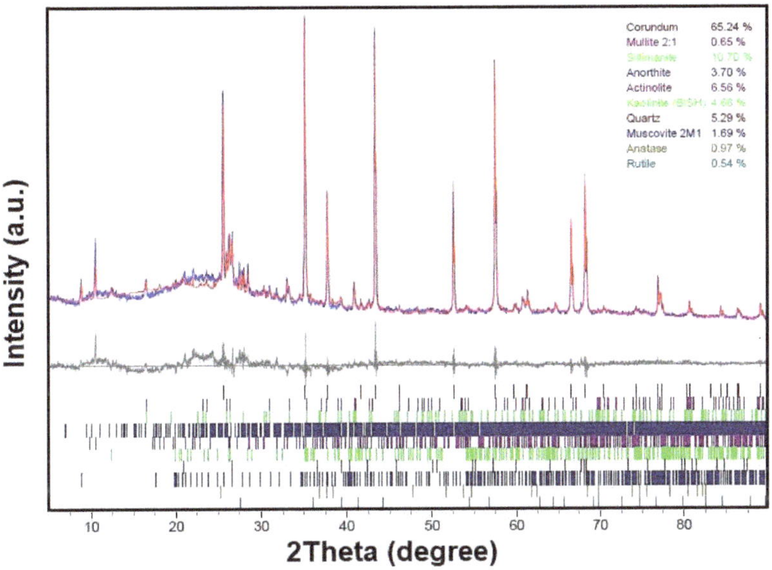

(a)

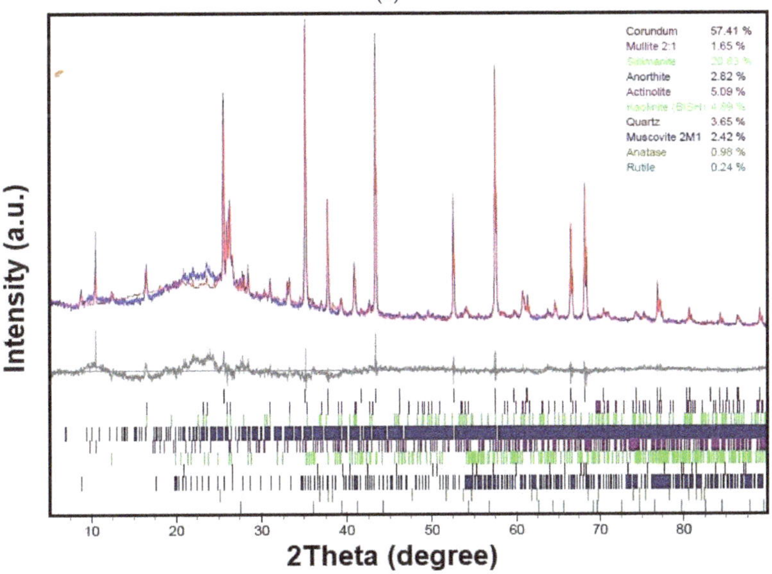

(b)

Figure 14. Rietveld refinement plots. (**a**) Rietveld refinement pattern of a normal porcelain insulator prototype; (**b**) Rietveld refinement pattern of a nano-TiO_2 coated porcelain insulator prototype.

4.3. Microstructural Analysis

Figures 15 and 16 depict the sintered microstructure that corresponds to the prototype porcelain insulator design based on international standards and the outdoor insulator composition used in industrial-scale railway electrification systems. Using the SEM and EDS methods in the back-scattered electron mode, and spectrum and quantity analysis were measured by TESCAN, Model: MIRA3. We compared the porcelain insulators with traditional coatings in Figure 15 and nano-TiO_2 coatings in Figure 16.

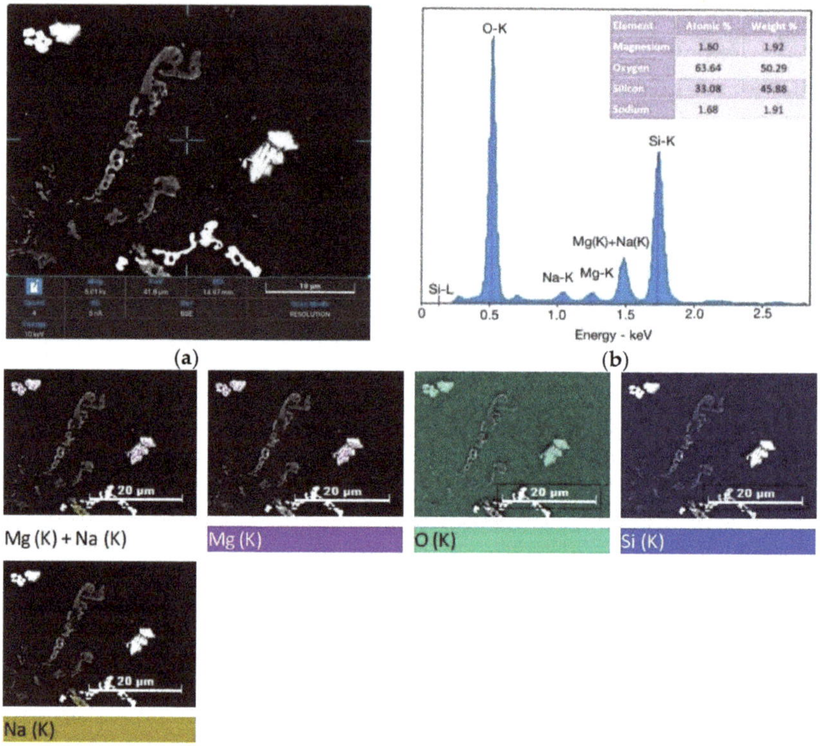

(c)

Figure 15. SEM and EDS of porcelain insulators with normal coatings. (**a**) SEM image of surface; (**b**) The EDS-spectrum and quantity analysis; (**c**) The element images.

In a glassy matrix with a consistent distribution of quasi-spherical pores, coarse quartz grains with variable shapes make up a typical siliceous porcelain microstructure. There are noticeable fissures all around the quartz grains. This occurrence is linked to the sample cooling stage after the prototype insulator has been burnt at 1300 °C for 60 h in a temperature-controlled kiln when the quartz grain volume falls by 2% because of the β-α quartz-phase transformation [7,31]. This shift may cause enough strain to cause both the glassy matrix and the quartz grains to crack because of the large difference in thermal expansion between the quartz grains, which have a coefficient of thermal expansion of 23×10^{-6} C^{-1}, and the glassy phase that surrounds them, which has a coefficient of thermal expansion of 3×10^{-6} C^{-1} (at temperatures between 20 and 700 °C). The size of the quartz particles and the rate at which they cool cause a reduction in the porcelain's mechanical and dielectric strength [31].

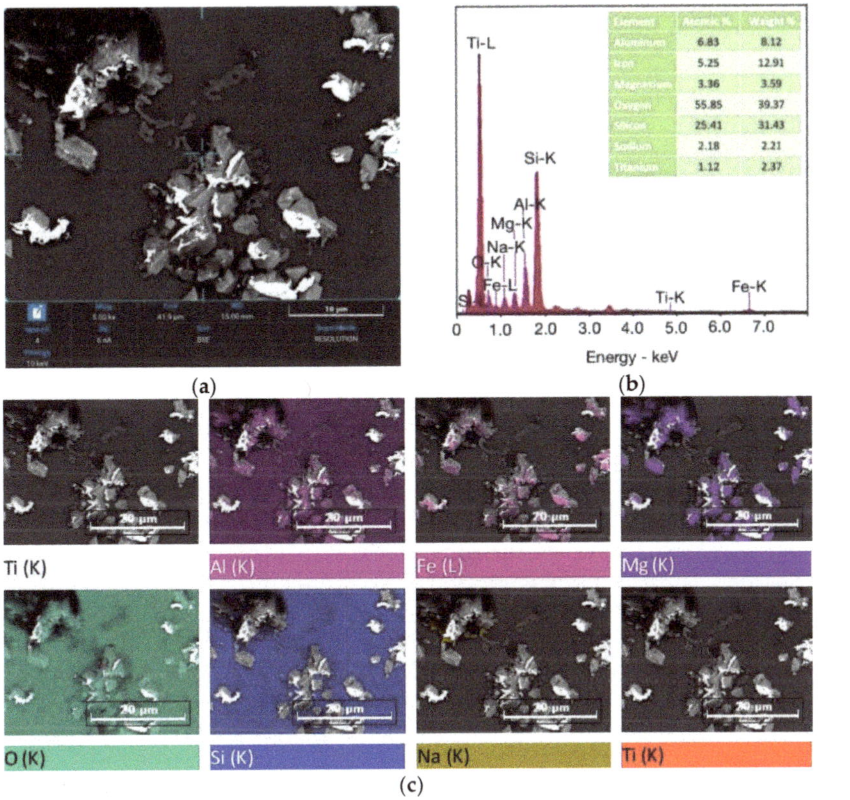

Figure 16. SEM and EDS of porcelain insulators with nano-TiO_2 coatings. (**a**) SEM image of surface; (**b**) The EDS-spectrum and quantity analysis; (**c**) The element images.

5. Conclusions

The test results concluded that when adding nano-TiO_2 at 0.1 wt.% of the amount of deionized water applications to outdoor insulators on railway electrification systems before processing coating and electrical testing, the performance of low-frequency flashover in dry and wet test conditions of the insulator coated with TiO_2 was higher than that coated with the normal coating. Furthermore, the test results of the lightning impulse critical-flashover voltage performance under normal test conditions showed that the insulator coated with nano-TiO_2 can cause a small change in the impulse breakdown voltage with both positive and negative polarity. Similarly, the positive polarity testing under contaminated test conditions, especially in the negative polarity of the impulse breakdown voltage, shows no improvement. However, it may be because the TiO_2 content is so low (only 0.1 wt.%) that it does not show a clear effect on the electrical properties of the coated insulators with improved coating as much as it should. In the future, we may add more TiO_2 to show higher electrical values. Finally, the results of this research showed that adding nano-TiO_2 to improve the electrical properties of the insulator will likely improve its performance in the future.

Author Contributions: Conceptualization, P.M.; methodology, P.M.; investigation, P.M.; writing—original draft preparation, P.M.; data curation, I.K.; resources, P.M. and I.K.; formal analysis, W.V. and I.K.; writing—review and editing, W.V. All authors have read and agreed to the published version of the manuscript.

Funding: This research project is supported by Rajamangala University of Technology Isan. Contract No. ENG 22/63.

Data Availability Statement: Not applicable.

Acknowledgments: The authors would like to acknowledge Asian Insulators Public Company Limited for technical support to develop the performance of insulators on the application of nanomaterials used to coat a thin TiO_2 film on the surface of porcelain insulators. Moreover, the authors are grateful to the Faculty of Engineering, Rajamangala University of Technology Isan, Khon Kaen Campus (RMUTI KKC), for providing the research support facility. A special thanks also to the high-voltage research team at the High Voltage Research Laboratory, RMUTI KKC.

Conflicts of Interest: The authors declare no conflict of interest.

References

1. Contreras, J.E.; Rodriguez, E.A.; Taha-Tijerina, J. Nanotechnology applications for electrical transformers—A review. *Electr. Power Syst. Res.* **2017**, *143*, 573–584. [CrossRef]
2. Eisawy, E.A. The Effect of Radiation Environment on Electrical Insulation Materials. *J. Nucl. Radiat. Phys.* **2019**, *14*, 11–20.
3. Montoya, G.; Hernández, R.; Montoya, J. Failures in outdoor insulation caused by bird excrement. *Electr. Power Syst. Res.* **2010**, *80*, 716–722. [CrossRef]
4. McAfee, R.; Heaton, R.; King, J.; Falster, A. A study of biological contaminants on high voltage porcelain insulators. *Electr. Power Syst. Res.* **1997**, *42*, 35–39. [CrossRef]
5. Gubanski, S.M. Outdoor high voltage insulation. *IEEE Trans. Dielectr. Electr. Insul.* **2010**, *17*, 325. [CrossRef]
6. Contreras, J.E.; Gallaga, M.; Rodriguez, E.A. Effect of Nanoparticles on Mechanical and Electrical Performance of Porcelain Insulator. In Proceedings of the 2016 IEEE Conference on Electrical Insulation and Dielectric Phenomena (CEIDP), Toronto, ON, Canada, 16–19 October 2016; pp. 583–586.
7. Meng, Y.; Gong, G.; Wu, Z.; Yin, Z.; Xie, Y.; Liu, S. Fabrication and microstructure investigation of ultra-high-strength porcelain insulator. *J. Eur. Ceram. Soc.* **2012**, *32*, 3043–3049. [CrossRef]
8. Hackam, R. Outdoor HV composite polymeric insulators. *IEEE Trans. Dielectr. Electr. Insul.* **1999**, *6*, 557–585. [CrossRef]
9. Carty, W.M.; Senapati, U. Porcelain-raw materials, processing phase evolution, and mechanical behavior. *J. Am. Ceram. Soc.* **1998**, *81*, 3–20. [CrossRef]
10. Morocutti, T.; Berg, T.; Muhr, M.; Gödel, G. Developments of High Voltage Porcelain Post Insulators. In Proceedings of the 2012 IEEE International Symposium on Electrical Insulation, San Juan, PR, USA, 10–13 June 2012; pp. 395–398.
11. Monire, T.; Mostafa, S.; Nasim, N.; Alireza, S.F. Reliability assessment of RTV and nano-RTV-coated insulators concerning contamination severity. *Electr. Power Syst. Res.* **2021**, *191*, 106892.
12. Hossain, S.S.; Roy, P.K. Sustainable ceramics derived from solid wastes: A review. *J. Asian Ceram. Soc.* **2020**, *8*, 984–1009. [CrossRef]
13. Zhuang, J.; Liu, P.; Dai, W.; Fu, X.; Li, H.; Zeng, W.; Liao, F. A novel application of nano nticontamination technology for outdoor high-voltage ceramic insulators. *Int. J. Appl. Ceram. Technol.* **2010**, *7*, E46–E53. [CrossRef]
14. Guo, C.; Liao, R.; Yuan, Y.; Zuo, Z.; Zhuang, A. Glaze icing on superhydrophobic coating prepared by nanoparticles filling combined with etching method for insulators. *J. Nanomater.* **2015**, *2015*, 404071. [CrossRef]
15. IEC/TS 60815-1. *Selection and Dimensioning of High-Voltage Insulators Intended for Use in Polluted Conditions*; International Electrotechnical Commission: Geneva, Switzerland, 2008.
16. Inframat Advanced Materials LLC. Available online: http://www.advancedmaterials.us (accessed on 18 May 2022).
17. Wen-Jen, L.; Yu-Ting, W.; Yi-Wei, L.; Yen-Ting, L. Graphite Felt Modified by Atomic Layer Deposition with TiO_2 Nanocoating Exhibits Super-Hydrophilicity, Low Charge-Transform Resistance, and High Electrochemical Activity. *Nanomaterials* **2020**, *10*, 1710.
18. Lee, W.J.; Hon, M.H.; Chung, Y.W.; Lee, J.H. A three-dimensional nanostructure consisting of hollow TiO_2 spheres fabricated by atomic layer deposition. *Jpn. J. Appl. Phys.* **2011**, *50*, 06GH06. [CrossRef]
19. Moya, A.; Kemnade, N.; Osorio, M.R.; Cherevan, A.; Granados, D.; Eder, D.; Vilatela, J.J. Large area photoelectrodes based on hybrids of CNT fibres and ALD-grown TiO_2. *J. Mater. Chem. A.* **2017**, *5*, 24695–24706. [CrossRef]
20. Li, M.; Zu, M.; Yu, J.; Cheng, H.; Li, Q.; Li, B. Controllable synthesis of core-sheath structured aligned carbon nanotube/titanium dioxide hybrid fibers by atomic layer deposition. *Carbon* **2017**, *123*, 151–157. [CrossRef]
21. Surfactants, Emulsifiers and Polyglycols High Performance Products. Available online: https://www.dow.com/en-us/product-technology/pt-surfactants-emulsifiers-polyglycols.html (accessed on 24 September 2022).
22. ANSI/NEMA C29.7-2015; American National Standard for Wet-Process Porcelain Insulators-High-Voltage Line Post-Type. National Electrical Manufacturers Association (NEMA): Rosslyn, Virginia USA, 2015.
23. TIS 2623-2560; Electrical power insulators Part 1 Test methods. Thai Industrial Standards: Bangkok, Thailand, 2017.
24. IEEE Std 4; IEEE Standard for High-Voltage Testing Techniques. IEEE Power and Energy Society: New York, NY, USA, 2013.
25. TIS 1077-2535; Line-Post Type Porcelain Insulators. Thai Industrial Standards: Bangkok, Thailand, 1992.

26. IEC/TS 60507; Artificial Pollution Tests on High Voltage Insulators to be Used on A. C. Systems. International Electrotechnical Commission: Geneva, Switzerland, 2013.
27. Baker, A.C.; Farzaneh, M.; Gorur, R.S.; Gubanski, S.M.; Hill, R.J.; Karady, G.G.; Schneider, H.M. Insulator selection for AC overhead lines with respect to contamination. *IEEE Trans. Power Deliv.* **2009**, *24*, 1633–1641. [CrossRef]
28. Ashwini, A.; Ravi, K.; Vasudev, N. Experimental Study on Aging of Polymeric Insulators by Dip Method. In Proceedings of the 2019 International Conference on High Voltage Engineering and Technology (ICHVET), Hyderabad, India, 7–8 February 2019; pp. 1–3.
29. Mohammad, R.A.V.; Mohammad, M.; Reza, S. Reliability assessment of aged SiR insulators under humidity and pollution conditions. *Int. J. Electr. Power Energy Syst.* **2020**, *117*, 105679.
30. Bowen, W.; Jiazheng, L.; Zhen, F.; Zhenglong, J.; Jianping, H. Development of Antithunder Composite Insulator for Distribution Line. *IEEJ Trans. Electr. Electron. Eng.* **2019**, *15*, 100–107.
31. Contreras, J.E.; Taha-Tijerina, J.; Lopez-Perales, J.F.; Banda-Munoz, F.; Díaz-Tato, L.; Rodríguez, E.A. Enhancing the quartz-clay-feldspar system by nano-Al_2O_3 addition for electrical insulators: From laboratory to prototype scale. *Mater. Chem. Phys.* **2021**, *263*, 124289. [CrossRef]
32. Chaudhuri, S.P.; Sarkar, P.; Chakraborty, A.K. Electrical resistivity of porcelain in relation to constitution. *Ceram. Int.* **1999**, *25*, 91–99. [CrossRef]
33. Srivastava, K.D.; Zhou, J.P. Surface Charging and Flashover of Spacers in SF6 Under Impulse Voltages. *IEEE Trans. Electr. Insul.* **1991**, *26*, 428–442. [CrossRef]
34. Miller, H.C.; Furno, E.J. The Effect of Mn/Ti Surface Treatment on Voltage-Holdoff Performance of Alumina Insulators in Vacuum. *J. Appl. Phys.* **1978**, *49*, 5416–5420. [CrossRef]
35. Iqbal, Y.; Lee, W. Fired porcelain microstructures revisited. *J. Am. Ceram. Soc.* **1999**, *82*, 3584–3590. [CrossRef]
36. Xu, R. The compositions and properties of electric porcelain materials in China. In Proceedings of the Second International Conference on Properties and Applications of Dielectric Materials, Beijing, China, 12–16 September 1988; pp. 256–259.
37. Mohsen, M. *Advanced Ceramic Materials*; IntechOpen: London, UK, 2021; pp. 267–268. ISBN 9781838812041.
38. Conconi, M.S.; Gauna, M.R.; Serra, M.F.; Suarez, G.; Aglietti, E.F.; Rendtorff, N.M. Quantitative firing transformations of a triaxial ceramic by X-ray diffraction methods. *Cerâmica* **2014**, *60*, 524–531. [CrossRef]
39. Bish, D.L.; Post, J.E. Quantitative mineralogical analysis using the Rietveld full-pattern fitting method. *Am. Mineral.* **1993**, *78*, 932–940.

energies

MDPI

Review

Deep Learning in High Voltage Engineering: A Literature Review

Sara Mantach [1], Abdulla Lutfi [2], Hamed Moradi Tavasani [1], Ahmed Ashraf [1], Ayman El-Hag [2] and Behzad Kordi [1,*]

[1] Department of Electrical & Computer Engineering, University of Manitoba, Winnipeg, MB R3T 5V6, Canada; mantachs@myumanitoba.ca (S.M.); moradith@myumanitoba.ca (H.M.T.); ahmed.ashraf@umanitoba.ca (A.A.)

[2] Department of Electrical & Computer Engineering, University of Waterloo, Waterloo, ON N2L 3G1, Canada; a2lutfi@uwaterloo.ca (A.L.); ahalhaj@uwaterloo.ca (A.E.-H.)

* Correspondence: behzad.kordi@umanitoba.ca

Abstract: Condition monitoring of high voltage apparatus is of much importance for the maintenance of electric power systems. Whether it is detecting faults or partial discharges that take place in high voltage equipment, or detecting contamination and degradation of outdoor insulators, deep learning which is a branch of machine learning has been extensively investigated. Instead of using hand-crafted manual features as an input for the traditional machine learning algorithms, deep learning algorithms use raw data as the input where the feature extraction stage is integrated in the learning stage, resulting in a more automated process. This is the main advantage of using deep learning instead of traditional machine learning techniques. This paper presents a review of the recent literature on the application of deep learning techniques in monitoring high voltage apparatus such as GIS, transformers, cables, rotating machines, and outdoor insulators.

Keywords: high voltage apparatus; deep learning; classification; localization; partial discharge; faults; outdoor insulators

Citation: Mantach, S.; Lutfi, A.; Moradi Tavasani, H.; Ashraf, A.; El-Hag, A.; Kordi, B. Deep Learning in High Voltage Engineering: A Literature Review. *Energies* **2022**, *15*, 5005. https://doi.org/10.3390/en15145005

Academic Editors: Zhijin Zhang and Hualong Zheng

Received: 14 June 2022
Accepted: 5 July 2022
Published: 8 July 2022

1. Introduction

Diagnosis of electrical insulation degradation is essential for monitoring the integrity of an electric power system. A well-known diagnostic method, which has been employed for a number of decades, is the measurement of localized discharges known as partial discharge (PD) [1]. Detecting fault or PD in electric apparatus, such as transformers, rotating machines, cables, gas insulated switchgear (GIS) and outdoor insulators, has always required the knowledge of expertise who are able to characterise and differentiate the different sources of fault, PD, defect, or degradation. Throughout the years, different parameters had to be extracted manually from recorded patterns or signals. The aim has been to use the manually-extracted parameters in order to implement a classifier that would be able to perform the task of differentiation and characterization of fault, PD, defect, or degradation. Though the process is partially automated, the fact that experts have to select the features presented a problem since different features might result in different outcomes. This influences the performance of the classifier due to its dependence on the manually-selected features.

Deep learning allows the feature selection stage to be integrated with the learning process, thus making the process all automated. In high voltage (HV) applications, the aim has mostly been to classify or localize faults, defects, or PD that occur in HV apparatus or determine the degradation of insulating material. The abundance of computational capabilities and the existence of big data has allowed researchers in different fields to take advantage of deep learning algorithms. Other than the main purpose of classifying and localizing the PD or fault in HV apparatus, a deep learning algorithm, namely the Generative Adversarial Network (GAN), allows researchers to generate more input data from a limited amount of experimental/simulation results (e.g., see [2]).

Classification refers to the process of differentiating between different sources of fault, defect, or PD or levels of degradation. Given that in real life scenarios, fault or PD can happen due to various sources, it is necessary to identify the source. When the source is identified, one can investigate techniques to eliminate that source from the high voltage system. Different sources or causes of fault, defect, PD, or degradation exhibit different characteristics that are unique to each source, making their classification (differentiation) possibly feasible. On the other hand, localization refers to the process of identifying the position of the fault or PD taking place in high voltage apparatus [3].

Extensive research has been done on the use of traditional machine learning techniques in high voltage applications, e.g., [4–7]. This paper only considers the literature that employs deep learning techniques. Furthermore, the focus of this review paper is solely on the application of deep learning in high voltage engineering and not on the deep learning algorithms themselves. Figure 1 gives an overview of the flow of this review paper.

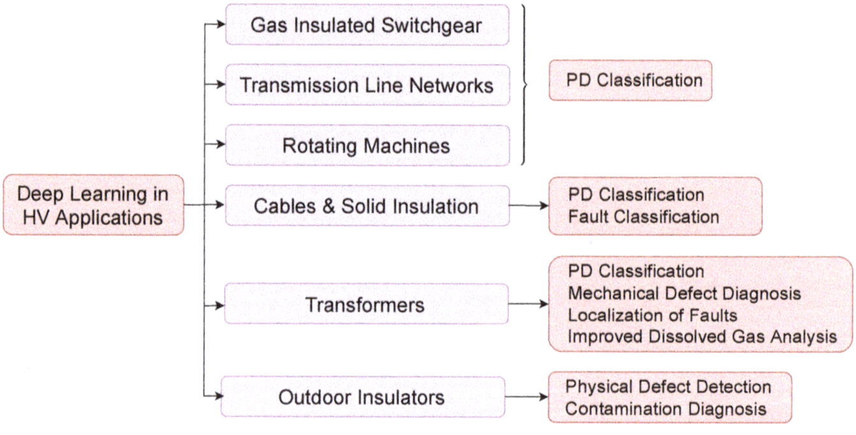

Figure 1. An overview of the topics covered in this review paper.

The majority of published literature focus on employing deep learning in PD classification and identification. Figure 2 gives an overview of the flow of the research done on the application of deep learning algorithms for classification of PDs in HV apparatuses.

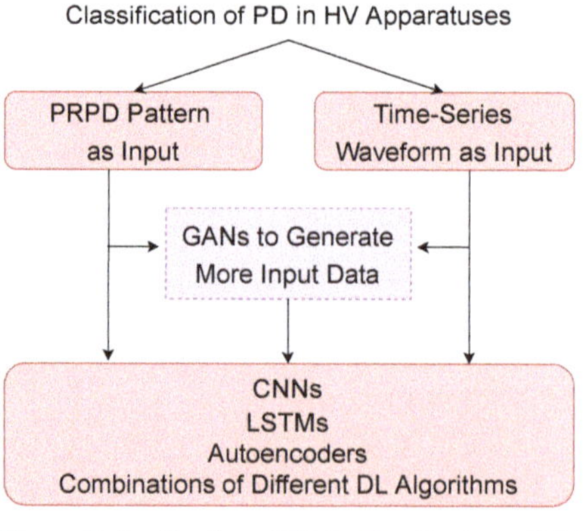

Figure 2. Application of DL in classification of PDs.

A summary of the papers on classification of PD using deep learning is shown in Table 1. Further to PD classification, deep learning has also been used for other applications in transformers and outdoor insulators that are also reviewed in this paper.

The organization of the paper is as follows: In Section 2, four deep learning techniques (namely convolutional neural networks, recurrent neural networks, autoencoders, and generative adversarial networks) are briefly reviewed. These are the commonly-used techniques in the HV application. In Sections 3–8, a review of papers on the application of deep learning in HV apparatus including GIS, transmission line networks, rotating machines, cables and solid insulators, transformers, and outdoor insulators is presented. Finally, the concluding remarks are presented in the last section of the paper, Section 9.

Table 1. Summary of deep learning algorithms used for classification of PD in various HV applications: specification of characteristics of collected data used and whether multiple-labeled sources of PDs are mentioned.

HV Application	Data Collected	Field/ Lab/ Simulations	Multiple Sources	DL Technique	References
GIS	PRPD	Lab	No	Stacked Sparse AE, LSTM, Siamese CNN network, multi-head self attention LSTM and self attention based neural network model	[8–11]
GIS	PRPD	Both	No	CNN(LeNet-5), Variational AE	[12,13]
Solid Insulation	PRPD	Lab	No	DBN , CNN	[14,15]
Transformer	PRPD	Lab	No	CNN-LSTM, CNN, lightweight attention mechanism Squeeze-and-Excitation (SE) module on top of CNN, CNN	[16–19]
Transformer	PRPD	Lab	Yes	Novel architecture based on CNN, LSTM	[20,21]
Transformer	PRPD	Lab	Yes *	ResNet	[22]
Cables	PRPD	Lab	No	Transfer Learning on CNN	[23]
Rotating Machines	PRPD	Field	Yes	ANNs incorporated in a hierarchical fashion	[24]
Cables	T-S waveforms	Simulation	No	CNN	[25]
Cables	T-S waveforms	Lab	No	CNN, DBN	[26,27]
Power Cables	T-S waveforms	Field	No	ensemble of deep learning algorithms (CNN, convolutional RNN, LSTM and bidirectional LSTM)	[28]
Hydro-generators	T-S waveforms	Field	No	Variational Autoencoder	[29]
GIS	T-S waveforms	Lab/ simulation	No	Conditional Variation Autoencoder and CNN, Convolutional Autoencoder, CNN(AlexNet), depth-wise CNN, 1D-CNN model where a multiple scale convolution kernel, CNN-LSTM	[30–35]
GIS	T-S waveforms	Lab/ simulation/ Field	No	domain adaptive deep transfer learning (DADTL) CNN	[36]
GIS	T-S waveforms	Simulation	No	CNN-LSTM	[37]
Power Lines	T-S waveforms	Field	No	1D-CNN with Global Average Pooling layer, Dual Cycle-consistency network	[38,39]
Conductors	T-S waveforms	Field	No	time-series decomposition and LSTM, CNN-LSTM with attention layer	[40,41]
Insulators	T-S waveforms	Lab	No	CNN with Bayesian optimization for hyper-parameters tuning	[42]

* Treating multiple PDs as a new class.

2. Deep Learning

Deep learning is a branch of machine learning that enables data-driven learning of feature representations for input data originating in diverse application domains [43–46]. Unlike traditional machine learning algorithms, where features need to be extracted explicitly through pre-defined hand-crafted rules, deep learning has the advantage of using raw data, and learn to extract features depending on the task [47]. This is appreciated especially in complex systems where such features are not necessarily known for a given dataset. As a result, deep neural networks subsume the feature extraction step within the learning phase, thereby computing intrinsic representations of the raw input data in an automatic manner.

Similar to traditional machine learning, deep learning also has the following three key paradigms: supervised, unsupervised, and reinforcement learning.

For the supervised setting, a labeled dataset is required. The type of output can either be continuous (used in a regression problem) or discrete/categorical (used for classification). For unsupervised systems, data with no labels are given, and the objective is either to cluster the data according to their intrinsic characteristics or learn representations which can be later used for downstream supervised or unsupervised settings [48–50]. In scenarios involving agent based learning, exhaustive collection of supervised data is often prohibitively difficult. In such situations, reinforcement learning is a powerful paradigm which allows data collection through interaction with the environment [51]. The agent's goal is to learn policies based on the environment in order to maximize long-term expected rewards [51]. In recent years, research in deep reinforcement learning has gained significant traction wherein the agent policies are learnt through deep neural networks [52–54] (see Figure 3). Depending on the inputs and the desired outputs for most of the high voltage application, a handful of mainly supervised deep learning algorithms have been of interest in this area of research. In the next section of this review paper, a brief introduction on major supervised deep learning algorithms is presented.

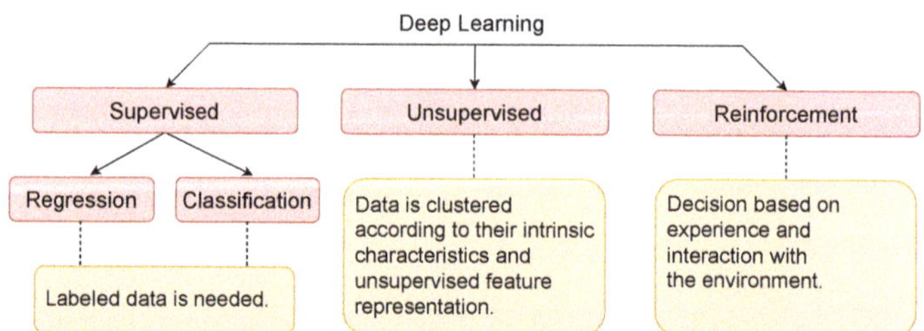

Figure 3. Different deep learning branches: supervised, unsupervised, and reinforcement learning.

2.1. *Convolutional Neural Networks*

Convolutional neural networks (CNNs) represent a class of deep learning architectures which were originally designed for processing data represented in a grid-like topology, e.g., images [47]. A CNN has four main components: convolutional layer, activation function layer, pooling layer, and fully connected layers. Typically, the output of the convolutional layer is passed to an activation function layer where the output of the latter is passed to the pooling layer. In a deep network, this set of the three components are often cascaded multiple times thereby constituting multiple layers and making the network progressively deeper [43,55,56]. While the initial layers usually end up learning low-level features, the deeper layers tend to learn more complex features. The cascade of these layers constitutes the automatic feature extraction stage, and the fully connected layers constitute the classification stage [57]. More details on each of these components are presented below:

Convolutional layer: This layer consists of a bank of learnable linear 1D, 2D, or 3D filters, which are also called kernels [58]. In the high voltage applications, usually 1D and 2D CNNs are used. The 1D-CNN, for example, is used with time-series waveforms, whereas in problems involving phase resolved partial discharge (PRPD) patterns or spectrograms, a 2D-CNN is used. Some of the researchers have employed a 2D-CNN for time series waveforms as well, where they considered an image of the signal as an input rather than the 1D data. These filters are convolved with the input data or the output from a previous layer. The output is a set of feature maps, where the number of feature maps is equal to the number of the filters.

Activation function layer: The purpose of adding activation layers is to introduce nonlinearity in the input-to-output mapping being learned by the neural network. This is desired because complex data include nonlinear features that need to be detected. Most frequently employed activation functions include sigmoid, ReLU [59] and tanh [60].

Pooling layer: The aim of pooling layer is to subsample the output feature maps so that wider receptive fields can be spanned during convolution without increasing the size of the filter kernel. Another advantage of this layer is to provide positional invariance or shift-invariance to the network [61]. Commonly-used pooling operations are maximum pooling and average pooling.

Fully connected layers: In a fully connected (FC) layer, every neuron in one layer is connected to every neuron in the next layer. FC layers are also referred to as dense layers in the literature [47]. In a CNN, the input to the first fully connected layer is the output of the last set of the first three components mentioned above, where the corresponding features maps are flattened into 1D vectors. For classification problems, the architecture is appended by FC layers and ends with a classification layer where the number of neurons is equal to the number of classes.

A typical CNN architecture is shown in Figure 4. The main advantage of CNNs compared to traditional neural networks is the weight sharing when training the learnable kernels, which reduces the learnable parameters in the network [62].

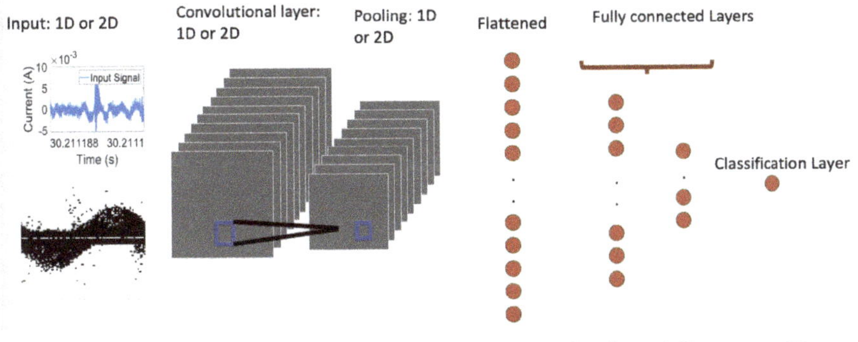

Figure 4. A simple CNN architecture: a convolutional layer, pooling layer, fully connected layers followed by the classification layer.

2.2. Recurrent Neural Networks

Recurrent Neural Network (RNN) is another family of deep learning architectures which are intended for the processing of sequential data [63,64]. A simple illustration of an RNN model is shown in Figure 5.

RNN process data from each time point in a sequential manner. However, the output is not just influenced by data at the current time, but also by the entire history of inputs that have been fed into the RNN previously. This is reflected by the cycles in the architecture, which are maintained in the hidden unit as a state vector including the history of the previous time points. RNN cells have one common set of weights, and when backpropagation runs, data from different time points contribute in updating the same set of weights.

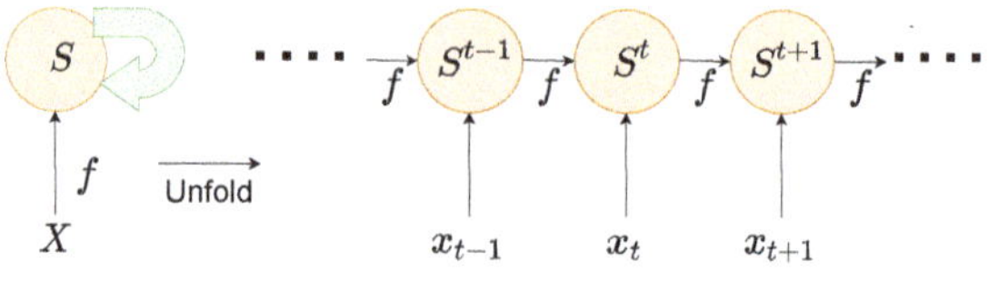

Figure 5. RNN architecture with no output: the network has feedback connections which can be unfolded in time and trained using back-propagation. The input X is processed by incorporating it into the state S that is passed forward through time.

The mathematical representation of the RNN is shown as

$$\mathbf{h}_t = f_{\mathbf{w}}(\mathbf{h}_{t-1}, \mathbf{x}_t) \tag{1}$$

where $\mathbf{h}_t$ represents the updated state vector, $\mathbf{h}_{t-1}$ is the hidden state vector from the previous time step, $\mathbf{x}_t$ represents the input vector at time t, and $f_{\mathbf{w}}$ represents a given function corresponding to learnable weight vector $\mathbf{w}$. The input can either be a vector or a sequence, and the output can either be a vector, sequence, or a value. For example, given a high voltage problem where classification of different PD pulses is required, the input is a vector of PD pulses and the output is a label corresponding to a PD source.

The drawback of a typical RNN is the long-term dependency where the current state depends on all the previous states, which causes the vanishing gradient problem [47]. The vanishing gradient emerges from the fact that, as RNN processes more time steps, repeated multiplication of small weights causes the gradients to approach zeros. To overcome this problem, long short-term memory (LSTM) architecture is used [65]. The main difference in an LSTM architecture is that, instead of computing the hidden state directly from the previous one, LSTM computes additional states, and this structure allows alternative paths to gradients to flow during the backpropagation avoiding repeated matrix multiplications [66]. An LSTM cell has two hidden states $\mathbf{c}_t$ corresponding to the cell state and $\mathbf{h}_t$ corresponding to the hidden state which are calculated as

$$\begin{aligned} \mathbf{c}_t &= \mathbf{f} \odot \mathbf{c}_{t-1} + \mathbf{i} \odot \mathbf{g} \\ \mathbf{h}_t &= \mathbf{o} \odot \tanh(\mathbf{c}_t) \end{aligned} \tag{2}$$

where $\mathbf{i}$ is the input gate, $\mathbf{f}$ is the forget gate, $\mathbf{o}$ is the output gate, and $\mathbf{g}$ is the gate. The operator $\odot$ is element-wise multiplication operation. The input gate decides what new information will be stored in the cell state, the forget gate decides what information will be removed from the cell state, and the output gate decides what information from the cell state will be used in the output [67].

2.3. *Autoencoder*

Autoencoders (AE) were introduced in 1980s [68] in order to learn useful representations in an unsupervised fashion by the use of the input data on its own [69]. They were then reintroduced in 2006 with the booming of the deep learning architectures [70]. The idea behind an autoencoder is to train a neural network such that the model learns a latent intrinsic representation of the original input. An autoencoder consists of an encoder–decoder architecture, wherein the role of the encoder is to transform the input to a latent representation, while the decoder is responsible for transforming the latent representation back to the original data. The two parts (encoder and decoder) are learned jointly so as to minimize the reconstruction error between the decoder's output and the network input. A simple illustration of an autoencoder model is shown in Figure 6.

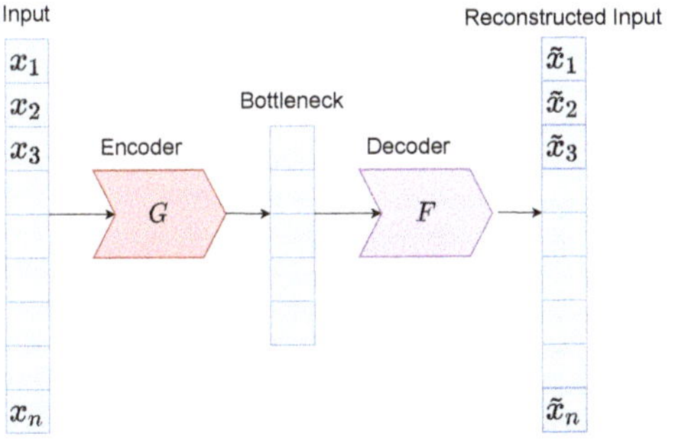

Figure 6. Autoencoder architecture: the output $\tilde{\mathbf{x}}$ is the reconstructed input where a bottleneck enables computing a latent representation of the original input $\mathbf{x}$. $\mathbf{x}$ is a vector consisting of n elements from x_1 to x_n.

Assuming that the input is $\mathbf{x}$ and the reconstructed output is $\tilde{\mathbf{x}}$, the model is trained to minimize the reconstruction error $\mathcal{L}(\mathbf{x}, \tilde{\mathbf{x}})$. The encoder and decoder can be fully connected layer networks or any deep learning architecture. The encoder is expressed as a function G such that

$$\mathbf{b}_i = G(\mathbf{x}_i) \tag{3}$$

where $\mathbf{b}_i$ represents the latent feature representation (bottleneck) of a single observation sample $\mathbf{x}_i$. The decoder F accepts $\mathbf{b}_i$ as input and produces $\tilde{\mathbf{x}}_i$ is. This is shown in

$$\tilde{\mathbf{x}}_i = F(\mathbf{b}_i) = F(G(\mathbf{x}_i)). \tag{4}$$

The goal is then to find F and G that would minimize

$$\arg\min_{F,G} \sum_i [\mathcal{L}(\mathbf{x}_i, F(G(\mathbf{x}_i))] \tag{5}$$

where the summation is over all the observations during training.

2.4. Generative Adversarial Networks

Generative models allow sampling data from the probability distribution of a given data. There is a long tradition of learning data distributions including methods for density estimation [71,72]. For high-dimensional continuous data, classical density estimation techniques become intractable. Generative Adversarial Networks (GANs) enable generative models for data stemming from unknown probability distributions [73]. The key idea is to learn a neural network to map a datapoint sampled from a simple distribution (such as normal) to data from the training data distribution. To assess if the generated data have modeled training distribution, another network, known as the discriminator, is trained to distinguish between the generated examples and examples from the original dataset. The goal of the GAN is to train the generator network such that the best discriminator performs as worse as possible. In other words, when the generator parameters have been updated in a way that it becomes difficult to train a classifier for distinguishing between generated and real samples, it implies that the generator is producing outputs that resemble examples from training data distribution. A simple illustration of a GAN model is shown in Figure 7.

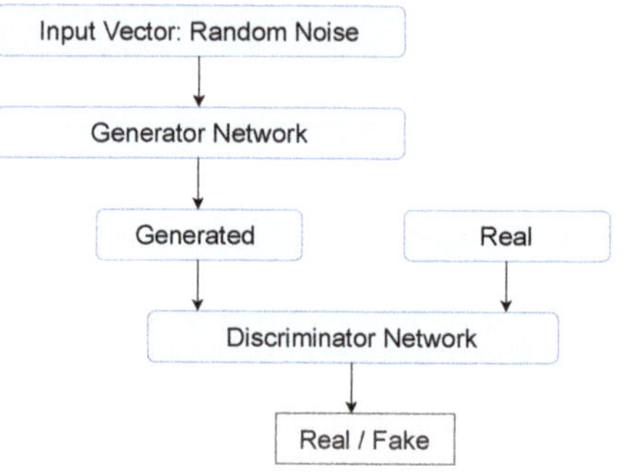

Figure 7. GAN architecture: discriminator network learns a classifier to distinguish between generated and real samples. Generator network updates its parameter so that the discriminator's task is as difficult as possible.

3. Gas Insulated Switchgear

Classification of PDs has been a standard procedure in the maintenance of high voltage assets. Various distinctive parameters extracted from PD measurements have been introduced for the PD classification application. Starting with current or voltage signals, time series waveforms have been proven to acquire unique behavior for each source of fault or PD [74]. Phase resolved partial discharge (PRPD) patterns have also been used to differentiate between different PD sources [75].

Gas insulated switchgear (GIS) is widely used in the industry [76]. A GIS platform is shown in Figure 8.

Figure 8. Gas insulated switchgear.

GIS has its components close to each other which makes the fault occurrence in one component transfer to other components easily. The components of the GIS are power conducting components and the control system. The power conducting components are responsible for ensuring the flow of the electric current in the system, and the control systems work on monitoring the behavior of the conducting components. Thermal, mechanical, and electrical faults comprise the main faults that take place in a GIS platform. This section reviews the application of deep learning to solve the problem of classification of only PDs (electrical faults) in GIS. For the condition monitoring of GIS, different detectors have been used in literature including antennas, coupling capacitors, high frequency current transformer (HFCT), UV sensor, thermal sensor and Rogowski coils.

There are different sources of PD that take place in a GIS platform. Whether using PRPD patterns or time-series waveforms, we can divide the literature into the following two main groups.

3.1. PD Classification Using the PRPD Pattern

In 2017, the authors of [8] simulated four sources of PDs that take place in a GIS platform. The four sources of PDs are: protrusion, contamination, gap, and particle defects. For each of the sources of PDs, four severity states of the PD were collected: normal state, attention state, serious state, and dangerous state. The authors proposed a stacked sparse autoencoder (AE) model, where the output of the middle layer (bottleneck) of the preceding AE is the input to the next AE. The output of middle layer of the final AE is the input to a softmax layer which decided on the assigned severity level label for each sample. The effect of changing different hyperparameters, such as the number of stacked AE or number of nodes in the middle layer, were examined. The proposed model was compared with support vector machine (SVM), where nine statistical characteristics were extracted from the PRPD patterns.The study reports enhanced average classification accuracy of the PD severity compared to SVM.

The authors of [12] published in 2018 collected PRPD data from experimental setup and more than 30 live GIS substations. The five defects studied in this work are floating electrode discharge, surface discharge, corona discharge, insulation void discharge, and free metal particle discharge. A known CNN architecture, LeNet-5, was employed and compared with back propagation neural network (BPNN) and SVM, where statistical features were extracted for the latter two and raw data were fed to the CNN. The statistical features extracted from the PRPD patterns include skewness, steepness, asymmetry and cross correlation coefficient of the PD amplitude and rate in both positive and negative half cycles of the applied voltage. In order to optimize the weights of the CNN, the authors trained an autoencoder and used the weights as an initialization for the CNN training. The authors reported an improved average accuracy of the CNN compared to that of SVM and BPNN.

A long short-term memory (LSTM) recurrent neural network (RNN) has been used to classify PRPDs in a GIS in [9]. In this work, four different types of defects have been simulated in a controlled lab environment, where the PRPD patterns have been collected. The PD sources simulated are: protruding electrodes, floating electrodes, free particles, and void defects. In addition to that, the authors simulated noise by using an air purifier and the noise signals were obtained using the external UHF sensor. The authors compared the classification accuracy with SVM and a fully dense artificial neural network. The study reported that, although the proposed model takes more training time, the classification accuracy is superior to the other two machine learning models.

In 2019, the authors of [13] simulated PDs using a laboratory setup and collected PD data from a live substation. The four PD sources considered in this work that take place in a GIS platform are: floating electrode defects, metallic protrusion defects, insulation void discharge defects, and free metal particle discharge defects. A variational autoencoder (VAE) was trained to extract the eigenvalues corresponding to the PRPD data. The training set included a mix of both the laboratory and substation data. For the test dataset, a matching algorithm based on cosine distance was used in order to decide to what class the test PRPD belongs to. The proposed method was compared with statistical features, deep belief networks (DBN) [77] and CNNs. The authors reported that the eigenvalues extracted from the VAE feature vector have improved results over the other methods used.

Four sources of PD that usually occur in GIS (protruding electrodes, floating electrodes, void defects, and free particles) were simulated in [10]. The authors used a Siamese network where the raw input data are pairs of PRPDs. The motivation behind using Siamese network is that PDs usually result in small datasets. Two identical independent CNN models are trained, and the distance between the embedded features resulting from the two CNN models is calculated. As a result, a decision is made whether the pair belong to the same or a different class. The authors compared their proposed architecture with SVM and a CNN model. They reported that the proposed method performed better compared to the latter two.

Ref. [11] presents the simulation of artificial defects that take place in a GIS platform, where the PRPD patterns corresponding to each source of PD were recorded. The four PD sources in this work are: corona, floating electrode, particle, and void. The authors proposed a multi-head self attention LSTM based model for PD (LSANPD) and a self attention based neural network model for PD (SANPD). They compared the classification of the two models with their previous published work which used an LSTM-RNN based model. They reported that SANPD and LSANPD are better in terms of classification accuracy and that SANPD is better than the LSANPD and LSTM-RNN model in terms of complexity.

3.2. PD Classification Using Time-Series Waveform

Some researchers have used sensors, such as HFCT or UHF sensors, to record voltage or current waveforms induced by PD. The waveforms are in the time domain and referred to as time-series waveforms. This section reviews the papers that have used time-series waveforms instead of PRPD patterns to train and optimize a deep learning algorithm.

In 2018, five sources of PD were generated in a GIS tank model in a laboratory setup that includes a floating electrode, a metal protrusion on the conductor and the tank, surface contamination, and free metal particles [35]. Four planar spiral antennas were installed at different locations on the tank. For each signal collected, the authors calculated three different short time Fourier transform (STFT) by changing the window lengths. The different window lengths correspond to high time resolution, high frequency resolution and medium resolution. The proposed model was a CNN-LSTM based model. The three different STFTs calculated from each signal were used to train three different CNN models, where the three outputs of the CNN models are combined by a fully connected layer. The output of the fully connected layer is the input to the LSTM. Since there are four sensors, the model is comprised of four fully-connected layers which are the input to four separate LSTMs. The outputs of the LSTMs collectively decide on the label of each input sample. The authors compared the model performance with other baseline models and with the case where a single window length was used for the STFT. The model showed improved results compared to the other models.

In 2019, the authors of [30] investigated GIS PD data which were collected from laboratory experiments and finite difference time domain simulations. A conditional variational autoencoder (CVAE) was used to generate more training data. A seven-layer CNN model was used for the classification of four different sources of PD, free metal particle, metal tip defects, floating electrode defects, and insulation void. The authors also reported a visualization of the feature maps from the first two convolutional layers. They compared their results with support vector machine (SVM), decision trees (DT), back propagation neural network (BPNN), and a few CNN architecture (LeNet5, AlexNet and VGG16). The proposed CNN model outperformed the above mentioned models.

Using a laboratory setup, four defects that take place in medium voltage switchgear were replicated [31]. The four sources of PDs included: cable termination floating earth, earth cable in contact with cable termination insulation, voltage presence indicating systems (VPIS) bushing screen disconnected, and earth grounding spring missing on bus bar connector. The spectrogram of the PD signals collected using a coupling capacitor is generated by applying the continuous wavelet transform (CWT). In addition, spectrograms from noises and other HF signals are generated. The authors proposed a convolutional autoencoder (CAE) that is able to reconstruct the spectrograms of the different sources of PDs and noise. After the CAE is trained, the decoder part of the autoencoder is removed and substituted by a fully connected layer followed by the classification layer. This model is trained using a labeled dataset, where the model is able to output the percentage of belonging of a tested spectrogram to each of the four PD sources and the noise/HF signals classes. The study reported high performance ability of the proposed model.

In another work [32], four sources of PDs in a gas-insulated switchgear were simulated in a lab setup. An existing CNN architecture, AlexNet, was the method used in this work where the inputs are the time series waveforms treated as images. The results of the proposed method are compared with the fractal method and mean discharge method. The time series waveforms were transformed into a PRPD plot for the sake of applying the fractal method and mean discharge method. Both the fractal and mean discharge methods provide features which are considered as input to two fully connected neural networks. The study included reporting the average classification accuracy of the three models with different percentage of noise added to the signals. The proposed method showed improved results compared with the other two methods especially with a high noise percentage. In addition, the author reported that the time consumed for PD classification was the least using the proposed CNN-based method.

In 2021, four types of PDs that take place in a GIS platform (free metal particle defects, metal tip defects, floating electrode defects and insulation void defects) were experimentally simulated and time-series waveforms corresponding to each type of PD were recorded using UHF sensors (butterfly antennas) [33]. The variability in the dataset was introduced by randomly changing the position of the defect. The deep learning model proposed by the authors was based on a depth-wise CNN model where the convolution is divided into two parts: the first part is composed of convolving one channel at a time with the convolution kernel (i.e., depth-wise convolution) and the second part is to mix the feature map using a 1×1 convolution kernel (i.e., point-wise convolution). A generative adversarial network (GAN) was also used in order to generate more data. The proposed model was compared with other CNN-based models such as MobileNetV1, MobileNetV2, Xception, ResNet, and LeNet models. The proposed model reported enhanced classification accuracy compared to the other models. Visualization of the feature maps of some layers is presented as well, which highlight what each layer was capable of learning.

In the same year, a research group simulated four different sources of PDs in a GIS controlled lab environment [34]. Varying the defect location for each of the artificial defects ensured the variability in the collected dataset. The authors proposed a 1D-CNN model where a multiple scale convolution kernel is used instead of a single scale convolution kernel. Channel shuffling was used on the outputs of the two feature maps produced by the multiple scale convolutional kernel in order to have a unified feature map. Since having labelled data is time consuming and needs expertise knowledge, the authors proposed domain adversarial transfer strategy (DATS), which is inspired by the GAN. Four different unbalanced datasets were acquired from an actual GIS in order to test the performance of the proposed model. The 1D-CNN that was trained on the experimental data are used to classify the on-site GIS PD data where some data had no labels. For the proposed 1D-CNN model, the authors compared the results of the proposed model with traditional 1D- and 2d-CNN models. With regard to the performance of DATS, the authors compared the results with other transfer learning (TL) techniques such as fine-tuning TL and domain adaptation TL. The study reports enhanced results using the proposed framework. The authors suggested that future work will focus on the automatic optimization of hyperparameters and on trying the platform in an online monitoring system.

In [36], the aim of the research work was to investigate transfer learning, especially domain adaptive deep transfer learning (DADTL) CNN, for GIS PD diagnosis. The authors used four different datasets for the training of their model. The datasets consisted of measured and simulated data. Dataset A included field data from three types of fault (rolling element, inner ring, and outer ring). Dataset B corresponded to the GIS PD simulation data using the finite difference time domain (FDTD) technique, where four sources of PDs (metal particle, tip, floating electrode, and insulator air gap defect) were simulated. Dataset C corresponded to a 252 kV GIS experimental platform, where signals were captured corresponding to the four defects mentioned in Dataset B. Finally, dataset D corresponded to the PD samples collected from a provincial power company's GIS failure.

By the use of maximum mean discrepancy to minimize the sliced Wasserstein distance (SWD), the authors aimed to ensure that transferable features have minimal discrepancy. The proposed model starts with data pre-processing, where samples from the larger datasets are classified as source domain samples, and samples from smaller dataset are classified as target domain samples. The aim is to minimize the differences between the features learned from both source and target datasets. The authors used residuals units in the CNN based architecture. The authors compared the results with traditional CNNs (LeNet and AlexNet) with the same number of layers. The proposed model reported improved results compared to the other deep learning models especially when the dataset is small.

Another work which used simulated data are presented in [37], where a CNN-LSTM network is proposed for the classification of PD sources in a GIS system. The model consists of two blocks of convolutional layers followed by pooling layers. The output of the second block is fed to an LSTM layer which is followed by a fully connected layer and ending with the classification layer. In order to generate the dataset, simulation software XFDTD is used. The four sources of PDs are metal tip defect, insulator air gap defect, floating electrode defect, and free metal particle defect. The authors reported the precision, recall, and F1-score of the four sources of PDs. They compared the performance of the proposed model with other models like SVM, LSTM and CNN. The proposed model reported high average classification accuracy compared to the other models.

Research has been focused on using different DL techniques for the purpose of identifying PD sources in a GIS platform varying from autoencoders, CNN, LSTM, or a combination of the above techniques. Authors have compared their proposed models with other DL models or traditional machine learning models. Despite the advantages of using different DL techniques, future work should focus on the quality of the input data with regard to the interference, noise, and the fact that multiple PDs can take place at the same time. The integration of the developed models in real-life systems would present its own challenges especially when it comes to developing industrial standards or regulations, and implementing condition-based maintenance asset management policies depending on the severity of the situation.

4. Transmission Line Networks

Transmission line networks are used to enable the long distance transmission of power. A few research groups have developed deep learning models to identify PD from non-PD signals collected from a publicly-available dataset. ENET Centre in Czech Republic developed a meter to measure the voltage signal induced by the stray electric field along covered conductors that contained PD or fault signals. The dataset contains noisy real world measurements from high-frequency voltage sensors, where the objective is to identify damaged three-phase, medium-voltage overhead power lines [78].

In 2020, the authors of [39] performed pre-processing of the raw data in order to remove noise and low-frequency components of the signals. The output of this process was the time and frequency representation of the signal by applying short-term Fourier transform. The time and frequency domain positive and negative half cycle signals are the input to the proposed deep learning algorithm. The proposed model is a Dual Cycle-Consistency network. Both time and frequency domain branches consist of three blocks. Each block contains a 2D convolutional layer, a Rectified Linear Unit (ReLU), and a batch normalization layer. The output from block-3 of the time-domain and frequency-domain branches is passed through a global average pooling layer, a shared fully connected layer, and a sigmoid layer. In order to calculate the cycle-consistency loss, the outputs are fed to the dual-domain attention module block (DDAM) for joint learning. The prediction is then based on the weighted average of the output from the fully connected layers and the output from the DDAM block. The results are compared with other models such as Random Forest, Resnet18 + VggNet11, and LSTM. The performance metric used is the Matthews Correlation Coefficient (MCC), in addition to precision, recall, and F1-score. The authors reported better results compared to the other approaches.

The authors of [40] developed a model based on time-series decomposition and LSTM for to classify PD from non-PD signals from the same public dataset. Seasonal-Trend decomposition using Loess (STL) was used to decompose each raw signal into three parts: trend, seasonal, and residual. PD is mostly reflected in the residual part. Four different STL modules with different seasonal window lengths were used to generate four different residual components. Feature engineering was then applied on the residual parts where a sequential feature vector is extracted. As a result, many-to-one sequential data are generated and are considered as the input to the long short-term memory network (LSTM) classifier. The proposed model was compared with other classifiers such as fully connected layers, SVM, XGBoost, and Multivariate Logistic Regression (MLR). The proposed model showed enhanced classification accuracy compared to the other models.

In 2021, the authors of [38] aimed to classify PD versus no-PD signals using the same publicly-available dataset of damaged power lines as the previous paragraph. The proposed model was a traditional 1D CNN model where a Global Average Pooling (GAP) layer is employed before the fully connected layer. Each sample in the dataset is compromised of voltage of the three phases over one period. For each phase, a highpass filter is used to remove the power frequency, after which a maximum filter is used to extract a set of pulses. Each set of pulses from each phase is the input to the trained 1D CNN. Finally, the decision on the label of the power line is based on the three outputs of each phase. In order to visualize what the model is looking at, in order to decide on the label, a pulse activation map (PAM) was used. The evaluation metrics used are Matthews Correlation Coefficient, precision, recall, and accuracy. The authors compared their results with other publicly reported results where models such as LSTM were used. The proposed model showed enhanced results, and the authors suggested that a larger dataset will be more compatible for hyperparameter tuning.

In [41], the authors aimed to classify PD versus no PD using the same dataset as the previous papers, which included the three-phase voltage signals. FFT noise reduction algorithms were used on the raw data. The proposed model was a CNN-LSTM model with an attention layer before the classification layer. Starting with two blocks of convolutional and max-pooling layers, the output is fed to a fully connected layer which is considered as the input to the LSTM layer. The output of the LSTM is the input to an attention layer, where multiplication of the feature vector obtained from the LSTM is done with learnable weight coefficients. The output of the attention layer is fed to to a sigmoid which decides on PD versus no PD label. The performance metrics used are precision, recall, and F1-score. The proposed model is compared with other traditional models such as SVM, CNN, and bidirectional LSTM. The study reported higher average accuracy compared to the other models.

The papers discussed in this section used the same publicly-available dataset, where different DL approaches were adopted to detect PD pulses from non-PD. This serves as proof that different DL algorithms can give good results; however, the deployment of such algorithms in real-life systems would definitely give a better perspective of what models to use. This section was also a good example to show how various DL techniques can be evaluated using a common dataset.

5. Rotating Machines

In rotating machines, voltages are generated due to time-varying magnetic fields, which is the result of the the change in the flux [79]. The change in the flux results from the mechanical motion of the rotating machine. Rotating machines consist of stator and rotor structures which are made of thin laminations of electrical steel, insulated from each other in order to reduce losses and prevent discharges and faults to take place. Various types of stress, such as thermal, electrical, ambient, or mechanical stress, can affect the insulation system of rotating machines. Statistical data show PD activities have preceded a large number of stator failures [80] and, as such, PD detection in rotating machines has been attracting attention. Extensive research has been done on using traditional machine learning techniques, such as Naïve Bayes-, SVM-, and kNN-based techniques [81], for rotating machine electrical insulation diagnosis. This section mentions the work done using deep learning for the rotating machine PD diagnosis.

The PPRDs of a number of sources of PD found in rotating machines through online PD measurements in hydro-generators operating in real-world conditions have been collected in [24]. The PD sources are: internal void, internal delamination, delamination between conductors and insulation, slot, corona, surface tracking, and gap discharges. The authors proposed a methodology to de-noise the PRPDs first and use an image pre-processing technique to separate different clouds in the PRPD patterns. The output of this stage was denoised sub-PRPDs which represent different sources of PD. Three features which are extracted for each sub-PRPD are the input to four different artificial neural networks (ANN). These different ANNs were incorporated in a hierarchical fashion in order to perform the final classification. The authors reported a good overall classification accuracy for all the PD sources.

A framework was proposed using visual data analysis for PD source classification in hydrogenerators with a minimum of labeled data [29]. A convolutional encoder was used to project the PD signals acquired from the generator stators to a 2D-visualization latent space. This serves as a visual aid for the expert to analyze the distribution of the training dataset. After being labeled by the experts, the labeled data are trained by a neural network classifier. Other unlabeled data are tested using the already trained classifier, and if any conflict area appeared on the 2D latent space, the human experts will have to label by conflict area sample data. The new labeled data are then added to the dataset, and this procedure is done in an iterative manner until the area of conflicted data is minimized. This study reported a base that integrates both expert knowledge and the advantages of deep learning in order to have a correctly-labeled dataset of PD sources.

Although performing preprocessing of the input data is crucial to help the models learn the intrinsic characteristics of patterns (i.e., denoising in HV applications), future work should focus on applying different denoising techniques. Investigating the effect of different denoising techniques on the effectiveness of the DL models yields better understanding of the learning process. Moreover, it is observed that more research is being directed towards exploring the problem of unlabeled data, which is a crucial step for the deployment of any algorithm in a real-life diagnosis system.

6. Cables and Solid Insulation

Electric power can be transmitted by underground cables or by overhead transmission lines. The main advantage of underground cables compared to overhead lines is the low maintenance cost. This is linked to the fact that overhead lines are exposed to environmental factors such as storms or lightning. An underground cable consists of one or more conductors which are covered with suitable insulation and the external component is the protecting cover [82]. The major disadvantage of using underground cables though is the problem of degradation and failure of the insulation under high voltage stress. Hence, detecting PDs/faults is crucial for assessing the health of the system. This section reviews the application of deep learning to solve the problem of classification of faults/PDs in cable insulation and solid dielectric using PRPD patterns or time-series waveforms as the input to train the classification model.

6.1. PD/Fault Classification Using the PRPD Pattern

Different sources of PD have been classified in a solid insulation in addition to the prediction of the aging stage of the insulation in [14]. The authors of this paper classified three different sources of PD (corona, surface, and internal discharges) which are simulated in a lab environment. They compared the deep belief network (DBN) output with three other machine learning approaches. The input to the DBN was the raw PRPDs, whereas the inputs to the other approaches were features extracted using statistical and vector-norm-based operators. Classification accuracy is the performance metric used in this study. The authors show that the DBN learns distinguishable features without any pre-processing of the PRPDs.

The performance of a CNN model was evaluated on the prediction of the ageing stage of high voltage insulation material using PRPD data [15]. Three classes of start, middle, and end as well as noise/disturbance were defined for the electrical insulation degradation process representing the ageing that occurs in an insulation specimen under electrical stress. Precision, recall, and F1 score were the metrics used for the evaluation of the CNN model. The author reported that the performance is consistent even with changes in the CNN hyper-parameters' values.

The effect of noise in PRPD patterns on the classification accuracy of different artificial defects in a 11 kV cross linked polyethylene (XLPE) cable joints has been investigated in [23]. There are a total number of five PD sources considered in this study. After training a CNN architecture using noise-free PRPD patterns, transfer learning was performed where the authors used this model to start training another CNN architecture but this time with noisy PRPDs. The results were compared with those obtained using traditional machine learning classifiers where hand crafted features were extracted. The authors reported that the CNN-based model was able to outperform the models that use manual feature extraction with an increase of 16.9% in the classification accuracy.

In [83], the authors simulated five different defects that take place in a 36 kV cross-linked polyethylene (XLPE) cable terminations such as protrusion, void, and corona discharge. The authors used a commercial PD detector in order to capture the PRPD patterns. A CNN architecture was proposed where the authors investigated the effect of different hyperparameters such as pooling and kernel size on the classification accuracy percentage. The authors compared their results with off-the-shelf CNN architectures such as AlexNet, VGG, ResNet and GoogleNet. They reported higher classification accuracy of their proposed model compared to the other models.

6.2. PD/Fault Classification Using Time-Series Waveform

In 2019, a traditional CNN model was used to differentiate between synthetic PD pulses in power cables [25]. The variability in the synthetic dataset was introduced by the signal-to-noise ratio (SNR) and the position at which the PD initiated. The model was compared with a support vector machine (SVM), where the study reported enhanced results using the proposed algorithm.

In the same year, five types of artificial defects in ethylene-propylene-rubber cables in a high voltage laboratory were collected to generate signals containing PD data [26]. Seventeen features were extracted from the time-series waveforms corresponding to characteristics such as pulse width, rise time, fall time, peak voltage, pulse polarity, mean voltage, and root mean square (RMS) voltage. In addition, 16 wavelet features were extracted from the transient signals using Wavelet Transform. In total, 33 features constituted the input corresponding to each signal to the proposed CNN model. Analysis was performed on the effect of the change in the hyperparameters of the CNN architecture such as the number of layers and the convolution kernel sizes. The results were compared with those obtained using SVM and back propagation neural network (BPNN) models. The study reported better classification accuracy when compared with the other two models.

In 2020, four common DC insulation faults were simulated during the operation of XLPE cables that include conductor burrs, external semi-conductive layer residue, internal air gap, and scratch on the insulation surface [27]. A modified Canny edge detection operator was used in order to extract the part of the time series signal which includes the PD. A deep belief network is proposed by the authors where the ADAM optimizer is used. The authors compared their model performance with naive Bayes, K-nearest neighbor (kNN), support vector machine (SVM), and back propagation neural networks (BPNN). The authors reported the confusion matrices of the models and compared the classification accuracy. The proposed model outperformed the other methods, and it was reported that, as the training dataset increases, the classification accuracy increases as well.

An ensemble of deep learning algorithms was used to differentiate between PD signals and noise signals in medium voltage power cables [28]. The samples were collected from offline in-service cables. The idea behind using ensemble learning is to allow more than one neural network to make the classification decision. CNN, convolutional RNN, LSTM, and bidirectional LSTM (BILSTM) were used in the ensemble frame, but two of these models were used at a time. In this paper, two scenarios were considered: In one scenario where there is difference in the prediction of a sample between two deep learning (DL) models, a human expert will have to decide on the label of that sample, and in the other, the output from the activation function of two different models were added together. Five different cables were tested on the trained models. Adaption training was done for each of the five cables where the classifier layer is re-trained with the measured calibration pulse specific to each cable. The authors reported the results of the five cables for the two ensemble scenarios and with different binary selection of the above-mentioned DL algorithms. They also reported the results when each DL model is used alone. It was reported that the CNN paired with the BILSTM gave the best results.

In Ref. [84], the authors aimed to target the problem of losing the voltage signal information that is used for plotting PRPD patterns. Different datasets consisting of three types of cable joint defects were generated in a lab environment, where the signals were recorded using an oscilloscope. Pulse sequence analysis (PSA) was performed by using the change in the magnitude of the PD pulses resulting in a magnitude difference heat map image that was as the input to the proposed CNN model. Investigation was carried out to optimize the CNN hyperparameters, in addition to investigating the effect of the different image features including the size, type, color, and marker size of the images. The authors compared the accuracy of the model with having PRPDs as input versus the PSA as input. The authors claimed that using PSA instead of PRPD yields a higher classification accuracy.

The authors in [85] classified and localized ten faults in an 11-kV, three-phase underground cable consisting of various combinations of phase to phase or to ground faults. The dataset was generated by simulation using PSCAD/EMTDC software with varying different system parameters such as fault inception angle and fault location. Additive Gaussian white noise was added to the signals. The authors proposed a CNN-LSTM architecture along with the application of a sliding window technique. The input to the architecture is the current and voltage signals, and the outputs are the fault location, fault inception time, and fault type. The authors compared their results with other deep learning architectures such as CNN and LSTM. The authors reported better performance for their model compared with other models.

A simulation model using Matlab was developed in [86] to generate fault signals of an underground cable distribution system consisting of sixteen cables. The fault types are ground fault, short-circuit fault, or open-circuit fault for each of the three phases of the sixteen cables. The authors proposed a deep belief network for this purpose and they compared the classification accuracy with a shallow neural network. The authors reported better results for the proposed deep learning model.

The authors of [87] located aged cable segments in underground power distribution systems labeled as even ageing, uneven ageing, and terminal ageing patterns. The signals were captured using an HFCT. The authors proposed a combined stacked autoencoder and CNN architecture for detecting aged segments. When an aged segment is detected, another CNN model was developed to indicate the location and severity of the aged segment. The authors compared the results with other machine learning models such as support vector regression and deep belief network. The proposed model performed better compared with other models for both detection and localization of the aged segments.

Ref. [88] proposed a technique to detect the inception faults that take place in cables. The authors used PSCAD/EMTDC in order to simulate the inception fault signals, over-current disturbance signals, and normal current signals. The authors proposed an architecture that includes a sparse autoencoder followed by a deep belief network. They compared the classification accuracy with support vector machine and K-nearest neighbor, where they reported better results performed by the proposed model.

The authors of [89] aimed to detect inception faults as well. They used PSCAD/EMTDC as well, where variability in the generated signals was introduced by changing some parameters in the simulation model such as fault impedance and fault location. The authors proposed restricted Boltzmann machine to compress the signals. This was the input to the stacked autoencoder. The authors investigated the trained model performance with simulated and measured data. In addition, they compared the classification accuracy with other models such as CNN, deep belief network, and random forest. The proposed model outperformed the other models.

In [90], the authors detected phase-to-ground faults in a typical 10 kV resonant grounding distribution system, which was simulated using PSCAD/EMTDC. The signals were generated under different fault conditions including different fault locations, different grounding resistances, and different fault initial phase angles. Continuous wavelet transform (CWT) was applied to the signals to generate 2D images. The images were transformed to grey-scale, and this was the input to the CNN. The authors investigated the robustness of the model to different parameters such as interference. The authors compared the results with SVM and adaBoost, and they reported that the proposed model gave better results.

Although the investigation of different DL techniques is essential for the sake of completeness of any work, it is observed that having common, publicly-available datasets can help in focusing on the generalization of any developed algorithm. In addition, as mentioned in previous sections, the investigation of the effect of any preprocessing technique is necessary in order to have a deeper understanding of the intrinsic characteristics that the DL model is learning.

7. Transformers

Power transformers play a significant role in power systems, so any failure in this apparatus may interrupt the power supply and cause outages and loss of profit. A photo of a power transformer is shown in Figure 9. One of the beneficial methods for preventing the failure in the power transformers and raising the reliability of these systems is detecting faults in power transformers accurately and promptly.

Figure 9. (**a**) A photo of a power transformer, (**b**) acoustic sensors installed on the tank of a transformer.

Whether the target is to classify/localize the sources of PDs/faults taking place in a transformer, or to identify overheat or vibration, deep learning has been used for this purpose. The following section summarizes the literature on the use of deep learning for the transformer application.

7.1. PD Classification Using the PRPD Pattern

Four typical transformer insulation defects were simulated in [16] that include metal protrusion, oil paper void, surface discharge, and floating potential defects. The authors developed a CNN-LSTM based model where the input is the PRPD data. They compared the results with a CNN-only model and an LSTM-only model. The evaluation metric used was the classification accuracy where the authors reported that CNN-LSTM has better overall recognition accuracy than CNN and LSTM alone.

In 2020, the authors of [17] simulated six types of PDs that take place in power transformers using artificial cells in a laboratory setup. They collected the PRPDs of the six PDs which include protruding electrode, moving particle, floating object, surface discharges, bad contact between windings, and void. In order to reduce the input size of the PRPD, the authors used the phase-amplitude (PA) response that is extracted from PRPDs. The authors proposed a CNN model for classification of PD sources. Comparing the classification accuracy of the proposed architecture versus other machine learning classifiers, such as linear and nonlinear SVM, the authors reported a better performance. They also reported that using the PA response as an input increases the accuracy by 1.46% compared to using the raw PRPDs as the input to the CNN model.

The PRPD data of different PD sources in a transformer were collected in a laboratory-controlled setup and reported in [18]. The PD sources included tip discharge, surface discharge, air gap discharge and suspended discharge. The squeeze-and-excitation (SE) module that is a lightweight attention mechanism and the nonlinear function hard-swish (h-swish) were used in addition to a CNN model in order to decrease the accuracy loss of the model further. The authors performed image pre-processing such as segmentation, binarization and enhancement of the data before feeding it to the training model. They compared the results of their model versus other models such as AlexNet, ResNet-18, and VGG16. They reported enhanced average accuracy versus the other models, in addition to less weight storage and reduction of parameters.

In the same year, an investigation of a transformer bushing insulation quality, which was affected by poor drying and impregnation, was reported in [19]. The authors used a simple CNN (i.e., 3300 parameters) for the identification of four types of dry impregnation defects using PRPDs as the input to the proposed CNN. The performance metrics used for the evaluation of the model are the precision rate, recall rate, and F1 score. The authors reported 97.1% average accuracy rate and indicated that their model can be used for online monitoring as it is a small model.

A novel convolutional architecture for single and multiple source PD classification, where the model is trained on single-source PDs, was proposed in [20]. The dataset included PRPDs of single and multiple sources of PD taking place in air, oil, and SF_6 which mimic common sources of PD. The six single PD sources of floating electrode in SF_6, moving particle in SF_6, fixed protrusion in SF_6, free particle in transformer oil, needle electrode in transformer oil, and corona in air were simulated in a laboratory setup. The proposed architecture has a convolutional backbone feeding into multiple fully connected neural networks (FCNs). The performance metrics used are the arithmetic mean of recall and precision in addition to the classification accuracy and false negative rate. The authors compared their results with one-versus-all CNN and reported that their model has better results than the traditional single-branch CNN architecture.

Adam et al. [21] simulated six artificial PD sources in a controlled lab environment that mimics PDs in a power transformer. The PD sources include two discharge sources in air and four discharge sources in mineral oil. The time at which the discharge takes place, the apparent discharge in pC, and the phase angle are recorded for each PD event. In addition, 100 PD events constituted a sample. Superimposed patterns were created by using the single sources patterns, where 30 different combinations of samples with two class labels are formed. The authors proposed an LSTM model which is able to classify multiple and single sources of PDs, where the training was done just on single sources of PDs. The study reported the multi-label accuracy in addition to the single-label accuracy. The multi-label accuracy is defined as the proportion of the correctly predicted labels to the total number of labels for each sample. The model showed a 99% average accuracy for single PD sources and 43% for the average multi-label classification problem.

Ref. [22] outlined the same objective as the previous two papers. Single and multiple sources of corona discharge in a controlled lab environment were simulated. The four single sources were: sphere–plane, sphere–sphere, needle–plane, and needle–needle. The three multiple sources were: needle–needle and sphere–sphere, needle–plane and sphere–sphere, and sphere–plane and needle–needle. The PRPD patterns were collected, and pre-processing was performed by filtering discharges that have small magnitude. The input to the deep learning models were greyscale images of 75 by 75 pixels. The classes were labeled from 0 for the first single class to 6 for the double-sourced configuration that is considering the multi-source classes as a new class. The authors proposed an optimized ResNet model which they compared with other DL models such as AlexNet, Inception-V3, residual network (ResNet), and DenseNet. The study reported enhanced classification accuracy and least computational cost.

7.2. Dissolved Gas Analysis Using Deep Learning

The dissolved gas analysis (DGA) is an established method for detecting internal faults in transformers. In recent years, the application of deep learning in DGA has received more attention. This section reviews the literature where the researchers have used deep learning to improve the accuracy of the DGA.

The DGA of insulating oil was conducted for transformer fault diagnosis by Dai et al. in 2017 [91]. To improve the efficiency of diagnosis, the authors proposed a novel transformer fault diagnosis approach based on deep belief networks (DBN), which outperforms power transformer fault diagnosis using support vector machine (SVM), back-propagation neural network (BPNN), and ratio methods. A variety of sources were used to collect the input DGA data, including data provided by the State Grid Corporation of China and previous publications. The proposed model was trained using different combinations of DGA ratios associated with fault patterns (the so-called non-code ratios). The training and testing accuracy of 96.4% and 95.9% were observed for the DBN with non-code ratios in this study, respectively.

In 2020, a semi-supervised autoencoder with an auxiliary task (SAAT) was introduced by Kim et al. to extract a health feature space for power transformer fault diagnosis considering DGA [92]. The DGA dataset was provided by Korea Electric Power Corporation. The proposed SAAT achieved an accuracy of over 90% in both fault detection and fault identification. The same group has also developed a framework that bridges Duval's method with a deep neural network (DNN) technique for power transformer fault diagnosis employing DGA [93]. The dataset employed contains 4000 unlabeled and 117 labeled DGA data. The obtained results emphasize the superiority of the proposed method compared with the existing AI-based methods in terms of accuracy.

Wu et al. introduced a CNN-LSTM deep parallel diagnostic method for transformer DGA employing its ability to extract nonlinear features [94]. The authors showed that this method has a better anti-interference ability compared to the other techniques studied in the paper. In this parallel CNN-LSTM based diagnostic method, the input was in the form of an image derived from the DGA numerical data. The issue of insufficient data was overcome by using the transfer learning technology. The results obtained in this study indicate that the diagnostic accuracy rate is 96.9% without complicated feature extraction.

In 2021, Taha et al. presented a CNN model to precisely diagnose a variety of transformer faults using DGA data and considering different noise levels [95]. The results obtained from applying the proposed method on 589 dataset samples, collected from utilities and literature with various noise levels up to ±20%, indicate that the CNN model with combined input ratios improves the prediction accuracy. The obtained accuracy was compared to traditional machine learning methods as well.

To improve fault diagnosis in transformers, Hu et al. [96] proposed a method based on refined deep residual shrinkage network (DRSN). The input dataset was based on the amount of gas in the transformer oil, the temperature data, and the number of collected data points based on the timing sequence. The recognition results indicate that the average accuracy of refined DRSN is around 99.67% for the training set and 97.82% for the test set.

On average, the proposed method could improve the recognition accuracy by 2% compared to the existing fault diagnosis methods.

A probabilistic neural network (PNN)-based fault diagnosis model was presented by Zhou et al. for power transformers [97]. This model optimizes the smoothing factor of the pattern layer of PNN using an optimization technique, improved gray wolf optimizer (IGWO), to enhance the classification accuracy of the PNN. Different fault types data from a real transformer were collected using smart sensors. The obtained results indicate a high diagnostic accuracy of 99.71% achieved by the IGWO-PNN model.

7.3. Detect Mechanical Defects in Winding

In 2021, Rucconi et al. [98] analyzed the vibration data measured by transformer sensors, such as accelerometers, installed on the transformer tank. These sensors record time series waveforms to build a dataset. An ensemble of fully-connected, feedforward deep neural networks was employed to classify the transformer winding condition (tight or loose). The robustness of the models was investigated by testing them with data collected by sensors at locations other than those used for training. The authors reported a high accuracy in the results.

Li et al. used the frequency response analysis (FRA) for detecting the mechanical defects of power transformers in 2021 [99]. The authors employed a lumped-parameter transformer model since creating actual faults experimentally on a real transformer was not practical. The proposed deep learning approach was based on a decision tree classification model and a fully connected neural network that used the FRA data for training. Fifty-five FRA samples were generated as the input to the proposed model by simulating a variety of transformer fault types and levels. The mean absolute error (MAE) and mean square error (MSE) of the validation set were both at a low level, which were employed to reflect the accuracy of the model.

A fault diagnosis technique was proposed in [100] by Hong et al. using the vibration analysis. The vibration samples were collected in more than 100 operating transformers and were divided into the categories of normal, degraded, and anomalous. Next, the vibration monitoring data were converted into an image. A deep learning method based on a CNN was employed to classify the images of various input sizes, which indicated an overall accuracy of 98.3%.

A fault diagnosis method based on a deep learning model was presented by Wang et al. applied on a 110 kV three-phase oil-immersed transformer in 2018 [101]. The model used self-powered radio-frequency identification (RFID) sensors and employed the stacked denoising autoencoder (SDA) to learn features. Based on experimental results, the highest accuracy was achieved by the proposed methods and in the shortest time in comparison with other existing methods.

In 2021, a method was proposed by Moradzadeh et al. for analyzing a transformer FRA using image processing and a deep learning method, graph convolutional neural network (CNN) [102]. The obtained results using simulation data indicate that the normal mode of CNN (without considering visual images) and with considering the visual images have an accuracy of 97.28% and 98.33%, respectively. Using experimental data, an accuracy of 98.01% and 100% was reported, respectively.

7.4. Detect Electrical Faults in Winding

The authors of [103] proposed a CNN model for the identification and localization of faults in transformer winding. The dataset was collected by generating single/multiple disc-to-disc faults of winding insulation in a transformer model at different winding positions, where the current waveforms were recorded. The faults were generated in an analog model of a 33 kV winding of a 3 MVA transformer. The training dataset was generated using an EMTP (Electromagnetic Transient Programming) based digital model, and the test dataset included the data collected from the analog model of the transformer. The results of the CNN model were compared with other methods such as self-organizing maps, fractal features aided SVM, and wavelet-aided SVM. The CNN showed improved classification results compared to the other methods.

In 2019, Duan et al. [104] presented an inter-turn fault diagnosis technique to diagnose 15 types of an inter-turn short circuit fault. A multi-channel signal matrix that contains voltage and current waveforms of a simulated transformer was generated as the input of a deep learning-based model. An autoencoder was employed for feature extraction followed by a classifier consisting of convolutional and pooling layers. The proposed model could achieve a recognition accuracy of 99.5%.

An online continuously-operating fault monitoring system for cast-resin transformers was presented by Fanchiang et al. in [105]. They presented an overheating fault diagnosis approach with a maximum accuracy of 99.95%. The model used infrared thermography (IRT) images as input provided by a thermal camera monitoring system. The images were used to train a Wasserstein autoencoder reconstruction (WAR) model and a differential image classification (DIC) model to classify a number of faults such as inter-turn short circuit or poor contact of primary and secondary sides.

Wu et al. introduced an approach to obtain an optimal identification of the operation state of a converter transformer based on vibration detection technology and a deep belief network optimization algorithm [106]. The fused feature extraction technique considered in this study accurately extracted the eigenvectors of the vibration signals. This deep belief network optimization algorithm has offered a high classification accuracy.

Similar to the application of DL in any area, the prerequisite of developing any DL algorithm to assess the health of a transformer is the availability of reliable training data. Some data acquisition techniques for transformers are easy to implement even with the transformer in operation (such as DGA or acoustic signals) while some (such as FRA) require an outage. Equipping transformers with sensors that can collect internal data will enhance the application of DL techniques. Most of the papers in the literature either employ simulated lab data or even numerically-generated data. Training a DL algorithm with such data will make the performance of the algorithm in the field require proper validation. Furthermore, research on the application of DL techniques that employ polarization methods data, such as recovery voltage method (RVM), polarization and depolarization currents (PDC), and frequency-domain dielectric spectroscopy (FDS) will enhance the ageing diagnosis of the transformer dielectric materials.

8. Outdoor Insulators

Outdoor insulators play an important role in distribution and transmission overhead lines. They mechanically support the high voltage conductors and electrically insulate the high voltage lines from the grounded tower structure. Although they account for approximately 5–8% of the total capital cost of transmission lines, they are responsible for more than 70% of power line outages [107]. Therefore, it is crucial to continuously inspect them to avoid any risks of premature failure. A photo of a 220-kV overhead transmission line is shown in Figure 10.

Outdoor insulators are classified into ceramic and non-ceramic insulators. Despite the differences in their characteristics, both are prone to aging due to the combined effect of electrical, mechanical, and environmental stresses. The main problems associated with outdoor insulators can be categorized into physical defects and pollution related issues. Physical defects like cracks in any parts of the insulator, air voids in the housing material or in the interface between various insulator materials and metallic sharp edges of insulator fittings can cause localized partial discharge (PD) activities to occur. These discharge activities can contribute to the insulation degradation. On the other hand, the accumulation of pollutants on the surface of insulators in the presence of moisture can reduce the leakage resistance and allow leakage currents (LCs) to flow on the surface. Therefore, heat is dissipated, evaporating part of the moisture, and forming dry bands. Since dry bands possess a relatively higher resistance compared to the wet surfaces, voltage stress will concentrate across these bands resulting in the formation of dry band arcing which may lead to complete flashover. As a result, estimating and forecasting pollution levels is critical for utilities to plan their washing schedule to avoid power interruption. Both the Equivalent Salt Deposit Density (ESDD) and the Non-Soluble Deposit Density (NSDD) are commonly used to assess pollution severity.

Figure 10. Photos of (**a**) a 220-kV overhead transmission line where outdoor insulators are employed to electrically isolate the conductors from the grounded towers, and (**b**) is a 12.47-kV wooden post using polymer insulators for isolating the conductors from the pole.

Nowadays, it is estimated that approximately 150 million ceramic insulators are deployed in North American overhead transmission and distribution networks [108]. A significantly high portion of them had either approached or exceeded their lifetime. As a result, utilities are increasingly favoring defective insulator detection systems that are fast, reliable, and cost-effective. To achieve this objective, it is crucial to select the proper sensors and measurement techniques, as well as to use effective machine/deep learning tools [109,110].

Condition monitoring techniques of outdoor insulators can be classified into intrusive and non-intrusive techniques. Since intrusive techniques are not safe, costly and may require the removal of the insulators from the field for further examination, they are time consuming and are not field inspection friendly. Non-intrusive techniques, on the contrary, are faster methods for assessing the health conditions of outdoor insulators and are therefore more favored in field inspections.

One of the most common non-intrusive inspection techniques deployed in the field involves the use of manned helicopters equipped with several sensors (like Cameras, IR Cameras, UV Cameras, etc.) for the purpose of recording inspection data. However, this possesses some risk since helicopters require hovering very closely to the electric power transmission line to obtain a better quality of the inspection data. To address this, several alternative solutions were proposed, and they fall mostly under two main categories: Unmanned Aerial Vehicles (UAVs) [111] and Rolling on Wire robots (ROW) [112]. Most of the work in the literature is moving towards UAVs because they have a slight advantage over ROW in the sense that their design does not need to adapt to different physical structures. However, UAVs also exhibit some restrictions in terms of the flying duration, which introduces some challenges when it is required to inspect a large number of insulators in the same trip. The massive volumes of data acquired throughout the inspection process are examined by an experienced crew of human inspectors, hence, the procedure may be time consuming, and the decisions made can be very subjective. Therefore, it is crucial to fuse artificial intelligence modules with UAVs for faster and better inspection performance.

In the recent years, machine leaning methodologies have evolved towards the use of deep learning techniques, which have proved to deliver excellent results for pattern recognition problems in a variety of applications. As a result, several authors have used deep learning models to assess the insulator condition. One of the drawbacks of deep learning models is its requirement of a large amount of training data. Despite this limitation, proper use of augmentation techniques can resolve this issue by altering the existing data to create more data for the model training process.

The aim of this section is to review deep learning applications along with non-intrusive condition monitoring techniques to assess both ceramic and non-ceramic insulators. In general, the work in the literature can be classified into two main categories, i.e., using deep learning models to detect different physical defects and/or predict the pollution severity level. Both categories relied primarily on either image processing or radiation-based approaches for classification. In the next sections, each category will be discussed along with the most recent research findings.

8.1. Physical Defect Detection

In ceramic insulators, researchers focused on detecting cracks, broken and missing discs in using UAVs. Most of the proposed methods based on deep learning algorithms share the same concept, i.e., the insulator in each image is located using object detection techniques; then, the defect is identified by a pre-trained deep neural network. The existing deep learning algorithms for object detection can be classified into two main categories: one stage and two stage networks [113]. Two stages networks consist of one stage for object detection and another stage for classification, while one-stage networks are end-to-end methods which can predict the position information and classification probability simultaneously in a rapid manner. Generally, two-stage networks possess a higher detection accuracy compared to one-stage networks; however, they have a relatively lower detection speed and thus may not be the best option for real-time operations. Some examples of two stage networks include: regions with convolutional neural networks (RCNN), Fast R-CNN, Region based fully convolutional neural network (R-FCN) and Mask R-CNN and examples of one-stage networks include algorithms like YOLO and single multi-box detector (SSD).

An example of two-stage networks is proposed in [114]. It is based on a novel deep CNN cascading architecture. The cascaded architecture is composed of two networks: the first network is responsible for detecting all the insulators in the images by confining them inside detection boxes and cropping them while the rest of the image is discarded. On the other hand, the second network detects the missing caps from the cropped images. The scarcity of the defective images for training was addressed by different data augmentation methods. The precision and recall of the proposed method were found to be 91% and 96%, respectively.

To address the issues of slow detection speeds, the authors of [115] proposed a one-stage network using a YOLOv3 deep learning model to recognize and classify images. Moreover, their proposed system combines deep learning with Internet of Things (IoT) through a Raspberry Pi. The work also considered the motion blur in aerial images by implementing a super resolution CNN to reconstruct the blurry images to a high-resolution image before classification. The results show that the proposed system obtains rapid and high accuracy of 95.6% in the identification and classification of insulators' defects. One of the early signs of surface damage of non-ceramic insulators is the loss of their hydrophobicity. Hence, measuring the hydrophobicity is crucial for assessing the insulator surface condition. According to IEC 62073, there are three methods to estimate the hydrophobicity level of insulators, i.e., the contact angle, surface tension and the spray methods [116]. Among these three methods, the spray method is the one that can be applied in the field. The method involves spraying distilled water on the non-ceramic insulator surface; then, the surface can be classified from HC1 (highly hydrophobic) to HC7 (Highly hydrophilic) as shown in Figure 11. The classes are determined based on the size of the wetted area and the contact angle of the droplets. Unfortunately, the main drawback of this method is the subjectivity of human judgment. To overcome this issue, numerous researchers have proposed digital image processing methods to analyze and quantify the hydrophobicity class.

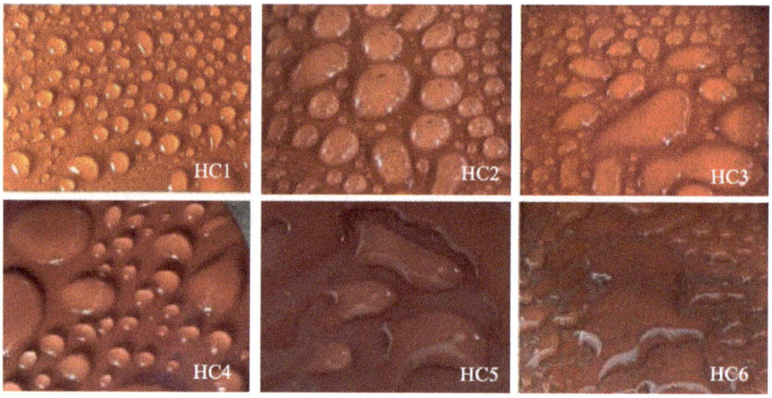

Figure 11. Hydrophobic classes from HC1 (highly hydrophobic) to HC6 (highly hydrophilic) [117].

In [118], the spray method was used to generate a huge amount of data images which were fed to a deep convolutional neural network model (AlexNet) for the purpose of wettability classification. Compared to other machine learning algorithms, deep learning overcomes the manual dependency on feature extraction and involves less training time due to the transferred learning approach that was used in the article. The algorithm's performance was very promising when compared to other networks like ResNet50, VGGNet16, VGGNet19 and GoogleNet with an overall accuracy of approximately 96%. However, this method may require the removal of the insulator from the field which can be impractical.

To resolve this problem, the authors of [119] proposed a method to detect the hydrophobicity of composite insulators using a UAV technology. The drone is equipped with a camera and water spray device in addition to an embedded artificial intelligence (AI) module for non-intrusive classification. Initially, the You Only Look Once version3 (YOLOv3) is used to locate the wet umbrella skirt area of the composite insulator in the complex aerial image, then VGGNet16 was used to classify the Hydrophobicity of the images. An overall classification accuracy of 92.57% was achieved.

Other approaches included using image data and deep learning algorithms to assess the material surface degradation in composite insulators. Tracking and Erosion is one of the irreversible physical defects that occurs in non-ceramic insulators which can lead to insulator failure. In [120,121], the authors used transfer learning to train CNN to estimate the severity of erosion in silicone rubber insulators. The algorithm showed robust performance against different lighting conditions which shows the potential of their proposed model in practical applications.

To the best of our knowledge, there seems to be a gap in the literature that involves training deep learning models to detect internal and external physical defects using radiation-based measurements like RF antenna and ultrasonic sensors. All the work that has been done on radiation-based techniques involves the use of feature extraction and machine learning techniques [122–126].

8.2. Contamination Diagnosis

Several methods have utilized image processing techniques to classify the pollution severity. In [127], for example, a total of 4500 images of ceramic and silicone rubber post insulators were captured under different surface conditions, i.e., clean dry surface, clean with water droplets, contaminated surfaces with cement, contaminated surfaces with soil, wet surface contaminated with soil and wet surface contaminated with cement. Deep CNNs were employed for classification, and a brute-force model selection was introduced to identify and optimize the structure of the CNN classifiers. It was demonstrated that this model selection has achieved a highly accurate architecture. Furthermore, a complexity reduction technique was then applied to achieve lighter architectures. This considers the potential of implementing the CNN classifier in resource limited embedded devices. The results show that this proposed model reduction technique corresponds to a three times lighter architecture at the expense of a slight reduction in the classification accuracy (6.5% only). This is intended to reduce memory usage and flop counts when implemented using embedded devices.

In [128], the pollution severity was estimated using UV images. First, Insulator samples were uniformly contaminated with an ESDD level of 0.1 mg/cm^2, 0.2 mg/cm^2 and 0.4 mg/cm^2. After applying the voltage, a UV camera was used to capture the discharge activities on the surface of contaminated porcelain insulators. The images were then preprocessed by first graying the image, then changing the pixels to 0 or 255. When doing so, the light spot becomes white, while the rest of the image becomes black, thus highlighting the regional characteristics of the discharge spot in the image. Finally, CNN was used to evaluate the pollution severity of the insulators. It has been found that there is a positive correlation between the pollution level and the severity of the discharge activities under the same voltage level.

Other approaches used deep learning models to estimate the LC and an indirect method to estimate the contamination level. For example, the authors of [129] proposed an online monitoring system that uses real-time weather data to predict and classify the LC using bidirectional long short-term memory (Bi-LSTM) model. The sequential weather data consist of parameters like humidity, temperature, rainfall, dew point, solar illumination, wind speed, air pressure and wind direction. They are measured hourly and transferred to data servers. Besides the meteorological data, the LCs are also measured for the purpose of training and validation of the networks using the current transformer. The LC is classified into one of eight groups (levels): i.e., 100 µA–500 µA, 500 µA–1 mA, 1 mA–5 mA, 5 mA–10 mA, 10 mA–100 mA, 100 mA–1 A, 1 A–10 A, and greater than 10 A. Grid search is used to tune the hyperparameters involved in the Bi-LSTM model. The results show that the model achieved an improvement by 12.8% in accuracy compared to other models like LSTM, GRU and RNN.

Seven PD sources pertinent to artificially damaged insulator sheds in a controlled lab experiment were simulated in [42]. The first three sources corresponded to damage in one shed of an HV insulator. The other four correspond to damage in two or all sheds in the HV insulator. For this matter, the CNN was used, however, in order to tune the hyperparameters in the CNN architecture, the authors used Bayesian optimization. To generate the training data, the scalogram pattern of the PD signal was generated and transformed using wavelets. Three different mother wavelets (Morse, Amor, and Bump) were used. In addition, different training optimizers' (including stochastic gradient descent with momentums (SGDM), RMSprop, Adam, and Nadam) were used. The authors compared the Bayesian-CNN (B-CNN) with the traditional CNN with no Bayesian optimization in addition to other off-the-shelf deep learning architectures such as VGG19, Resnet 50, and Googlenet. The average classification accuracy was used as the performance metric. The study reported enhanced results for the B-CNN compared to the other architectures with the Bump mother wavelet. The authors also tested the model on another 15-kV porcelain insulator dataset, and the average classification accuracy showed optimistic results which reflect the generalization capabilities of the B-CNN.

In general, the literature shows an inadequate amount of work that has been devoted to train deep network architectures to classify and predict pollution levels non-intrusively. All the work done is either focused on applying deep learning models to intrusive measurement techniques [130,131] or applying classical machine learning using non-intrusive techniques [132–135]. More research is needed to combine deep learning models with non-intrusive approaches for monitoring, particularly those based on radiation type sensors. Furthermore, the majority of the publications were focused on employing one type of sensor for their diagnostics, although this may yield satisfactory results, but could be further improved using multiple sensors. For example, ultrasonic sensors can detect both low and high frequency surface discharges, but it might be difficult to detect internal discharges. On the other hand, RF antennas can be utilized to detect internal and external high frequency discharges but cannot be used to detect low frequency discharges. Thus, combining ultrasonic sensors and RF antenna can be used to detect and classify a wider range of defects.

9. Conclusions

Monitoring of electrical insulation of high voltage apparatus is crucial for the reliable operation of power systems. Such a high voltage apparatus includes but is not limited to gas-insulated switchgear (GIS), transformers, cables, rotating machines, and outdoor insulators. Extensive research has been done on the classification of sources of partial discharge (PD), detection and localization of faults that take place in such apparatus, and the quality and remaining lifetime of insulating material. Modern techniques have been based on machine learning methods, where the input to such methods is composed of manually-extracted features, i.e., feature extraction has required the intervention of human experts. Deep learning, which is a branch of machine learning, has been used to enhance the performance of PD classification, fault and defect detection, contamination diagnosis of outdoor insulators, etc. This enhancement is attributed to the capability of deep learning techniques to use raw data as the input to the classification model. In other words, instead of using manually-extracted features, raw data such as PRPD patterns, time-series waveforms, or images are used as the input to the deep learning systems. This allows the classification model to be fully automated where the feature extracting stage is integrated into the learning stage.

In this article, the potential of applying deep learning in assessing the health conditions of different power system assets is highlighted. The following shortcomings/future needs are identified:

1. Most published research employs training data generated in a laboratory environment or by computer simulation that leads to achieving high classification accuracy. A limitation is always presented when data are collected in a controlled lab environment due to the fact that acquiring real data is expensive, intrusive, and time-consuming. Hence, integrating research work in real online or offline systems is always appreciable in order to incorporate all the uncertainties of such systems in the learning process of deep-learning models. Moreover, future research should focus more on the utilization of the use of a generative adversarial network in order to generate more data that mimic real data instead of using lab data that present its own limitations. In addition, future directions should focus more on the utilization of DL techniques such as one shot learning [136] towards the issue of small datasets, which is a typical restriction in the HV application.
2. Prior knowledge of the defect types and/or knowing the exact location of the defect is far from the reality of the field conditions. Moreover, unknown sources and types of external noise may hinder the deep learning algorithms capabilities to identify and/or localize the defect type. Hence, future research needs to focus more on unsupervised learning when it comes to high voltage applications.
3. More work should focus on the occurrence of multiple, simultaneous PDs or faults. The reason is that, in real-life systems, multiple sources of faults or PDs can take place at the same time. Therefore, more focus should be directed towards this problem.
4. One of the limitations of the reported research is the utilization of single sensors like ultrasonic sensors, RF antenna, or IR camera. It is expected that the use of multiple sensors can improve the overall classification accuracy when sensor fusion is applied and different 1D and/or 2D signals are fed to the deep learning classifiers.
5. Integrating the state-of-the-art deep learning algorithms along with promising technologies like drones can improve the inspection efficiency of outdoor insulation systems. With the current improved computational power of micro-controllers, real-time condition monitoring and diagnostics of different defects are feasible using drones and deep learning algorithms.

6. Future research directions should focus on developing electrical insulation ageing models using DL techniques that employ polarization methods data, such as RVM, PDC, and FDS.
7. Using deep learning techniques in the high voltage application is still in the starting stage. More work should be done on deciding on the best standard to specify the optimal architecture per application. One aspect of this is utilizing the use of already established hyperparameter optimization techniques such as the Bayesian optimization technique. In addition, the industrial deployment of the DL algorithms should be addressed, since this requires a different action for each scenario. If the deployment is taking place on a local server, the aim would be to maximize the performance of the algorithm while taking advantage of high-speed and high-end hardware resources. This is different in case the deployment is to take place on a portable monitoring device, where the restriction of space and speed will be presented.
8. With the emerging of the digital twin technologies, deep learning should be utilized for different digital twins of assets, such as transformers or rotating machines. Digital twins are virtual representations of the interactions and behavior that assets can undergo in the physical world. More information on the application of digital twins in power system assets can be found in [137–139].

Author Contributions: Conceptualization: S.M., A.L., H.M.T., A.A., A.E.-H., B.K.; formal analysis: S.M., A.L., H.M.T., A.A., A.E., B.K.; funding acquisition: A.E.-H., B.K.; funding acquisition: S.M., A.L., H.M.T., A.A., A.E.-H., B.K.; supervision: A.A., A.E.-H., B.K.; Writing—original draft: S.M., A.L., H.M.T.; writing—review and editing: S.M., A.L., H.M.T., A.A., A.E.-H., B.K. All authors have read and agreed to the published version of the manuscript.

Funding: Financial support from the Natural Sciences and Engineering Research Council of Canada (NSERC) for both the University of Manitoba and University of Waterloo, and the Faculty of Graduate Studies, University of Manitoba is acknowledged.

Institutional Review Board Statement: Not applicable.

Informed Consent Statement: Not applicable.

Data Availability Statement: Not applicable.

Conflicts of Interest: The authors declare no conflict of interest.

References

1. Luo, Y.; Li, Z.; Wang, H. A review of online partial discharge measurement of large generators. *Energies* **2017**, *10*, 1694. [CrossRef]
2. Ardila-Rey, J.A.; Ortiz, J.E.; Creixell, W.; Muhammad-Sukki, F.; Bani, N.A. Artificial generation of partial discharge sources through an algorithm based on deep convolutional generative adversarial networks. *IEEE Access* **2020**, *8*, 24561–24575. [CrossRef]
3. Sinaga, H.; Phung, B.; Blackburn, T. Partial discharge localization in transformers using UHF detection method. *IEEE Trans. Dielectr. Electr. Insul.* **2012**, *19*, 1891–1900. [CrossRef]
4. Lu, S.; Chai, H.; Sahoo, A.; Phung, B. Condition monitoring based on partial discharge diagnostics using machine learning methods: A comprehensive state-of-the-art review. *IEEE Trans. Dielectr. Electr. Insul.* **2020**, *27*, 1861–1888. [CrossRef]
5. Long, J.; Wang, X.; Zhou, W.; Zhang, J.; Dai, D.; Zhu, G. A comprehensive review of signal processing and machine learning technologies for UHF PD detection and diagnosis (I): Preprocessing and Localization Approaches. *IEEE Access* **2021**, *9*, 69876–69904. [CrossRef]
6. Wu, M.; Cao, H.; Cao, J.; Nguyen, H.L.; Gomes, J.B.; Krishnaswamy, S.P. An overview of state-of-the-art partial discharge analysis techniques for condition monitoring. *IEEE Electr. Insul. Mag.* **2015**, *31*, 22–35. [CrossRef]
7. Hussain, M.R.; Refaat, S.S.; Abu-Rub, H. Overview and partial discharge analysis of power transformers: A literature review. *IEEE Access* **2021**, *9*, 64587–64605. [CrossRef]
8. Tang, J.; Jin, M.; Zeng, F.; Zhang, X.; Huang, R. Assessment of PD severity in gas-insulated switchgear with an SSAE. *IET Sci. Meas. Technol.* **2017**, *11*, 423–430. [CrossRef]
9. Nguyen, M.T.; Nguyen, V.H.; Yun, S.J.; Kim, Y.H. Recurrent neural network for partial discharge diagnosis in gas-insulated switchgear. *Energies* **2018**, *11*, 1202. [CrossRef]
10. Tuyet-Doan, V.N.; Do, T.D.; Tran-Thi, N.D.; Youn, Y.W.; Kim, Y.H. One-Shot learning for partial discharge diagnosis using ultra-high-frequency sensor in gas-insulated switchgear. *Sensors* **2020**, *20*, 5562. [CrossRef]
11. Tuyet-Doan, V.N.; Nguyen, T.T.; Nguyen, M.T.; Lee, J.H.; Kim, Y.H. Self-attention network for partial-discharge diagnosis in gas-insulated switchgear. *Energies* **2020**, *13*, 2102. [CrossRef]

12. Song, H.; Dai, J.; Sheng, G.; Jiang, X. GIS partial discharge pattern recognition via deep convolutional neural network under complex data source. *IEEE Trans. Dielectr. Electr. Insul.* **2018**, *25*, 678–685. [CrossRef]
13. Dai, J.; Teng, Y.; Zhang, Z.; Yu, Z.; Sheng, G.; Jiang, X. Partial discharge data matching method for GIS case-based reasoning. *Energies* **2019**, *12*, 3677. [CrossRef]
14. Karimi, M.; Majidi, M.; MirSaeedi, H.; Arefi, M.M.; Oskuoee, M. A novel application of deep belief networks in learning partial discharge patterns for classifying corona, surface, and internal discharges. *IEEE Trans. Ind. Electron.* **2019**, *67*, 3277–3287. [CrossRef]
15. Florkowski, M. Classification of partial discharge images using deep convolutional neural networks. *Energies* **2020**, *13*, 5496. [CrossRef]
16. Zhou, X.; Wu, X.; Ding, P.; Li, X.; He, N.; Zhang, G.; Zhang, X. Research on transformer partial discharge UHF pattern recognition based on CNN-lSTM. *Energies* **2019**, *13*, 61. [CrossRef]
17. Do, T.D.; Tuyet-Doan, V.N.; Cho, Y.S.; Sun, J.H.; Kim, Y.H. Convolutional-neural-network-based partial discharge diagnosis for power transformer using UHF sensor. *IEEE Access* **2020**, *8*, 207377–207388. [CrossRef]
18. Sun, Y.; Ma, S.; Sun, S.; Liu, P.; Zhang, L.; Ouyang, J.; Ni, X. Partial discharge pattern recognition of transformers based on mobileNets convolutional neural network. *Appl. Sci.* **2021**, *11*, 6984. [CrossRef]
19. Liu, Y.; Hu, M.; Dai, Q.; Le, H.; Liu, Y. Online recognition method of partial discharge pattern for transformer bushings based on small sample ultra-micro-CNN network. *AIP Adv.* **2021**, *11*, 045221. [CrossRef]
20. Mantach, S.; Ashraf, A.; Janani, H.; Kordi, B. A convolutional neural network-based model for multi-source and single-source partial discharge pattern classification using only single-source training set. *Energies* **2021**, *14*, 1355. [CrossRef]
21. Adam, B.; Tenbohlen, S. Classification of superimposed partial discharge patterns. *Energies* **2021**, *14*, 2144. [CrossRef]
22. Borghei, M.; Ghassemi, M. A deep learning approach for discrimination of single-and multi-source corona discharges. *IEEE Trans. Plasma Sci.* **2021**, *49*, 2936–2945. [CrossRef]
23. Raymond, W.J.K.; Xin, C.W.; Kin, L.W.; Illias, H.A. Noise invariant partial discharge classification based on convolutional neural network. *Measurement* **2021**, *177*, 109220. [CrossRef]
24. Araújo, R.C.; de Oliveira, R.; Barros, F.J. Automatic PRPD image recognition of multiple simultaneous partial discharge sources in online hydro-generator stator bars. *Energies* **2022**, *15*, 326. [CrossRef]
25. Khan, M.A.; Choo, J.; Kim, Y.H. End-to-end partial discharge detection in power cables via time-domain convolutional neural networks. *J. Electr. Eng. Technol.* **2019**, *14*, 1299–1309. [CrossRef]
26. Peng, X.; Yang, F.; Wang, G.; Wu, Y.; Li, L.; Li, Z.; Bhatti, A.A.; Zhou, C.; Hepburn, D.M.; Reid, A.J.; et al. A convolutional neural network-based deep learning methodology for recognition of partial discharge patterns from high-voltage cables. *IEEE Trans. Power Deliv.* **2019**, *34*, 1460–1469. [CrossRef]
27. Li, Z.; Xu, Y.; Jiang, X. Pattern recognition of DC partial discharge on XLPE cable based on ADAM-DBN. *Energies* **2020**, *13*, 4566. [CrossRef]
28. Yeo, J.; Jin, H.; Mor, A.R.; Yuen, C.; Tushar, W.; Saha, T.K.; Ng, C.S. Identification of partial discharge through cable-specific adaption and neural network ensemble. *IEEE Trans. Power Deliv.* **2021**, *37*, 1598–1607. [CrossRef]
29. Zemouri, R.; Levesque, M.; Amyot, N.; Hudon, C.; Kokoko, O.; Tahan, S.A. Deep convolutional variational autoencoder as a 2d-visualization tool for partial discharge source classification in hydrogenerators. *IEEE Access* **2019**, *8*, 5438–5454. [CrossRef]
30. Wang, Y.; Yan, J.; Yang, Z.; Liu, T.; Zhao, Y.; Li, J. Partial discharge pattern recognition of gas-insulated switchgear via a light-scale convolutional neural network. *Energies* **2019**, *12*, 4674. [CrossRef]
31. Barrios, S.; Buldain, D.; Comech, M.P.; Gilbert, I. Partial discharge identification in MV switchgear using scalogram representations and convolutional autoEncoder. *IEEE Trans. Power Deliv.* **2020**, *36*, 3448–3455. [CrossRef]
32. Gu, F.C. Identification of partial discharge defects in gas-insulated switchgears by using a deep learning method. *IEEE Access* **2020**, *8*, 163894–163902. [CrossRef]
33. Wang, Y.; Yan, J.; Yang, Z.; Zhao, Y.; Liu, T. Optimizing GIS partial discharge pattern recognition in the ubiquitous power internet of things context: A MixNet deep learning model. *Int. J. Electr. Power Energy Syst.* **2021**, *125*, 106484. [CrossRef]
34. Wang, Y.; Yan, J.; Yang, Z.; Wang, J.; Geng, Y. A novel 1DCNN and domain adversarial transfer strategy for small sample GIS partial discharge pattern recognition. *Meas. Sci. Technol.* **2021**, *32*, 125118. [CrossRef]
35. Li, G.; Wang, X.; Li, X.; Yang, A.; Rong, M. Partial discharge recognition with a multi-resolution convolutional neural network. *Sensors* **2018**, *18*, 3512. [CrossRef] [PubMed]
36. Wang, Y.; Yan, J.; Yang, Z.; Jing, Q.; Qi, Z.; Wang, J.; Geng, Y. A domain adaptive deep transfer learning method for gas-insulated switchgear partial discharge diagnosis. *IEEE Trans. Power Deliv.* **2021**. [CrossRef]
37. Liu, T.; Yan, J.; Wang, Y.; Xu, Y.; Zhao, Y. GIS partial discharge pattern recognition based on a novel convolutional neural networks and long short-term memory. *Entropy* **2021**, *23*, 774. [CrossRef]
38. Michau, G.; Hsu, C.C.; Fink, O. Interpretable detection of partial discharge in power lines with deep learning. *Sensors* **2021**, *21*, 2154. [CrossRef]
39. Zunaed, M.; Nath, A.; Rahman, M. Dual-CyCon Net: A cycle consistent dual-domain convolutional neural network framework for detection of partial discharge. *arXiv* **2020**, arXiv:2012.11532.
40. Dong, M.; Sun, J. Partial discharge detection on aerial covered conductors using time-series decomposition and long short-term memory network. *Electr. Power Syst. Res.* **2020**, *184*, 106318. [CrossRef]

41. Li, Z.; Qu, N.; Li, X.; Zuo, J.; Yin, Y. Partial discharge detection of insulated conductors based on CNN-LSTM of attention mechanisms. *J. Power Electron.* **2021**, *21*, 1030–1040. [CrossRef]
42. Vigneshwaran, B.; Iruthayarajan, M.W.; Maheswari, R. Recognition of shed damage on 11-kV polymer insulator using Bayesian optimized convolution neural network. *Soft Comput.* **2022**, *26*, 6857–6869. [CrossRef]
43. Krizhevsky, A.; Sutskever, I.; Hinton, G.E. Imagenet classification with deep convolutional neural networks. *Adv. Neural Inf. Process. Syst.* **2012**, *25*, 1097–1105. [CrossRef]
44. Oord, A.v.d.; Dieleman, S.; Zen, H.; Simonyan, K.; Vinyals, O.; Graves, A.; Kalchbrenner, N.; Senior, A.; Kavukcuoglu, K. Wavenet: A generative model for raw audio. *arXiv* **2016**, arXiv:1609.03499.
45. Vaswani, A.; Shazeer, N.; Parmar, N.; Uszkoreit, J.; Jones, L.; Gomez, A.N.; Kaiser, Ł.; Polosukhin, I. Attention is all you need. *Adv. Neural Inf. Process. Syst.* **2017**, *30*.
46. Levy, J.J.; Titus, A.J.; Petersen, C.L.; Chen, Y.; Salas, L.A.; Christensen, B.C. MethylNet: An automated and modular deep learning approach for DNA methylation analysis. *BMC Bioinform.* **2020**, *21*, 1–15. [CrossRef] [PubMed]
47. Goodfellow, I.; Bengio, Y.; Courville, A. *Deep Learning*; MIT press: Cambridge, MA, USA, 2016.
48. Zhang, G.; Liu, Y.; Jin, X. A survey of autoencoder-based recommender systems. *Front. Comput. Sci.* **2020**, *14*, 430–450. [CrossRef]
49. Kiran, B.R.; Thomas, D.M.; Parakkal, R. An overview of deep learning based methods for unsupervised and semi-supervised anomaly detection in videos. *J. Imaging* **2018**, *4*, 36. [CrossRef]
50. Jaiswal, A.; Babu, A.R.; Zadeh, M.Z.; Banerjee, D.; Makedon, F. A survey on contrastive self-supervised learning. *Technologies* **2020**, *9*, 2. [CrossRef]
51. Sutton, R.S.; Barto, A.G. *Reinforcement Learning: An Introduction*; MIT press: Cambridge, MA, USA, 2018.
52. Fan, J.; Wang, Z.; Xie, Y.; Yang, Z. A theoretical analysis of deep Q-learning. In Proceedings of the Learning for Dynamics and Control, online, 10–11 June 2020; pp. 486–489.
53. Ning, B.; Lin, F.H.T.; Jaimungal, S. Double deep q-learning for optimal execution. *arXiv* **2018**, arXiv:1812.06600.
54. Gu, S.; Lillicrap, T.; Ghahramani, Z.; Turner, R.E.; Levine, S. Q-prop: Sample-efficient policy gradient with an off-policy critic. *arXiv* **2016**, arXiv:1611.02247.
55. Simonyan, K.; Zisserman, A. Very deep convolutional networks for large-scale image recognition. *arXiv* **2014**, arXiv:1409.1556.
56. He, K.; Zhang, X.; Ren, S.; Sun, J. Deep residual learning for image recognition. In Proceedings of the IEEE Conference on Computer Vision and Pattern Recognition, Las Vegas, NV, USA, 27–30 June 2016 ; pp. 770–778.
57. Gu, J.; Wang, Z.; Kuen, J.; Ma, L.; Shahroudy, A.; Shuai, B.; Liu, T.; Wang, X.; Wang, G.; Cai, J.; et al. Recent advances in convolutional neural networks. *Pattern Recognit.* **2018**, *77*, 354–377. [CrossRef]
58. O'Shea, K.; Nash, R. An introduction to convolutional neural networks. *arXiv* **2015**, arXiv:1511.08458.
59. Nair, V.; Hinton, G.E. Rectified Linear Units Improve Restricted Boltzmann Machines. In Proceedings of the 27th International Conference on Machine Learning, Haifa, Israel, 21–24 June 2010.
60. LeCun, Y.A.; Bottou, L.; Orr, G.B.; Müller, K.R. Efficient Backprop. In *Neural Networks: Tricks of the Trade*; Springer: Berlin/Heidelberg, Germany, 2012; pp. 9–48.
61. Yu, D.; Wang, H.; Chen, P.; Wei, Z. Mixed Pooling for Convolutional Neural Networks. In Proceedings of the International Conference on Rough Sets and Knowledge Technology, Shanghai, China, 24–26 October 2014; pp. 364–375.
62. Alzubaidi, L.; Zhang, J.; Humaidi, A.J.; Al-Dujaili, A.; Duan, Y.; Al-Shamma, O.; Santamaría, J.; Fadhel, M.A.; Al-Amidie, M.; Farhan, L. Review of deep learning: Concepts, CNN architectures, challenges, applications, future directions. *J. Big Data* **2021**, *8*, 1–74. [CrossRef] [PubMed]
63. Rumelhart, D.E.; Hinton, G.E.; Williams, R.J. Learning representations by back-propagating errors. *Nature* **1986**, *323*, 533–536. [CrossRef]
64. Zaremba, W.; Sutskever, I.; Vinyals, O. Recurrent neural network regularization. *arXiv* **2014**, arXiv:1409.2329.
65. Hochreiter, S.; Schmidhuber, J. Long short-term memory. *Neural Comput.* **1997**, *9*, 1735–1780. [CrossRef]
66. Graves, A.; Fernández, S.; Schmidhuber, J. Bidirectional LSTM Networks for Improved Phoneme Classification and Recognition. In Proceedings of the International Conference on Artificial Neural Networks, Munich, Germany, 17–19 September 2005; pp. 799–804.
67. Michalek, J.; Vaněk, J. A survey of recent DNN architectures on the TIMIT phone recognition task. In Proceedings of the International Conference on Text, Speech, and Dialogue, Brno, Czech Republic, 8–11 September 2018; pp. 436–444.
68. Rumelhart, D.E.; Hinton, G.E.; Williams, R.J. *Learning Internal Representations by Error Propagation*; Technical Report; University of California: Oakland, CA, USA, 1985.
69. Baldi, P. Autoencoders, unsupervised learning, and deep architectures. In Proceedings of the 2011 International Conference on Unsupervised and Transfer Learning Workshop, Bellevue, WA, USA, 2 July 2011; pp. 37–50.
70. Hinton, G.E.; Salakhutdinov, R.R. Reducing the dimensionality of data with neural networks. *Science* **2006**, *313*, 504–507. [CrossRef]
71. Yin, H.; Allinson, N.M. Self-organizing mixture networks for probability density estimation. *IEEE Trans. Neural Netw.* **2001**, *12*, 405–411. [CrossRef]
72. Hwang, J.N.; Lay, S.R.; Lippman, A. Nonparametric multivariate density estimation: A comparative study. *IEEE Trans. Signal Process.* **1994**, *42*, 2795–2810. [CrossRef]
73. Goodfellow, I.; Pouget-Abadie, J.; Mirza, M.; Xu, B.; Warde-Farley, D.; Ozair, S.; Courville, A.; Bengio, Y. Generative adversarial nets. *Adv. Neural Inf. Process. Syst.* **2014**, *27*.

74. Okamoto, T.; Tanaka, T. Novel partial discharge measurement computer-aided measuremnet systems. *IEEE Trans. Electr. Insul.* **1986**, *EI-21*, 1015–1019. [CrossRef]
75. Cachin, C.; Wiesmann, H.J. PD recognition with knowledge-based preprocessing and neural networks. *IEEE Trans. Dielectr. Electr. Insul.* **1995**, *2*, 578–589. [CrossRef]
76. Sahoo, A.; Subramaniam, A.; Bhandari, S.; Panda, S.K. A review on condition monitoring of GIS. In Proceedings of the 2017 International Symposium on Electrical Insulating Materials (ISEIM), Toyohashi, Japan, 11–15 September 2017; Volume 2, pp. 543–546.
77. Hinton, G.E. Deep belief networks. *Scholarpedia* **2009**, *4*, 5947. [CrossRef]
78. ENET-Centre. VSB Power Line Fault Detection. 2019. Available online: https://www.kaggle.com/c/vsb-power-line-fault-detection (accessed on 9 June 2022).
79. Fitzgerald, A.E.; Kingsley, C.; Umans, S.D.; James, B. *Electric Machinery*; McGraw-Hill: New York, NY, USA, 2003; Volume 5.
80. Lévesque, M.; Amyot, N.; Hudon, C.; Bélec, M.; Blancke, O. Improvement of a hydrogenerator prognostic model by using partial discharge measurement analysis. In Proceedings of the Annual Conference of the PHM Society, St. Petersburg, FL, USA, 2–5 October 2017; Volume 9.
81. Liu, R.; Yang, B.; Zio, E.; Chen, X. Artificial intelligence for fault diagnosis of rotating machinery: A review. *Mech. Syst. Signal Process.* **2018**, *108*, 33–47. [CrossRef]
82. Chen, G.; Hao, M.; Xu, Z.; Vaughan, A.; Cao, J.; Wang, H. Review of high voltage direct current cables. *CSEE J. Power Energy Syst.* **2015**, *1*, 9–21. [CrossRef]
83. Uckol, H.I.; Ilhan, S.; Ozdemir, A. Workmanship defect classification in medium voltage cable terminations with convolutional neural network. *Electr. Power Syst. Res.* **2021**, *194*, 107105. [CrossRef]
84. Chang, C.K.; Chang, H.H.; Boyanapalli, B.K. Application of Pulse Sequence Partial Discharge Based Convolutional Neural Network in Pattern Recognition for Underground Cable Joints. *IEEE Trans. Dielectr. Electr. Insul.* **2022**, *29*, 1070–1078. [CrossRef]
85. Swaminathan, R.; Mishra, S.; Routray, A.; Swain, S.C. A CNN-LSTM-based fault classifier and locator for underground cables. *Neural Comput. Appl.* **2021**, *33*, 15293–15304. [CrossRef]
86. Qin, X.; Zhang, Y.; Mei, W.; Dong, G.; Gao, J.; Wang, P.; Deng, J.; Pan, H. A cable fault recognition method based on a deep belief network. *Comput. Electr. Eng.* **2018**, *71*, 452–464. [CrossRef]
87. Wu, Y.; Zhang, P.; Lu, G. Detection and location of aged cable segment in underground power distribution system using deep learning approach. *IEEE Trans. Ind. Inform.* **2021**, *17*, 7379–7389. [CrossRef]
88. Liu, N.; Fan, B.; Xiao, X.; Yang, X. Cable incipient fault identification with a sparse autoencoder and a deep belief network. *Energies* **2019**, *12*, 3424. [CrossRef]
89. Wang, Y.; Lu, H.; Xiao, X.; Yang, X.; Zhang, W. Cable incipient fault identification using restricted Boltzmann machine and stacked autoencoder. *IET Gener. Transm. Distrib.* **2020**, *14*, 1242–1250. [CrossRef]
90. Guo, M.F.; Zeng, X.D.; Chen, D.Y.; Yang, N.C. Deep-learning-based earth fault detection using continuous wavelet transform and convolutional neural network in resonant grounding distribution systems. *IEEE Sens. J.* **2017**, *18*, 1291–1300. [CrossRef]
91. Dai, J.; Song, H.; Sheng, G.; Jiang, X. Dissolved gas analysis of insulating oil for power transformer fault diagnosis with deep belief network. *IEEE Trans. Dielectr. Electr. Insul.* **2017**, *24*, 2828–2835. doi: 10.1109/TDEI.2017.006727. [CrossRef]
92. Kim, S.; Jo, S.H.; Kim, W.; Park, J.; Jeong, J.; Han, Y.; Kim, D.; Youn, B.D. A semi-supervised autoencoder with an auxiliary task (saat) for power transformer fault diagnosis using dissolved gas analysis. *IEEE Access* **2020**, *8*, 178295–178310. doi: 10.1109/ACCESS.2020.3027830. [CrossRef]
93. Kim, S.; Park, J.; Kim, W.; Jo, S.H.; Youn, B.D. Learning from even a weak teacher: Bridging rule-based Duval method and a deep neural network for power transformer fault diagnosis. *Int. J. Electr. Power Energy Syst.* **2022**, *136*, 107619. doi: 10.1016/J.IJEPES.2021.107619. [CrossRef]
94. Wu, X.; He, Y.; Duan, J. A deep parallel diagnostic method for transformer dissolved gas analysis. *Appl. Sci.* **2020**, *10*, 1329. doi: 10.3390/APP10041329. [CrossRef]
95. Taha, I.B.; Ibrahim, S.; Mansour, D.E.A. Power transformer fault diagnosis based on DGA using a convolutional neural network with noise in measurements. *IEEE Access* **2021**, *9*, 111162–111170. doi: 10.1109/ACCESS.2021.3102415. [CrossRef]
96. Hu, H.; Ma, X.; Shang, Y. A novel method for transformer fault diagnosis based on refined deep residual shrinkage network. *IET Electr. Power Appl.* **2022**, *16*, 206–223. [CrossRef]
97. Zhou, Y.; Yang, X.; Tao, L.; Yang, L. Transformer fault diagnosis model based on improved gray wolf optimizer and probabilistic neural network. *Energies* **2021**, *14*, 3029. doi: 10.3390/en14113029. [CrossRef]
98. Rucconi, V.; de Maria, L.; Garatti, S.; Bartalesi, D.; Valecillos, B.; Bittanti, S. *Deep Learning for Fault Detection in Transformers Using Vibration Data*; IFAC-PapersOnLine; Elsevier: Amsterdam, The Netherlands, 2021; Volume 54, pp. 262–267. doi: 10.1016/j.ifacol.2021.08.369. [CrossRef]
99. Li, Z.; Zhang, Y.; Abu-Siada, A.; Chen, X.; Li, Z.; Xu, Y.; Zhang, L.; Tong, Y. Fault diagnosis of transformer windings based on decision tree and fully connected neural network. *Energies* **2021**, *14*, 1531. doi: 10.3390/en14061531 . [CrossRef]
100. Hong, K.; Jin, M.; Huang, H. Transformer winding fault diagnosis using vibration image and deep learning. *IEEE Trans. Power Deliv.* **2021**, *36*, 676–685. doi: 10.1109/TPWRD.2020.2988820 . [CrossRef]
101. Wang, T.; He, Y.; Li, B.; Shi, T. Transformer fault diagnosis using self-powered RFID sensor and deep learning approach. *IEEE Sens. J.* **2018**, *18*, 6399–6411. doi: 10.1109/JSEN.2018.2844799 . [CrossRef]

102. Moradzadeh, A.; Moayyed, H.; Mohammadi-Ivatloo, B.; Gharehpetian, G.B.; Aguiar, A.P. Turn-to-Turn short circuit fault localization in transformer winding via image processing and deep learning method. *IEEE Trans. Ind. Inform.* **2021**, *18*, 4417–4426. doi: 10.1109/TII.2021.3105932. [CrossRef]
103. Dey, D.; Chatterjee, B.; Dalai, S.; Munshi, S.; Chakravorti, S. A deep learning framework using convolution neural network for classification of impulse fault patterns in transformers with increased accuracy. *IEEE Trans. Dielectr. Electr. Insul.* **2017**, *24*, 3894–3897. [CrossRef]
104. Duan, L.; Hu, J.; Zhao, G.; Chen, K.; Wang, S.X.; He, J. Method of inter-turn fault detection for next-generation smart transformers based on deep learning algorithm. *High Volt.* **2019**, *4*, 282–291. doi: 10.1049/hve.2019.0067. [CrossRef]
105. Fanchiang, K.H.; Huang, Y.C.; Kuo, C.C. Power electric transformer fault diagnosis based on infrared thermal images using wasserstein generative adversarial networks and deep learning classifier. *Electronics* **2021**, *10*, 1161. [CrossRef]
106. Wu, Y.; Zhang, Z.; Xiao, R.; Jiang, P.; Dong, Z.; Deng, J. Operation state identification method for converter transformers based on vibration detection technology and deep belief network optimization algorithm. *Actuators* **2021**, *10*, 56. doi: 10.3390/ACT10030056 . [CrossRef]
107. Castro, J.; Aponte, G.; Sanchez, V.H.; Castro, M.; Espinosa, A.; Rosales, N. Colombian experience on insulation pollution level measurement applying the ESDD methodology. In Proceedings of the 2006 IEEE/PES Transmission Distribution Conference and Exposition: Latin America, Caracas, Venezuela, 15–18 August 2006; pp. 1–5. doi: 10.1109/TDCLA.2006.311620. [CrossRef]
108. Gorur, R.S.; Shaffner, D.; Clark, W.; Vinson, R. Utilities share their insulator field experience. *Transm. Distrib. World* **2005**, *57*, 17–27.
109. Álvarez, F.; Ortego, J.; Garnacho, F.; Sánchez-Urán, M. Advanced techniques for online PD measurements in high voltage systems. In Proceedings of the 2014 ICHVE International Conference on High Voltage Engineering and Application, Poznan, Poland, 8–11 September 2014; pp. 1–4.
110. Álvarez, F.; Garnacho, F.; Albarracín, R.; Granizo, R.; Dong, M.; Ortego, J. Practical experience of insulation condition evaluation in an on-site HV installation applying a PD measuring procedure. In Proceedings of the 2016 International Conference on Condition Monitoring and Diagnosis (CMD), Xi'an, China, 25–28 September 2016; pp. 509–513.
111. Nguyen, V.N.; Jenssen, R.; Roverso, D. Intelligent monitoring and inspection of power line components powered by UAVs and deep learning. *IEEE Power Energy Technol. Syst. J.* **2019**, *6*, 11–21. [CrossRef]
112. Morozovsky, N.; Bewley, T. SkySweeper: A low DOF, dynamic high wire robot. In Proceedings of the 2013 IEEE/RSJ International Conference on Intelligent Robots and Systems, Tokyo, Japan, 3–7 November 2013; pp. 2339–2344.
113. Miao, X.; Liu, X.; Chen, J.; Zhuang, S.; Fan, J.; Jiang, H. Insulator detection in aerial images for transmission line inspection using single shot multibox detector. *IEEE Access* **2019**, *7*, 9945–9956. [CrossRef]
114. Tao, X.; Zhang, D.; Wang, Z.; Liu, X.; Zhang, H.; Xu, D. Detection of power line insulator defects using aerial images analyzed with convolutional neural networks. *IEEE Trans. Syst. Man Cybern. Syst.* **2018**, *50*, 1486–1498. [CrossRef]
115. Sarkar, D.; Gunturi, S.K. Online health status monitoring of high voltage insulators using deep learning model. *Vis. Comput.* **2021**, 1–12. [CrossRef]
116. *IEC/TS Standard 62073 Ed 1.0*; Guide to the Measurement of Wettability of Insulator Surfaces. International Electrotechnical Commission (IEC): Geneva, Switzerland, 2003.
117. Huang, X.; Nie, T.; Zhang, Y.; Zhang, X. Study on hydrophobicity detection of composite insulators of transmission lines by image analysis. *IEEE Access* **2019**, *7*, 84516–84523. [CrossRef]
118. Chatterjee, S.; Roy, S.S.; Samanta, K.; Modak, S. Sensing wettability condition of insulation surface employing convolutional neural network. *IEEE Sens. Lett.* **2020**, *4*, 1–4. [CrossRef]
119. Yunpeng, L.; Ziru, M. Research on intelligent detection method of composite insulator hydrophobicity based on water spraying of drone. In Proceedings of the 2020 8th International Conference on Condition Monitoring and Diagnosis (CMD), Phuket, Thailand, 25–28 October 2020; pp. 145–148.
120. Ibrahim, A.; Dalbah, A.; Abualsaud, A.; Tariq, U.; El-Hag, A. Application of machine learning to evaluate insulator surface erosion. *IEEE Trans. Instrum. Meas.* **2019**, *69*, 314–316. [CrossRef]
121. El Haj, Y.; El-Hag, A.H.; Ghunem, R.A. Application of deep-learning via transfer learning to evaluate silicone rubber Material surface erosion. *IEEE Trans. Dielectr. Electr. Insul.* **2021**, *28*, 1465–1467. [CrossRef]
122. Brar, R.K.; El-Hag, A.H. Application of machine learning in discharge classification. In Proceedings of the 2020 IEEE Conference on Electrical Insulation and Dielectric Phenomena (CEIDP), East Rutherford, NJ, USA, 18–30 October 2020; pp. 43–46.
123. Anjum, S.; Jayaram, S.; El-Hag, A.; Jahromi, A.N. Detection and classification of defects in ceramic insulators using RF antenna. *IEEE Trans. Dielectr. Electr. Insul.* **2017**, *24*, 183–190. [CrossRef]
124. Anjum, S.; Jayaram, S.; El-Hag, A.; Naderian, A. Condition monitoring system for the detection of defects in ceramic insulators based on their radio frequency (RF) signatures. In Proceedings of the International Conference on Condition Monitroing and Diagnosis 2014 (CMD 2014), Lyon, France, 15–16 December 2014.
125. Polisetty, S.; El-Hag, A.; Jayram, S. Classification of common discharges in outdoor insulation using acoustic signals and artificial neural network. *High Volt* **2019**, *4*, 333–338. [CrossRef]
126. Sopelsa Neto, N.F.; Stefenon, S.F.; Meyer, L.H.; Bruns, R.; Nied, A.; Seman, L.O.; Gonzalez, G.V.; Leithardt, V.R.Q.; Yow, K.C. A study of multilayer perceptron networks applied to classification of ceramic insulators using ultrasound. *Appl. Sci.* **2021**, *11*, 1592. [CrossRef]

127. Serikbay, A.; Bagheri, M.; Zollanvari, A.; Phung, B. Accurate surface condition classification of high voltage insulators based on deep convolutional neural networks. *IEEE Trans. Dielectr. Electr. Insul.* **2021**, *28*, 2126–2133. [CrossRef]
128. Liu, Y.; Lai, T.; Liu, J.; Li, Y.; Pei, S.; Yang, J. Insulator contamination diagnosis method based on deep learning convolutional neural network. In Proceedings of the 2021 3rd Asia Energy and Electrical Engineering Symposium (AEEES), Chengdu, China, 26–29 March 2021; pp. 184–188.
129. Yeh, C.T.; Thanh, P.N.; Cho, M.Y. Real-time leakage current classification of 15 kV and 25 kV distribution insulators based on bidirectional long short-term memory networks with deep learning machine. *IEEE Access* **2022**, *10*, 7128–7140. [CrossRef]
130. Mussina, D.; Irmanova, A.; Jamwal, P.K.; Bagheri, M. Multi-modal data fusion using deep neural network for condition monitoring of high voltage insulator. *IEEE Access* **2020**, *8*, 184486–184496. [CrossRef]
131. Vigneshwaran, B.; Maheswari, R.; Kalaivani, L.; Shanmuganathan, V.; Rho, S.; Kadry, S.; Lee, M.Y. Recognition of pollution layer location in 11 kV polymer insulators used in smart power grid using dual-input VGG convolutional neural network. *Energy Rep.* **2021**, *7*, 7878–7889. [CrossRef]
132. Jahromi, A.N.; El-Hag, A.H.; Jayaram, S.H.; Cherney, E.A.; Sanaye-Pasand, M.; Mohseni, H. A neural network based method for leakage current prediction of polymeric insulators. *IEEE Trans. Power Deliv.* **2005**, *21*, 506–507. [CrossRef]
133. El-Hag, A.H.; Jahromi, A.N.; Sanaye-Pasand, M. Prediction of leakage current of non-ceramic insulators in early aging period. *Electr. Power Syst. Res.* **2008**, *78*, 1686–1692. [CrossRef]
134. Ferreira, T.V.; Germano, A.D.; da Costa, E.G. Ultrasound and artificial intelligence applied to the pollution estimation in insulations. *IEEE Trans. Power Deliv.* **2012**, *27*, 583–589. [CrossRef]
135. El-Hag, A.; Mukhopadhyay, S.; Al-Ali, K.; Al-Saleh, A. An intelligent system for acoustic inspection of outdoor insulators. In Proceedings of the 2017 3rd International Conference on Condition Assessment Techniques in Electrical Systems (CATCON), Rupnagar, India, 16–18 November 2017; pp. 122–125.
136. Vinyals, O.; Blundell, C.; Lillicrap, T.; Wierstra, D. Matching networks for one shot learning. *arXiv* **2016**, arXiv:1606.04080.
137. Yang, Y.; Chen, Z.; Yan, J.; Xiong, Z.; Zhang, J.; Yuan, H.; Tu, Y.; Zhang, T. State evaluation of power transformer based on digital twin. In Proceedings of the 2019 IEEE International Conference on Service Operations and Logistics, and Informatics (SOLI), Zhengzhou, China, 6–8 November 2019; pp. 230–235.
138. Moutis, P.; Alizadeh-Mousavi, O. Digital twin of distribution power transformer for real-time monitoring of medium voltage from low voltage measurements. *IEEE Trans. Power Deliv.* **2020**, *36*, 1952–1963. [CrossRef]
139. Liu, T.; Yu, H.; Yin, H.; Zhang, Z.; Sui, Z.; Zhu, D.; Gao, L.; Li, Z. Research and application of digital twin technology in power grid development business. In Proceedings of the 2021 6th Asia Conference on Power and Electrical Engineering (ACPEE), Chongqing, China, 8–11 April 2021; pp. 383–387.

energies MDPI

Article

Experimental Study and Modeling of the Effect of ESDD/NSDD on AC Flashover of SiR Outdoor Insulators

Mohammed El Amine Slama [1,*], Adnan Krzma [1,2], Maurizio Albano [1] and Abderrahmane Manu Haddad [1]

1 Advanced High Voltage Engineering Centre, School of Engineering, Cardiff University, Queen's Buildings, The Parade, Cardiff CF24 3AA, UK; salamaaek@elmergib.edu.ly (A.K.); albanom@cardiff.ac.uk (M.A.); haddad@cardiff.ac.uk (A.M.H.)
2 Department of Electrical and Computer Engineering, Faculty of Engineering, Elmergib University, Khoms P.O. Box 40414, Libya
* Correspondence: slamame@cardiff.ac.uk

Abstract: Pollution flashover occurs when soluble and nonsoluble materials cover the surface of an insulator, and this may ultimately cause a reduction in its performance. In this paper, the common type of sodium chloride (NaCl) was used as a soluble pollutant (ESDD) and kaolin as a nonsoluble pollutant (NSDD). Samples of silicone rubber (SiR) insulators were selected for this study and fabricated at the Advanced High Voltage Engineering Research Centre (AHIVEC) at Cardiff University. The samples were preconditioned and polluted according to standard specifications. Additionally, the AC voltage ramp technique was used to achieve flashover (FOV) voltage with different pollution levels. The aim of this work was to investigate the effect of nonsoluble materials on flashover characteristics to understand their interaction with dry-band arcs by using FOV electrical equations and experimental data. The test results show that the FOV voltage of the silicone rubber insulator substantially decreased with the increase in both ESDD and NSDD values. It was also identified from these results that the dry-band arcs were considerably influenced by both ESDD and NSDD levels. This impact can be quantified by determining the variation of discharge parameters (N, n). Based on the FOV equations and experimental data, a mathematical model was suggested, taking into account the effect of both ESDD and NSDD.

Keywords: pollution insulators; nonsoluble deposit density; flashover characteristics; arc parameters

Citation: Slama, M.E.A.; Krzma, A.; Albano, M.; Haddad, A.M. Experimental Study and Modeling of the Effect of ESDD/NSDD on AC Flashover of SiR Outdoor Insulators. *Energies* **2022**, *15*, 3782. https://doi.org/10.3390/en15103782

Academic Editors: Zhijin Zhang and Hualong Zheng

Received: 12 April 2022
Accepted: 13 May 2022
Published: 20 May 2022

1. Introduction

One of the major problems facing HV electrical grid engineers is the disturbance generated by insulator arcing due to pollution flashover (FOV), especially in harsh environments [1,2]. Pollution FOV of insulators is a complex phenomenon and has become a significant aspect in the design of HV overhead lines and substations [2,3]. This phenomenon can be affected by many parameters such as insulator material, insulator profile, pollution level, and atmospheric conditions (air pressure, temperature, humidity, fog, rain, snow, ice, etc.) [1,3–6].

The chemical constitution of the pollution layer is one of the key factors affecting the FOV characteristics and performance of insulators [7,8]. Sodium chloride (NaCl) is commonly utilized as the key reference indicator for conductive pollution materials [9,10]. The pollution degree is defined by the equivalent salt deposit density (ESDD) and represents the concentration of NaCl that, when dissolved in demineralized water, gives the same volumetric conductivity as that of the natural deposit removed from a given surface of the insulator divided by the area of the surface, generally expressed in mg/cm^2. However, in the real field, the pollution constitution depends on the local environment where NaCl is not the predominant salt, as established by many researchers [7,8,11–15]. On the other hand, the pollution layer contains insoluble materials consisting mainly of ferrite, silica, and alumina [4,9,13,15–18]. The pollution index for insoluble or nonsoluble materials is defined as nonsoluble deposit density (NSDD) and represents the amount of nonsoluble residue removed from a given surface of the insulator, divided by the area of the surface, expressed in mg/cm^2 [9]. Several studies have been performed to understand the effect of NSDD on the FOV performance of ceramic, glass, and polymeric insulators in artificial pollution tests, and the general conclusion is that FOV voltage decreases when NSDD increases for the same ESDD [14,16,17]. In the case of polymeric insulators, it has been reported that NSDD decreases hydrophobic performance and FOV voltage [19–22].

From a modeling point of view, it has been established that the variation of pollution FOV voltage with ESDD can be described as [4]

$$V_{\mathrm{FOV}} = \alpha \cdot \mathrm{ESDD}^{\beta} \tag{1}$$

where coefficients α and β depend on the test conditions, and $\beta < 0$.

On the other hand, the flashover equation is based on the Obenaus model [23,24]. The model consists of a cylindrical discharge (partial arc) of axial length X in series with a linear resistance r_p representing the pollution layer, supplied by a voltage V (Figure 1) and expressed as follows

$$V = XNI^{-n} + r_p(L - X)I \tag{2}$$

where I is the leakage current flowing across the pollution layer, r_p represents the linear resistance of the unabridged part of the pollution, and L is the total leakage length of the insulator. N and n are discharge characteristic parameters. Usually, the values of N and n are considered depending upon the experimental conditions in which the discharge burns. For example, Hampton [25] considers that the discharge burns in an environment rich in water vapor, while Wilkins [26] proposes that the discharge burns in a dry-air medium. Slama et al. [27] demonstrated that the parameters N and n are not constant and depend on the electrical equivalent circuit consisting of the polluted insulator, the discharge, and the thermal parameters of the discharge.

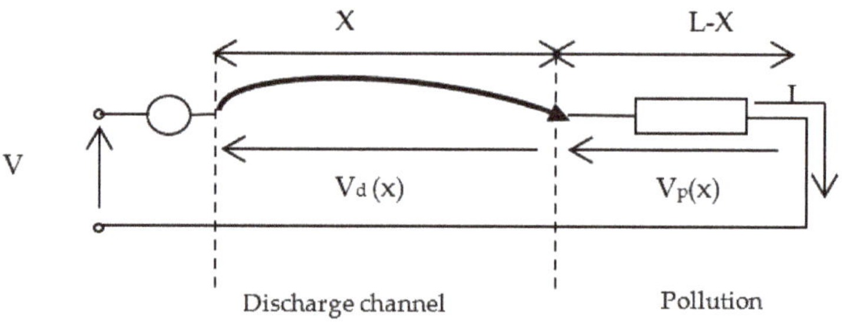

Figure 1. Equivalent electrical circuit based on Obenaus model. V: applied voltage, $V_d(X)$: discharge voltage, $V_p(X)$: pollution voltage, I: leakage current, X: discharge length, L-X: nonshunted insulator creepage length.

The critical flashover voltage is

$$V_{\mathrm{FOV}} = LN^{\frac{1}{1+n}} r_p^{\frac{n}{1+n}} \tag{3}$$

In this study, the relationship between the variation of FOV voltage, the coefficients α and β, the parameters N and n, and the amount of NSDD was investigated. For this purpose, AC clean fog tests were carried out on artificially polluted silicone rubber (SiR) insulators. The soluble pollutant was simulated by NaCl, and different amounts of NSDD (kaolin) levels were applied. The influence of NSDD values on FOV characteristics was investigated to establish a mathematical model of the influence of both ESDD/NSDD on FOV voltage. This study aims to improve the understanding of polluted FOV insulators to enable better selection of outdoor insulation used in high-voltage systems.

2. Experimental Arrangements

2.1. Insulator Preparation

Four-shed insulators were used in this investigation, which were fabricated at the AHIVEC by using the following facilities: room-temperature vulcanized (RTV-2), two-component SiR (herein referred to as x-A/B), a fiberglass core with metallic end fittings at both ends, a 3D-printed mold, and an oven. The details of the used facilities are well described in [28–30]. The electrical and mechanical properties of the liquid SiR used in this study are given in Table 1, and the fabrication process is the same as that used in [28–30]. Table 2 gives the geometrical parameters of the insulators. Figure 2 illustrates an example of the SiR insulator fabricated and used for the purpose of this study.

Table 1. Characteristics of x-A/B bicomponent SiR materials.

Property	Inspection Method	x-A/B
Permittivity	IEC 60250	2.9
Dielectric strength (kV/mm)	IEC 60243	23
Tracking resistance	IEC 60587	1 A 3.5
Dissipation factor	IEC 60250	3×10^{-4}
Tensile strength (N/mm^2)	ISO 37	6.50
Hardness shore	ISO 868	30
Tear strength (N/mm)	ASTM D 624 B	20
Elongation at break	ISO 37	500
Volume resistivity	IEC 60093	10^{15}

Table 2. Dimensions of tested insulators.

Parameters	Dimensions (mm)
Creepage distance	375
Fitting separation	175
Shed diameter	90
Shed separation	46
Trunk diameter	28
Inner core diameter	18
Form factor	2.7

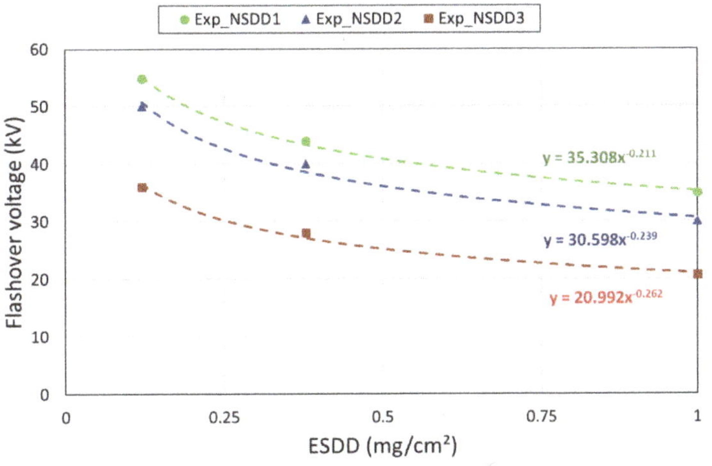

Figure 2. Variation of average flashover voltage vs. ESDD for different NSDD levels.

2.2. Experimental Setup

The experimental setup consists of a fog chamber where the polluted insulators are suspended and wetted with fog generated. High voltage is supplied with the Hipotronics AC Dielectric test (150 kVA, 75 kV, 50 Hz, 2 A). The applied voltage is controlled manually or by using a program to follow a predetermined voltage pattern with the panel by adjusting the primary voltage (0–840 V). A digital display control panel allows reading the RMS applied voltage, and an RC divider (10,000:1) records the voltage waveform. The measurement is realized through a shunt that is connected to a data acquisition system (DAQ). The DAQ is connected to a personal computer with an IEEE interface to monitor and record the voltage and leakage current signals using a developed LabVIEW program. Postprocessing software is used for data analysis. More details about the experimental setup are given in [28–32].

2.3. Artificial Pollution of Insulators

The pollutant was prepared according to the IEC 60507 solid-layer method [9], consisting of a Triton X-100 wetting agent (1 g), tap water (1 L), kaolin (40 g), and an adequate concentration of sodium chloride (NaCl) to achieve the volume conductivity value needed. The volume conductivity was selected to represent different salt deposit density (ESDD) levels. In this investigation, the ESDD values are 0.12 mg/cm^2, 0.38 mg/cm^2 and 1.0 mg/cm^2, and the nonsoluble (NSDD) values of kaolin are 0.11 mg/cm^2, 0.23 mg/cm^2, and 0.52 mg/cm^2.

Before the test, all insulators were carefully cleaned and allowed to dry naturally indoors to avoid any pollution or dust. Then, before application of the pollutant, the insulator was preconditioned with brushed dry kaolin according to the recommendations of Cigré WG C4.303 [33] and left for at least 1 h. This step is used to reduce the hydrophobicity of the silicone rubber. The insulator was then polluted by spraying the pollutant on the insulator surface, ensuring application to all parts of the insulator surface, and then the insulator was left to dry for 24 h. Next, the insulator was mounted in the fog chamber. The polluted insulator was wetted with a fog flow rate of 3 L/h, which is sufficient to achieve maximum conductance of the pollution layer within 10–40 min from the instant the fog is applied depending on the used ESDD [31].

2.4. Flashover Ramp Test Method

The used test technique, a high-voltage ramp, is the same as that proposed in previous work [28,29] with a predetermined voltage ramp of 6.5 kV/min. This rate allows exciting the FOV event within a 5 to 10 min time gap. Fog generation is applied at the same time as the applied voltage and kept running until the end of the test. When FOV occurs, the protection interrupts the voltage supply. The tested insulator is then left to cool naturally for 5 min, and then a new ramp test is started until the required number of FOVs is realized (5 tests). During the test, video and infrared cameras are used to monitor the flashover event and discharge activity on the insulator surface.

3. Experimental Results, Analysis, and Discussion

3.1. Flashover Test Results

Figure 2 illustrates the variation of V_{FOV} vs. ESDD for a different NSDD. According to this figure, for the same ESDD, V_{FOV} decreases with the increase in NSDD. The relationship between V_{FOV} and ESDD was fitted to determine the values of the coefficients α and β for each NSDD. Table 3 illustrates the values of α and β and shows that these coefficients vary with the amount of NSDD: α decreases while β increases.

In Figure 3, V_{FOV} is plotted as a function of NSDD for a different ESDD. According to this figure, NSDD drastically impacts V_{FOV} where the V_{FOV} ratio corresponding to the minimum and maximum NSDD (0.11 mg/cm^2 and 0.52 mg/cm^2) for each ESDD level is, respectively, 1.83 and 1.7.

Table 3. Values of parameters α and β.

NSDD	α	β
0.11	35.308	−0.211
0.23	30.598	−0.239
0.52	20.992	−0.262

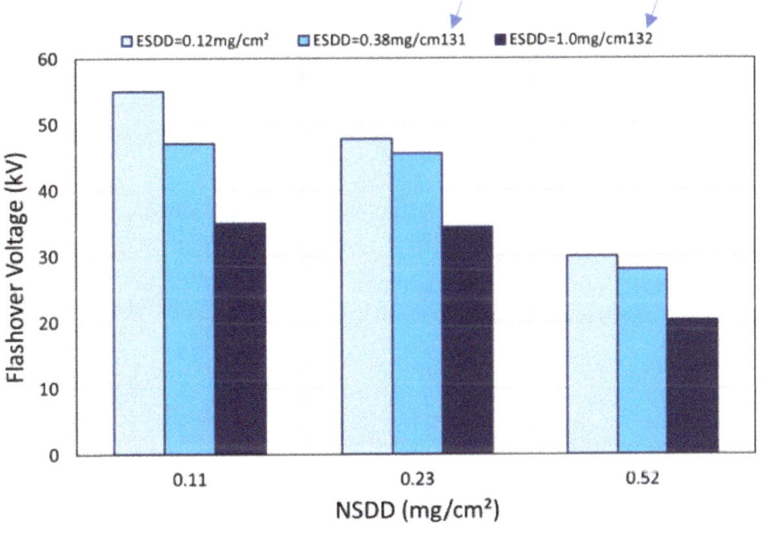

Figure 3. Effect of NSDD level on the averaged flashover voltage of the conventional insulator with different ESDD values.

Figures 4 and 5 illustrate examples of the instantaneous leakage current and voltage during the FOV test for two pollution levels ($ESDD_1 = 0.12$ mg/cm^2 and $ESDD_2 = 0.23$ mg/cm^2) and two NSDD levels ($NSDD_1 = 0.11$ mg/cm^2 and $NSDD_2 = 0.52$ mg/cm^2). The current and V_{FOV} related to $NSDD_1$ are lower than those corresponding to $NSDD_2$, whatever the ESDD. This can be explained by the fact that the greater the amount of insoluble materials, the greater the moisture absorption, and the greater the pollution conductive layer.

Figures 6 and 7 illustrate the infrared and thermal recording during the FOV tests related to Figure 4. The first observation is that the dry bands (DBs) appear at the region between the shed and the trunk. Next, dry-band discharges (DbDs) and dry-band arcs (DBAs) developed and elongated along the trunk and the sheds. The inception of DBs is not the same, and the ESDD is kept constant. In Figure 6, the DBs are incepted at the trunk region and are followed, after a certain time, by DbDs and DBAs. In Figure 7, DbDs and DBAs appear quickly just after the inception of DBs at the trunk. This is clearly observed in Figures 8 and 9, which show the visual DbD and DBA activity at the polluted insulator with $ESDD_1$, $NSDD_1$, and $NSDD_2$.

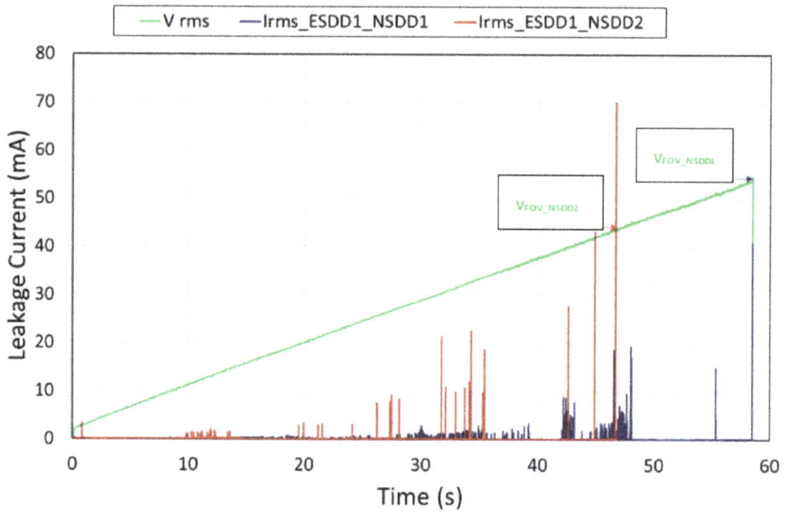

Figure 4. Leakage current and voltage during the FOV test with $ESDD_1$, $NSDD_1$, and $NSDD_2$.

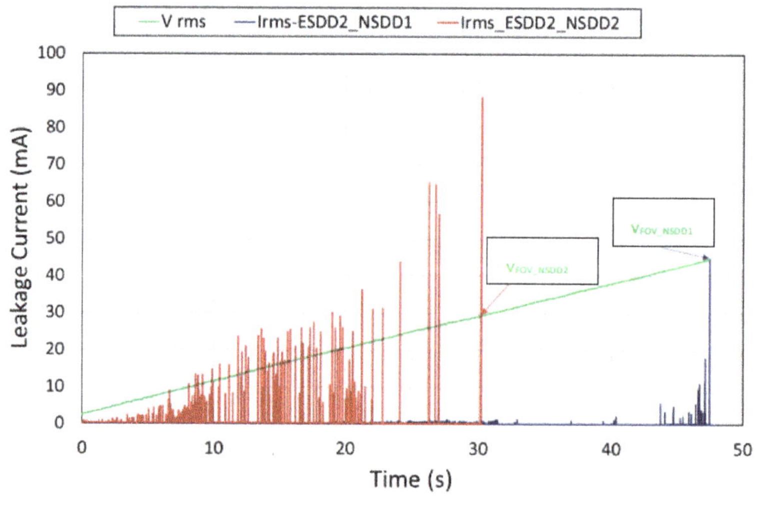

Figure 5. Leakage current and voltage during the FOV test with $ESDD_2$, $NSDD_1$, and $NSDD_2$.

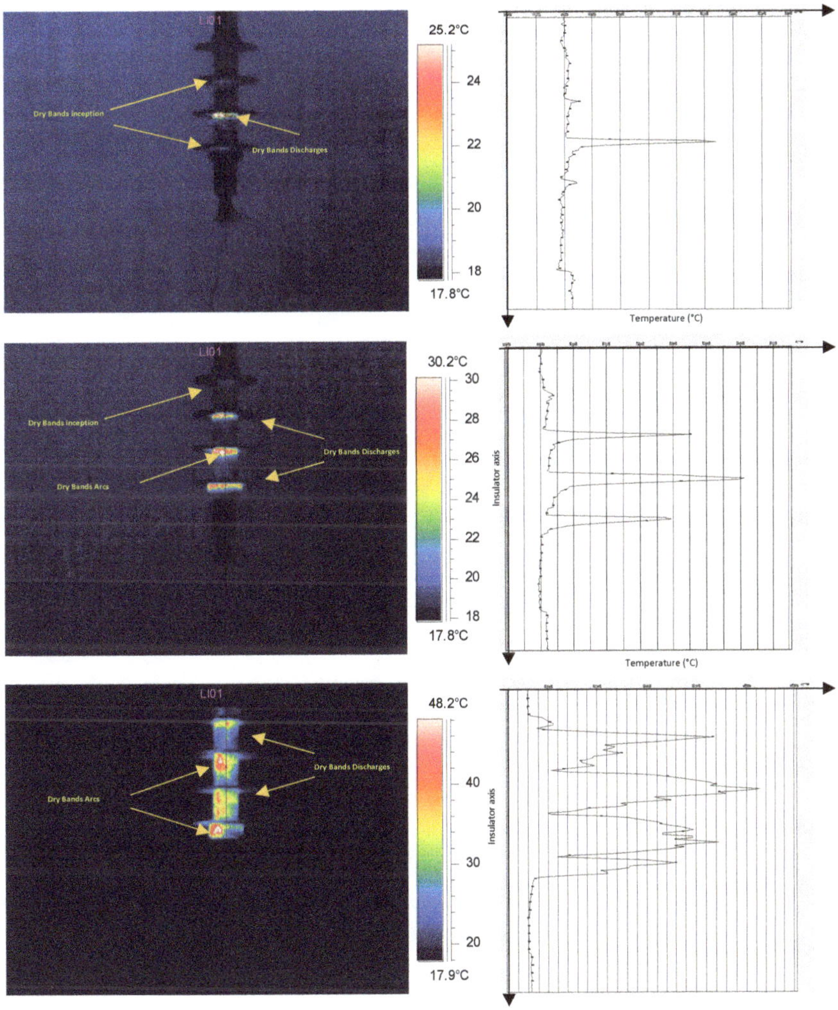

Figure 6. Infrared recording and thermal activity during the FOV test with $ESDD_1$ = 0.12 mg/cm^2 and $NSDD_1$ = 0.11 mg/cm^2.

Based on these observations, one can deduce that NSDD considerably affects the resistance of the pollution layer and can be considered a key parameter for the calculation of FOV voltage.

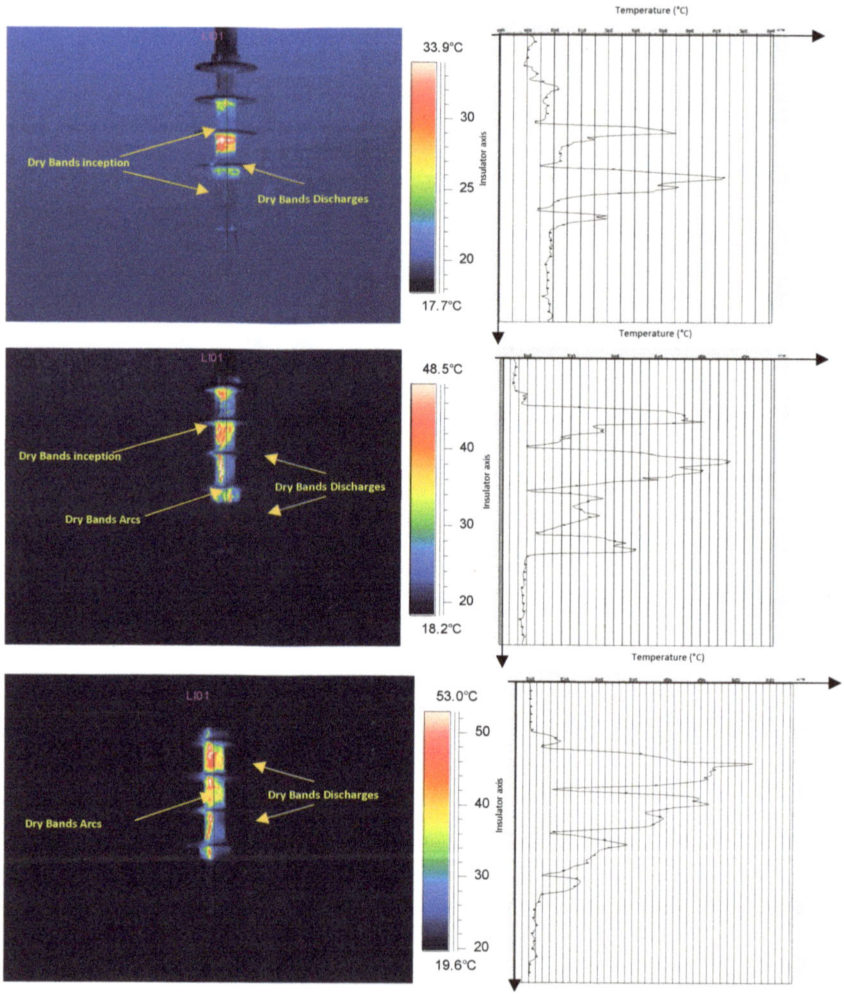

Figure 7. Infrared recording and thermal activity during the FOV test with $ESDD_1 = 0.12$ mg/cm^2 and $NSDD_2 = 0.23$ mg/cm^2.

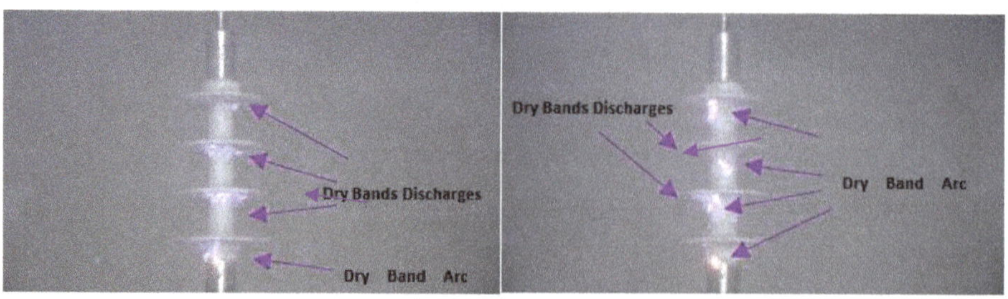

Figure 8. Visual observation of DbDs and DBAs activity with $ESDD_1 = 0.12$ mg/cm^2 and $NSDD_2 = 0.11$ mg/cm^2.

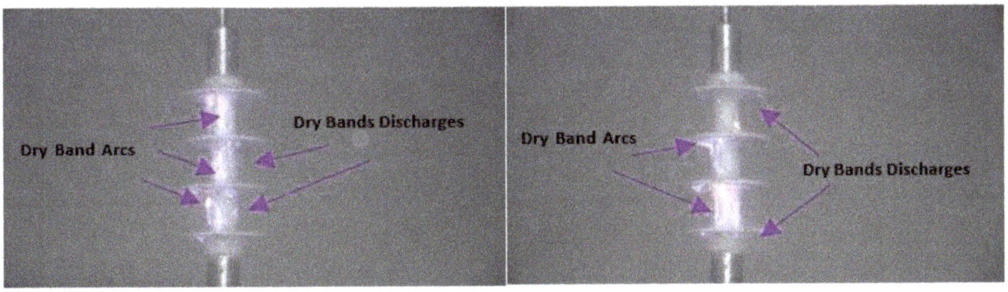

Figure 9. Visual observation of DbDs and DBAs activity with $ESDD_1$ = 0.12 mg/cm^2 and $NSDD_2$ = 0.23 mg/cm^2.

3.2. Effect of NSDD on the Leakage Resistance and Surface Resistivity

Based on the conclusion of the previous section, special attention is paid to the low-voltage leakage resistance of the different polluted insulators before DB creation related to the FOV tests. Figures 10 and 11 illustrate the corresponding instantaneous low-voltage leakage resistance before the FOV tests of the examples in Figures 4 and 5. The leakage resistances are different and depend on the NSDD level, even though the ESDD level is the same. For $ESDD_1$ = 0.12 mg/cm^2, the minimum resistances are: $R_{\_Min\text{-}NSDD1}$ = 10.75 MΩ and $R_{\_Min\text{-}NSDD2}$ = 4.77 MΩ. For $ESDD_2$ = 0.38 mg/cm^2, $R_{\_Min\text{-}NSDD1}$ = 10.63 MΩ, $R_{\_Min\text{-}NSDD2}$ = 2.325 MΩ. These examples show that the minimum leakage resistance is greatly influenced by NSDD, and this influence increases with both ESDD and NSDD. Indeed, the ratio of the leakage resistance corresponding to $NSDD_1$ and $NSDD_2$ is 1.01 for ESDD1, while it is 2.05 for $ESDD_2$.

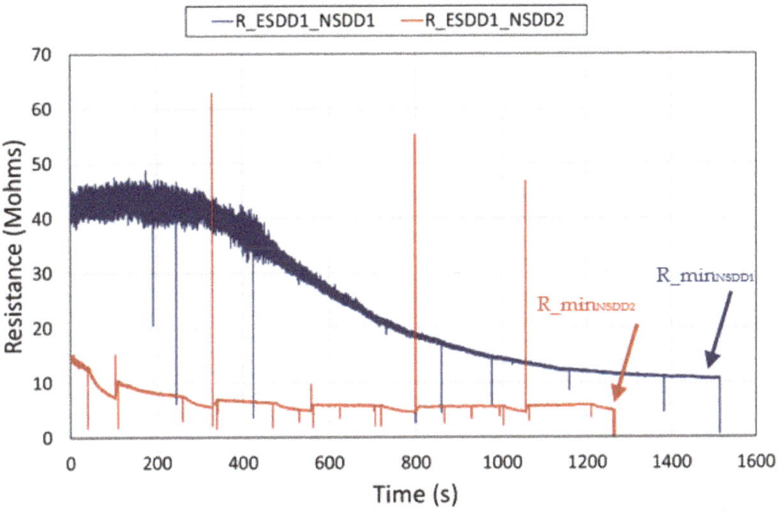

Figure 10. Leakage resistance during the low-voltage (840 V) test with $ESDD_1$.

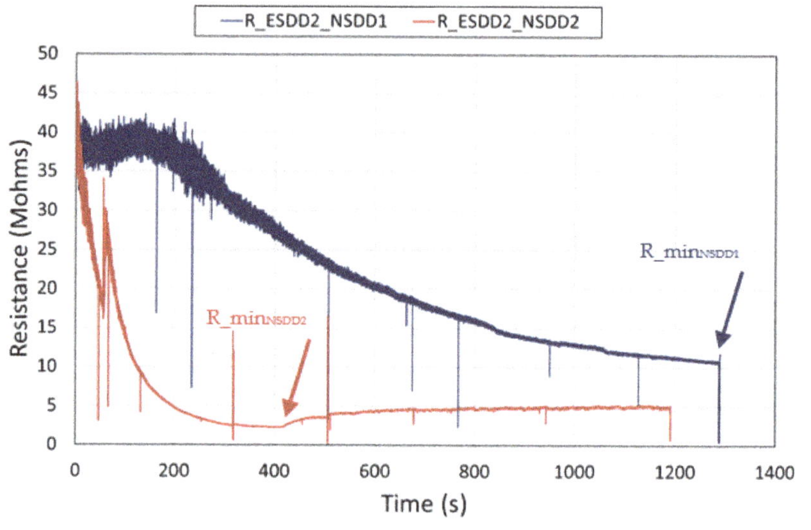

Figure 11. Leakage resistance during the low-voltage (840 V) test with $ESDD_2$.

Table 4 gives the average minimum resistance corresponding to each NSDD for the used ESDD. According to this table, the surface resistance depends on the amount of insoluble matter density in the pollution layer, even though the ESDD is the same. This can be related to the influence of the porosity of the layer that changes with the kaolin density. The greater the density, the greater the humidity absorption. The direct effect is that the resistance decreases with the increase in the moisture contained (trapped) in the pollution layer. On the other hand, the results of Table 4 show that the variation of the minimum resistance is clearly observable at the extremes: $ESDD_1$ = 0.12 mg/cm^2 and $ESDD_3$ = 1.0 mg/cm^2.

Table 4. Average minimum resistance corresponding to each NSDD for the used ESDD.

ESDD (mg/cm^2)	R_min_{NSDD1} (MΩ)	R_min_{NSDD2} (MΩ)	R_min_{NSDD3} (MΩ)
0.12	1.62	1.03	0.73
0.38	1.57	0.88	0.57
1.00	1.20	0.60	0.20

By considering that the pollution layer is uniform and continuous at the insulator surface, the resistance can be expressed as [9]

$$R_p(x) = \rho_s \int_0^L \frac{dx}{2\pi D(x)} = \rho_s \cdot K_f \tag{4}$$

where r_s is the surface resistivity, $D(x)$ is the partial diameter until the path dx, and K_f is the form factor. Then, the surface resistivity can be deduced from Equation (4).

Table 5 gives the average minimum surface resistivity deduced from Equation (4) corresponding to each ESDD with a different NSDD.

Table 5. Average minimum surface resistivity corresponding to each ESDD for various NSDD.

ESDD (mg/cm^2)	Average Minimum Surface Resistivity (MΩ)		
	NSDD$_1$ (mg/cm^2)	NSDD$_2$ (mg/cm^2)	NSDD$_3$ (mg/cm^2)
0.12	4.07	2.59	2.10
0.38	3.60	2.22	1.30
1.00	3.04	1.52	0.40

Based on the values in Table 5, the variation of the surface resistivity with ESDD can be approximated with a power function

$$\rho_s = A \cdot \mathrm{ESDD}^B \tag{5}$$

The values suggest that parameters A and B depend on the amount of NSDD and can be approximated with the following equations

$$\begin{gathered} A = 1.5897 \cdot \mathrm{NSDD}^{-0.429} \\ B = -1.3516 \cdot \mathrm{NSDD} - 1.0072 \end{gathered} \tag{6}$$

3.3. Modeling

The pollution resistance covering the insulator is an important parameter for the calculation of flashover parameters. In the case of long-rod insulators, an approximation of the form factor can be made by considering the insulator's equivalent diameter. Then, Equation (1) can be rewritten as

$$R_p = \rho_s \frac{L}{2\pi D_{eq}} \tag{7}$$

The form factor will be

$$K_f = \frac{L}{2\pi D_{eq}} \tag{8}$$

Equations (7) and (8) indicate that the insulator can be reduced to an equivalent cylinder with a length L and an equivalent diameter D_{eq}.

One can deduce that the per-unit resistance is

$$r_p = \frac{\rho_s}{D_{eq}} \tag{9}$$

By combining (3) and (9)

$$V_{\mathrm{FOV}} = L\left[N\left(\frac{\rho_s}{D_{eq}}\right)^n\right]^{\frac{1}{1+n}} \tag{10}$$

Equation (10) gives the flashover voltage as a function of the pollution surface resistivity and form factor.

By combining (5) and (10), it yields

$$V_{\mathrm{FOV}} = L \cdot \left\{N\left[\frac{A(\mathrm{ESDD})^B}{D_{eq}}\right]^n\right\}^{\frac{1}{1+n}} \tag{11}$$

Equation (11) shows that the flashover voltage depends on the pollution level, form factor, NSDD amount, and arc characteristics.

By comparing (1) and (11), one deduces

$$\begin{gathered} \alpha = L \cdot \left[\left(\frac{N \cdot A}{D_{eq}}\right)^n\right]^{\frac{1}{1+n}} \\ \beta = B\frac{n}{n+1} \end{gathered} \tag{12}$$

According to Equation (12), α depends on the geometrical characteristics of the insulator, the discharge parameters N and n, and the amount of NSDD, while β depends on parameter n and the amount of NSDD.

For the estimation of the variation of the discharge parameters N and n with NSDD, Equation (12) should be rewritten as a function of A, B, α, and β

$$n = \frac{\beta}{B-\beta}$$
$$N = \left[\left(\frac{\alpha}{L}\right)^{B} \cdot \left(\frac{D_{eq}}{A}\right)^{\beta}\right]^{\frac{1}{B-\beta}} \quad (13)$$

Equation (13) shows that the discharge parameters are closely related to the NSDD amount.

3.4. Estimation of Discharge Parameters

Based on Equations (12) and (13), the discharge parameters N and n are deduced and summarized in Table 6 for each used NSDD.

Table 6. Values of parameters n and N for fog rate 3 l/h.

NSDD	N	n
0.11	149.45	0.214
0.23	133.60	0.211
0.52	83.61	0.173

Based on Table 6, discharge parameters N and n vary with the insoluble amount: both N and n decrease with NSDD. The results are in accordance with the results observed in [8,34] and confirmed by [11,17]. This can be related to the effect of the pollution constituent on the discharge temperature, as proposed in [8,27].

Figure 12 illustrates the comparison between the measured average flashover voltages and the calculated flashover voltages based on Equation (11). The calculated values are close to the experimental values.

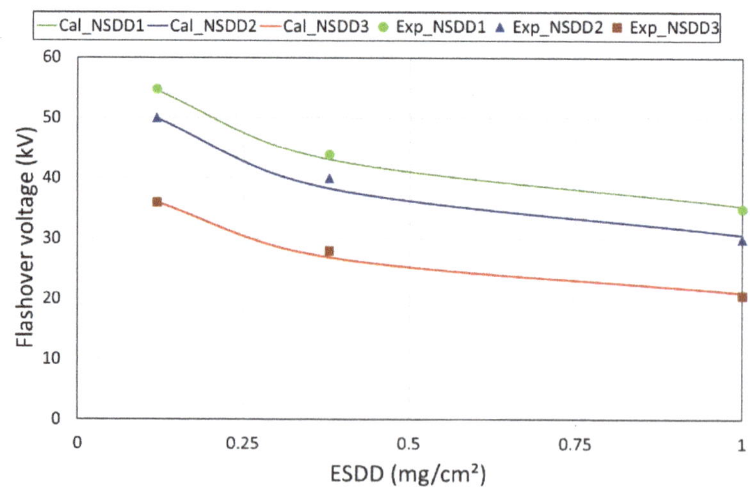

Figure 12. Comparison between calculated and measured flashover voltages vs. ESDD for various NSDD.

4. Conclusions

This paper was aimed at the experimental study of the effect of NSDD on the FOV performance of SiR insulators to establish a mathematical model of FOV voltage, taking into account the effect of both ESDD and NSDD. The experimental results showed that for the same ESDD, V_{FOV} decreases with the increase in NSDD. By plotting V_{FOV} and ESDD, it was found that coefficients α and β vary with the amount of NSDD.

The infrared and thermal recording during FOV tests showed that DBs appear at the region between the shed and the trunk. Next, the DbDs and the DBAs developed and elongated along the trunk and the sheds. The inception of DBs and the development of DbDs and DBAs are closely related to the NSDD amount.

The surface resistance depends on the amount of insoluble matter density in the pollution layer, even though the ESDD is the same. Then, the variation of the surface resistivity with ESDD and NSDD was proposed.

The link between FOV voltage equations based on the Obenaus model and the empirical equations based on ESDD was established. The equivalence between the discharge parameters N and n and the coefficients α and β was proposed. Based on the model and the experimental results, it was deduced that N and n are not constant and depend on the NSDD amount.

Author Contributions: Conceptualization, M.E.A.S. and A.K.; methodology, M.E.A.S. and A.K.; A.K.; validation, M.E.A.S., A.K., M.A. and A.M.H.; formal analysis, M.E.A.S. and A.K.; investigation, A.K.; resources, A.M.H.; data curation, M.E.A.S. and A.K.; writing—original draft preparation, M.E.A.S. and A.K.; writing—review and editing, M.E.A.S., A.K., M.A. and A.M.H.; visualization, M.E.A.S., A.K., M.A. and A.M.H. All authors have read and agreed to the published version of the manuscript.

Funding: This research received no external funding.

Institutional Review Board Statement: Not applicable.

Informed Consent Statement: Not applicable.

Data Availability Statement: Not applicable.

Conflicts of Interest: The authors declare no conflict of interest.

References

1. Haddad, A.; Warne, D. *Advanced in High Voltage Engineering*; IET Edition: London, UK, 2004.
2. Slama, M.E.; Hadi, H.; Flazi, S. Study on Influence of the Non-Uniformity of Pollution at the Surface of HVAC Lines Insulators on Flashover Probability. In Proceedings of the Conference on Electrical Insulation and Dielectric Phenomena CEIDP 2007, Vancouver, BC, Canada, 14–17 October 2007; pp. 562–566.
3. Looms, J.S.T. *Insulators for High Voltages*; Peter Pergrinus Ltd.: London, UK, 1988; no. 7.
4. Farzaneh, M.; Farokhi, S.; Chisholm, W.A. *Electrical Design of Overhead Power Transmission Lines*, 1st ed.; McGraw-Hill Professional: New York, NY, USA, 2013.
5. Rizk, F.M.; Rezazada, A.Q. Modeling of Altitude Effects on Ac Flashover of Polluted High Voltage Insulators. *IEEE Trans. Power Deliv.* **1997**, *12*, 810–822. [CrossRef]
6. Saleem, M.Z.; Akbar, M. Review of the Performance of High-Voltage Composite Insulators. *Polymers* **2022**, *14*, 431. [CrossRef] [PubMed]
7. Yang, Z.; Jiang, X.; Han, X.; Zhang, Z.; Hu, J. Influence of pollution chemical components on AC flashover performance of various types of insulators. *High Volt.* **2019**, *4*, 105–112. [CrossRef]
8. Slama, M.E.; Beroual, A.; Hadi, H. Influence of Pollution Constituents on DC Flashover of High Voltage Insulators. *IEEE Trans. Dielectr. Electr. Insul.* **2013**, *20*, 401–408. [CrossRef]
9. International Electrotechnical Commission. *Selection and Dimensioning of High-Voltage Insulators Intended for Use in Polluted Conditions—Part 1. IEC Technical Specification IEC/TS 60815-1*; IEC: Geneva, Switzerland, 2008.
10. IEC 60507:1991. *Artificial Pollution Tests on High-Voltage Insulators to Be Used on A.C. Systems*; IEC: Geneva, Switzerland, 2014.
11. Zhang, D.; Zhang, Z.; Jiang, X.; Yang, Z.; Zhao, J.; Li, Y. Study on Insulator Flashover Voltage Gradient Correction Considering Soluble Pollution Constituents. *Energies* **2016**, *9*, 954. [CrossRef]
12. Zhang, Z.; Zhang, D.; Jiang, X.; Liu, X. Effects of pollution materials on the AC flashover performance of suspension insulators. *IEEE Trans. Dielectr. Electr. Insul.* **2015**, *22*, 1000–1008. [CrossRef]

13. Ramos, N.G.; Campillo RM, T.; Naito, K. A study on the characteristics of various conductive contaminants accumulated on high voltage insulators. *IEEE Trans. Power Deliv.* **1993**, *8*, 1842–1850. [CrossRef]
14. Zhang, D.; Zhang, Z.; Jiang, X.; Yang, Z.; Liu, Y. Study on the Flashover Performance of Various Types of Insulators Polluted by Nitrates. *IEEE Trans. Dielectr. Electr. Insul.* **2017**, *24*, 167–174. [CrossRef]
15. Akbar, M.; Zedan, F. Performances of HV transmission line insulators on desert conditions—Part 3: Pollution measurement at a coastal site in the eastern region of Saudi Arabia. *IEEE Trans. Power Deliv.* **1991**, *6*, 429–438. [CrossRef]
16. Matsuoka, R.; Kondo, K.; Nako, K.; Ishii, M. Influence of Non-soluble Contaminants on the Flashover Voltages of Artificially Contaminated Insulators. *IEEE Trans. Power Deliv.* **1996**, *11*, 420–430. [CrossRef]
17. Lshii, M.; Komatsubara, M.; Matsuoka, R.; Matsumoto, T.; Kurokawa, M.; Naito, K. Behavior of Insoluble Materials in Artificial Contamination Tests. *IEEE Trans. Dielectr. Electr. Insul.* **1996**, *3*, 438.
18. Flazi, M.E.S.S.; Hadi, H.; Flazi, S.; Tchouar, N. Belakadi Etude du dépôt de pollution responsable du contournement des isolateurs des lignes aériennes du réseau électrique THT national. *Rev. Sci. Technol. B* **2007**, 43–50. (In French). Available online: https://www.asjp.cerist.dz/en/downArticle/407/0/25/58337 (accessed on 11 April 2022).
19. Jiang, X.; Yuan, J.; Zhang, Z.; Hu, J.; Sun, C. Study on AC Artificial-Contaminated Flashover Performance of Various Types of Insulators. *IEEE Trans. Power Deliv.* **2007**, *22*, 2567–2574. [CrossRef]
20. Sundhar, S. Influence of non-soluble contaminants on flashover performance of artificially contaminated polymer insulators. In Proceedings of the IEEE Conference on Electrical Insulation and Dielectric Phenomena—(CEIDP'94), Arlington, TX, USA, 23–26 October 1994; pp. 657–662.
21. Papailiou, K.O.; Schmuck, F. *Silicone Composite Insulators: Materials, Design, Applications*; Springer: Berlin/Heidelberg, Germany, 2013.
22. Swift, D.A.; Spellman, C.; Haddad, A. Hydrophobicity Transfer from Silicone Rubber to Adhering Pollutants and its Effect on Insulator Performance. *IEEE Trans. Dielectr. Electr. Insul.* **2006**, *13*, 820–829. [CrossRef]
23. Obenaus, F. Fremdschichtüberschlag und Kriechweglänge. *Deutsch. Electrotech.* **1958**, *4*, 135–136.
24. Rizk, F.A.M. Mathematical Models for Pollution Flashover. *Electra* **1981**, *78*, 71–103.
25. Hampton, B.F. Flashover Mechanism of Polluted Insulation. *Proc. Inst. Electr. Eng.* **1964**, *111*, 985–990. [CrossRef]
26. Wilkins, R. Flashover Voltage of HV Insulators with Uniform Surface Pollution Films. *Proc. Inst. Electr. Eng.* **1969**, *116*, 457–465. [CrossRef]
27. Slama, M.E.; Beroual, A.; Hadi, H. Analytical Computation of Discharge Characteristic Constants and Critical Parameters of Flashover of Polluted Insulators. *IEEE Trans. Dielectr. Electr. Insul.* **2010**, *17*, 1764–1771. [CrossRef]
28. Charalampidis, P.; Albano, M.; Griffiths, H.; Haddad, A.; Waters, R.T. Silicone rubber insulators for polluted environments, part 1: Enhanced artificial pollution tests. *IEEE Trans. Dielectr. Electr. Insul.* **2014**, *21*, 740–748. [CrossRef]
29. Albano, M.; Charalampidis, P.; Griffiths, H.; Haddad, A.; Waters, R.T. Silicone rubber insulators for polluted environents, part 2: Textured insulators. *IEEE Trans. Dielectr. Electr. Insul.* **2014**, *21*, 749–757. [CrossRef]
30. Krzma, A.; Albano, M.; Haddad, A. Comparative characterization of conventional and textured 11 kV insulators using the rotating wheel dip test. *IET High Volt.* **2020**, *5*, 739–746. [CrossRef]
31. Slama, M.E.A.; Albano, M.; Haddad, A.M.; Waters, R.T.; Cwikowski, O.; Iddrissu, I.; Knapper, J.; Scopes, O. Monitoring of dry bands and discharge activities at the surface of textured insulators with AC clean fog test conditions. *Energies* **2021**, *14*, 2914. [CrossRef]
32. Albano, M.; Waters, R.T.; Charalampidis, P.; Griffiths, H.; Haddad, A. Infrared Analysis of Dry-band FOV of Silicone Rubber Insulators. *IEEE Trans. Dielectr. Electr. Insul.* **2016**, *23*, 304–310. [CrossRef]
33. Working Group C4.303. *Artificial Pollution Test for Polymer Insulators. Results of Round Robin Test*; Cigré: Paris, France, 2013.
34. Slama, M.E.A.; Hadi, H.; Flazi, S. Investigation on influence of salts mixture on the determination of flashover discharge constant. Part I: A preliminary study. In Proceedings of the Annual Report—Conference on Electrical Insulation and Dielectric Phenomena, Quebec City, QC, Canada, 26–29 October 2008; pp. 674–677.

energies

MDPI

Article

Study on the Characteristics of Water Jet Injection and Temperature Spatial Distribution in the Process of Hot Water Deicing for Insulators

Qi Yang [1], Zhijin Zhang [2,*], Shenghuan Yang [2,3], Huarong Zeng [1], Xiaohong Ma [1], Huan Huang [1] and Guohui Pang [2]

[1] Electric Power Research Institute of Guizhou Power Grid Co., Ltd., Guiyang 550002, China; yangqi_cqu@163.com (Q.Y.); zenghuar@126.com (H.Z.); maxh1978gz@163.com (X.M.); huanghuan2005@163.com (H.H.)

[2] State Key Laboratory of Power Transmission Equipment & System Security and New Technology, Chongqing University, Chongqing 400044, China; yangshenghuan@cqu.edu.cn (S.Y.); victorpang@cqu.edu.cn (G.P.)

[3] State Grid Deyang Power Supply Company, Deyang 618000, China

* Correspondence: zhangzhijing@cqu.edu.cn; Tel.: +86-138-8320-7915

Abstract: For the benefit of knowing the method of insulator deicing using hot water, this paper investigated the physical aspects of propagation and the spatial temperature distribution of a hot water jet in the air. The numerical calculation model for the water jet flow field was established and validated, and the temperature spatial variation rule and influencing factors of the water jet were analyzed. The results indicate that the water column breaks as the distance increases. The numerical calculation results and the experimental results of the water jet flow field were approximated. As the distance increases, the mass entrainment rate of the water jet gradually increases, and the normalized axis velocity decreases approximately linearly. The jet temperature rises with the rise in the initial temperature of the hot water, and the rising rate decreases with the increase in distance. The temperature of the water jet falls as the jet distance increases. With the rise in the outlet pressure, the temperature of the water jet drops slightly. The influence of environmental wind speed on the water jet temperature is more significant than ambient temperature—the water jet temperature increases as the nozzle diameter increases. The thorough deicing efficiency is most outstanding when the hot water output temperature is around 86 °C and the jet pressure is ~3.5 MPa.

Keywords: insulator; deicing; hot water jet; numerical calculation; temperature spatial distribution

Citation: Yang, Q.; Zhang, Z.; Yang, S.; Zeng, H.; Ma, X.; Huang, H.; Pang, G. Study on the Characteristics of Water Jet Injection and Temperature Spatial Distribution in the Process of Hot Water Deicing for Insulators. *Energies* **2022**, *15*, 2298. https://doi.org/10.3390/en15062298

Academic Editor: Giorgio Ficco

Received: 12 February 2022
Accepted: 14 March 2022
Published: 21 March 2022

1. Introduction

Since the 20th century, the icing problem has posed a danger to the security of transmission line operations. Icing significantly downgrades transmission lines' mechanical and electrical performance, and threatens the power grid's stable operation [1,2]. The icing of wires and insulators has the most significant impact on the power grid. After the insulator is coated with ice, its electrical performance declines sharply. Ice flashover may lead to tripping in severe cases and interrupt the power supply [3–7].

At present, the deicing technology for transmission lines and equipment is mainly aimed at wires and substation equipment. According to the working principle [8–10], the deicing methods mainly include thermal deicing (relying on a deicing medium and ice layer heat exchange) [11], mechanical deicing (relying on external mechanical force to destroy the ice layers' bonding state) [12], and natural passive deicing (depending on ice layer gravity and other factors) [13]. Among them, the DC ice-melting technology [14] has the advantage that the ice-melting line does not consume reactive power. Still, the cost of the ice-melting device is high, the equipment utilization rate is low during the non-melting period, and it does not apply to ice-covered insulators. Wei et al. [15] controlled a deicing robot to move on a steel tube simulating wire at a certain speed, and deicing was achieved through the cutting of the deicing knife and vibration of the robot arm; however, this method is inefficient, and cannot remove the icing from insulators.

To protect the insulator from ice flashover and tripping, manual deicing was used in the past, with an excellent safety profile and low efficiency. At present, the existing deicing methods for insulators include laser or infrared deicing, hot air deicing, etc. [16–20]. In [16], Zhu used a CO_2 laser with a power of 35 W, and put forward an experimental scheme of deicing by combining laser thermal melting with ice gravity. The experiment indicated that removing 1 kg of ice consumes 90 kJ of energy and takes 26 min. However, a high-power laser and long-term irradiation may cause thermal damage and thermal stress damage on the surface of the porcelain insulator, so the laser operation needs to be meticulous and professional [17]. The authors of [18] carried out infrared deicing experiments; it was found that under the experimental conditions (the temperature of the infrared emission device and the ice distance) of 1073 K, 100 cm, and 2073 K, 40 cm, the melting of 6.5 kg ice took 1746 s and 1002 s, respectively. The research results of [19,20] show that hot air's temperature and outlet velocity are key factors affecting the deicing efficiency; its deicing efficiency is lower than that of infrared rays. Therefore, the critical power and irradiation time to avoid insulator damage are difficult to control when laser deicing is used. In comparison, the downside of infrared ray or hot air deicing is that the amount of deicing per unit of time is negligible.

The deicing methods above essentially take advantage of heat conduction and convection to complete the heat transfer between the ice and the deicing medium. Water has a larger specific heat capacity and thermal conductivity than air. The convective heat transfer coefficient between the water jet and ice is more significant than between hot air and ice at the same flow velocity [21]. Moreover, using a hot water jet can prevent insulator damage more easily. Regarding water jet deicing, the relationship between deicing efficiency using a water jet at room temperature and jet pressure and ice layer temperature was studied preliminarily in [22,23], which revealed that increasing the ice layer temperature helps deicing. At present, hot water deicing has been used in road deicing and other fields [24,25], and related numerical simulation and laboratory research has been carried out on hot water deicing for EMUs (electric multiple units) [26]. Therefore, a hot water jet can be used as a deicing medium for the insulator to study its deicing efficiency and influencing factors.

This research focuses on the physical properties and temperature distribution of hot water jet propagation in the air. Experiment and numerical calculations were used to investigate the effects of the initial temperature of hot water, jet distance, and outlet pressure on the spatial distribution of jet temperature. The numerical and experimental results of the water jet flow field are similar. According to the research findings, the melting process of the insulator surface under the action of hot water was also researched, as well as the deicing efficiency and safety distance of hot water deicing technology. The findings confirm theoretical study of hot water deicing, and indicate the mechanisms and parameters influencing insulator deicing.

2. Physical Process of Hot Water Deicing

The main principle of hot water deicing for the insulator is to transfer the heat and energy of hot water to the ice using a water jet device with high pressure. The ice gradually melts and falls off under hot water jet melting and impact force, as shown in Figure 1. The hot water jet arrives from the left, and the ice falls to the right, as marked by the red circle and lines. The physical process consists of two stages:

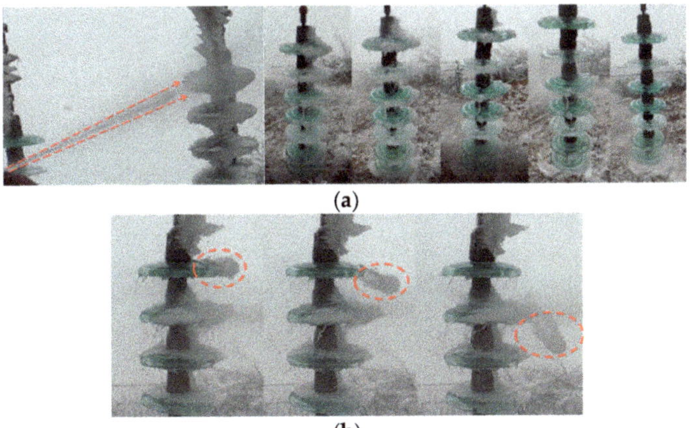

Figure 1. Hot water deicing process of an LXY-120 glass insulator. (**a**) Process from 0 s to 120 s at 30 s intervals; (**b**) Ice falling off.

(1) High-speed hot water jets from the nozzle outlet to the ice layer (Figure 1a). The reference temperature, wind speed, nozzle diameter, ambient temperature, injection distance, and outlet pressure on the spatial distribution of water jet temperature will determine the technical parameters of the hot water deicing device. This paper will focus on this process;
(2) The hot water causes the ice to melt and shed through convective heat transfer and impact force (Figure 1b). The insulation performance of the water jet needs to be considered in the actual deicing operation. It is necessary to investigate the influence of jet distance, water conductivity, and outlet pressure on the water jet's leakage current in order to validate the shortest deicing length under various combinations of water conductivity and outlet pressure.

3. Numerical Model of Water Jet Flow Field

The study of the temperature distribution of water jets in space was carried out via ANSYS FLUENT numerical calculation and the experimental verification of the Xuefeng Mountain Natural Icing Test Base of Chongqing University.

3.1. The Physical Structure of the Nozzle

According to the shape of the inner hole, the nozzles can be cylindrical, conical-straight, streamlined, etc. The conical-straight nozzle is the most commonly used water jet nozzle due to its significant flow coefficient [27–29]. The conical-straight nozzle was taken as the prototype, and its physical model was simplified and obtained, as shown in Figure 2. The parameters of the nozzle structure is shown in Table 1.

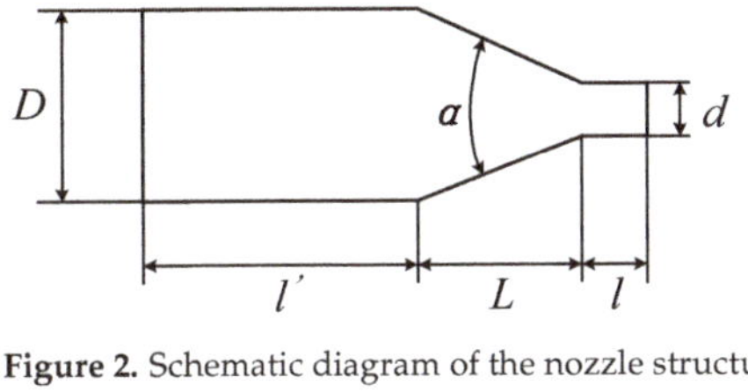

Figure 2. Schematic diagram of the nozzle structure.

Table 1. Parameters of the nozzle structure.

Parameters	D (mm)	l' (mm)	L (mm)	α (°)	d (mm)
	7	10	5	45	1–2

The nozzle structure parameters mainly include inlet diameter D, outlet diameter d, contraction angle α, length of converging section L, the ratio of smoothing section length to outlet diameter—that is, l/d—and access section length l'.

3.2. Geometric Model and Meshing of the Water Jet Flow Field

The size of the three-dimensional flow field was set to 5 m × 4 m × 4 m to investigate the effects of side wind on the properties of the jet flow field. When meshing, the mixed grid was applied in the computational domain (Figure 3), divided into two parts: the near-axis area through which the water jet flows, and the far-axis area away from the water jet. The scope of the paraxial area was 1 m. The above two areas had meshed independently. The near-axis area was meshed using a hexahedral grid with better calculation performance. The grid size increased gradually from the inside to the outside with a growth rate of 1.2, and the initial minimum edge length was 0.1 mm. The far-axis area was meshed using a tetrahedral–hexahedral mixed grid with a larger size, as shown in Figure 3.

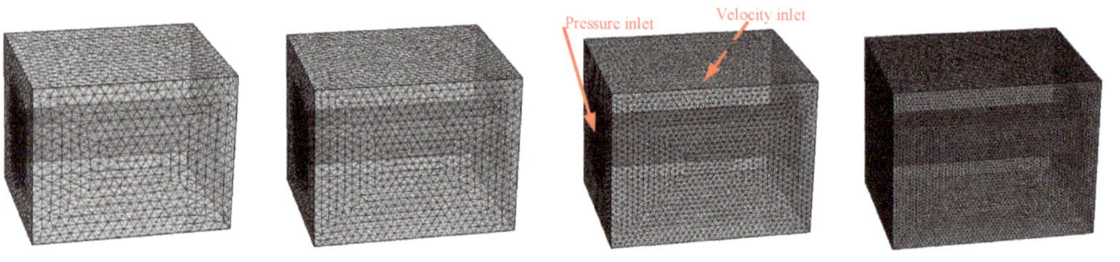

Figure 3. Different meshing and boundary condition settings of the water jet flow field.

At least four different grid calculations must be performed in order to analyze the grid convergence test, as show in Figure 3. As the mesh density increased, the velocity distribution on the axial section at the nozzle exit remained the same, and there was no significant difference. Furthermore, the change rate of the inlet and outlet pressure drop between adjacent grids was less than 0.9%. Therefore, the physical model and discrete format used in the calculation have spatial convergence and numerical stability, and the selection of a coarser grid number is sufficient to meet the accuracy requirements of the calculation.

3.3. Calculation Model and Governing Equation

After flowing through the contraction section of the nozzle, the high-speed water is turbulent, and turbulent diffusion and momentum exchange with the ambient air occurs violently [30]. Therefore, the nozzle jet belongs to the gas–liquid two-phase flow problem, and in the calculation of the jet flow field of the nozzle, the high-pressure water jet belongs to the state of high turbulence. Therefore, selecting a suitable turbulence model and multiphase flow model is necessary. The turbulence model selects standard k–ε governing equations.

The standard k–ε model is suitable for a flow field with low swirl intensity and regular wall shape, and it is a semi-empirical model based on model transport equations for the turbulent viscosity (μ_t), turbulent kinetic energy (κ), and its dissipation rate (ε), which are given by:

$$\mu_t = \rho C_\mu \frac{k^2}{\varepsilon} \tag{1}$$

$$\frac{\partial(\rho\kappa)}{\partial t}+\frac{\partial(\rho\kappa u_i)}{\partial x_i}=\frac{\partial}{\partial x_j}\left[\left(\mu+\frac{\mu_t}{\sigma_\kappa}\right)\frac{\partial\kappa}{\partial x_j}\right]+G_\kappa+G_b-\rho\varepsilon-Y_M+S_\kappa \tag{2}$$

$$\frac{\partial(\rho\varepsilon)}{\partial t}+\frac{\partial(\rho\kappa u_i)}{\partial x_i}=\frac{\partial}{\partial x_j}\left[\left(\mu+\frac{\mu_t}{\sigma_\varepsilon}\right)\frac{\partial\varepsilon}{\partial x_j}\right]+G_{1\varepsilon}\frac{\varepsilon}{\kappa}(G_\kappa+G_{3\varepsilon}G_b)+G_{2\varepsilon}\rho\frac{\varepsilon^2}{\kappa}+S_\varepsilon \tag{3}$$

where G_k represents the generation of turbulent kinetic energy due to the mean velocity gradients in these equations, G_b is the turbulent kinetic energy produced by buoyancy, and Y_M fluctuates due to excessive diffusion of incompressible turbulence. $C_{1\varepsilon}$, $C_{2\varepsilon}$, $C_{3\varepsilon}$, and C_μ are constants, σ_k and σ_ε are the turbulence Prandtl numbers for k and ε, respectively, and S_k and S_ε are user-defined. The model constants $C_{1\varepsilon}$, $C_{2\varepsilon}$, $C_{3\varepsilon}$, σ_k, and σ_ε were assumed to have the following values: $C_{1\varepsilon} = 1.44$, $C_{2\varepsilon} = 1.92$, $C_\mu = 0.09$, $\sigma_k = 1.0$, and $\sigma_\varepsilon = 1.3$.

Since the VOF (volume of fluid) model is applicable to solve the spatiotemporal distribution of gas–liquid interface in stratified flow, sloshing, dam breaking, jet attenuation, etc. [31], the VOF model was selected for the multiphase flow model.

In the VOF model, the content of each phase fluid in each calculation unit is characterized by the volume fraction α. For example, in the case of q kinds of fluids, the volume fraction of each phase is $\alpha_1, \alpha_2, \ldots, \alpha_q$, and each volume fraction satisfies $\alpha_1 + \alpha_2 + \ldots + \alpha_q = 1$. The governing equations of the VOF model mainly include the volume fraction equation, momentum equation, and energy equation (Appendix A).

Assuming that the density changes of water and air can be ignored, the viscous flow in the computational domain is incompressible. The solver adopts the transient and pressure-based implicit solution, and the energy equation is opened to solve the jet temperature field simultaneously. The wall function in the standard k–ε turbulence model is set as the standard near-wall function. In the VOF two-phase model, the primary phase is air, and the secondary phase is water. The relationship between the physical parameters—such as thermal conductivity, viscosity, and specific heat capacity—and air and water temperatures is input into the material property column.

3.4. Boundary Conditions

The setting of boundary conditions of the computational domain is shown in Figure 3. The pressure reflects the nozzle outlet speed, and the relationship between them can be obtained from the literature [32]. Therefore, the nozzle inlet was set as the pressure inlet, and the pressure direction was perpendicular to the nozzle inlet surface. The turbulence intensity I and hydraulic diameter D_H of the inlet can be calculated by Equation (4):

$$\begin{cases} I = 0.16\times(R_{eD})^{-1/8} \\ D_H = D \end{cases} \tag{4}$$

where R_{eD} is the Reynolds number of the flow in the nozzle, and D_H is equal to the ratio of four times the flow area to the wetting circumference. D_H is the nozzle inlet diameter D for flows in a circular tube.

The volume fraction of water at the nozzle inlet was kept at 1, and the inlet temperature was set as the initial temperature of the hot water. The inner wall of the nozzle was set to have no sliding wall surface, and the wall material was set as steel. The nozzle outlet–air interface was set as the internal boundary. The front and back boundaries were selected as the velocity inlet in the external flow field area, used to simulate the transverse wind speed. The velocity direction was perpendicular to the jet axis. The upper and lower boundaries and the left and right boundaries were the pressure outlets. A pressure value was 0.1 MPa, one barometric pressure, and the temperature value was the ambient temperature. Before starting the calculation, the computational domain was initialized by filling the domain with air and setting the initial temperature as the ambient temperature.

For non-steady-state numerical simulation, it is necessary to determine the number of grids used, the time-step size, and the non-correlation with the calculation results—that is, the grids should be verified for non-correlation and time-independence. The four grid numbers were tested for independence according to the time step of 0.001 s.

After selecting the fixed number of grids, we set the Courant number to four levels from small, and chose the time step according to the convergence and error of the iterative residuals. In general, the Courant number is essential in explicit differential method schemes, and it should be small in order to ensure stability. The maximum control residual is 0.0001 in the iterative calculation. The numerical values used for the thermal parameters of water and air are shown in Table 2.

Table 2. The numerical values used for the thermal parameters.

Properties	Density (kg/m^3)	Cp (J/(kg K))	Thermal Conductivity (W/(m K))	Viscosity (kg/(m s))	Reference Temperature (°C)
Air	1.225	1006.43	0.0242	1.7894e-05	25
Water	998.2	4182	0.65	0.001003	24.85

4. Numerical Calculation Results and Validation of the Water Jet Flow Field

4.1. Numerical Calculation Results of Water Jet Flow Field

Taking the outlet pressure of 3 MPa as an example, when the wind speeds are 0 m/s and 3 m/s, the temperature and velocity contours of the water jet cross-section are shown in Figures 4 and 5, respectively.

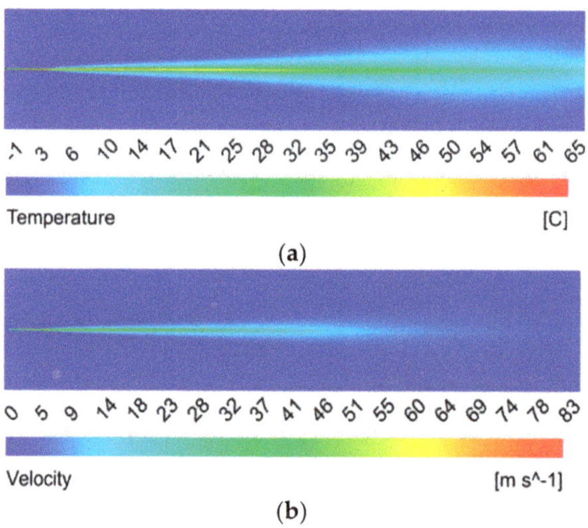

Figure 4. The numerical calculation results of the water jet when the ambient wind speed is 0 m/s: (**a**) The temperature contour of the water jet. (**b**) The velocity contour of the water jet.

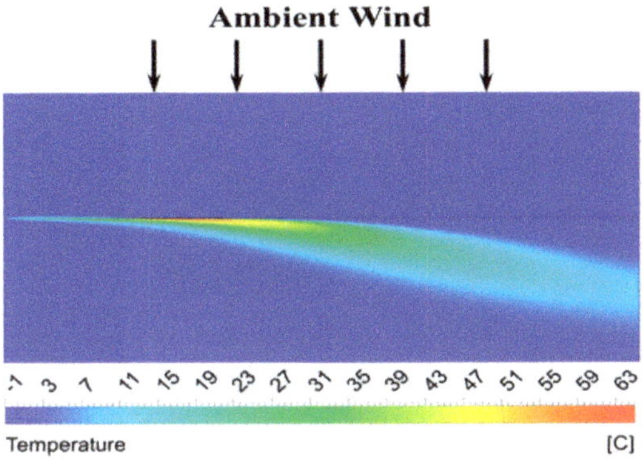

(a)

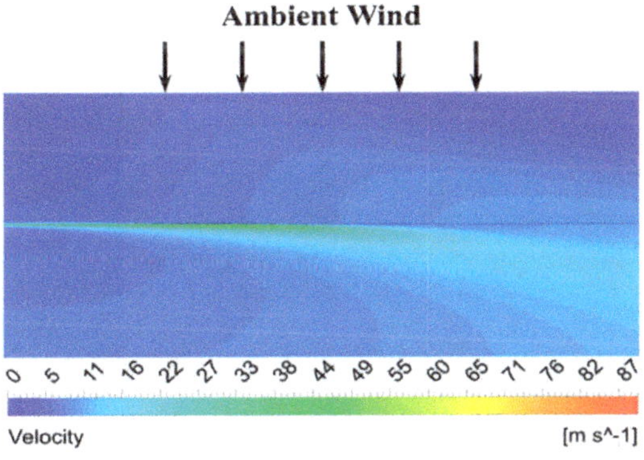

(b)

Figure 5. The numerical calculation results of the water jet when the ambient wind speed is 3 m/s: (**a**) The temperature contour of the water jet. (**b**) The velocity contour of the water jet.

It can be seen from Figures 4 and 5 that regardless of the presence of ambient wind, after the water jet sprays from the nozzle, the jet temperature and velocity gradually decrease along the axial direction, and the radial diffusion width gradually increases. Compared with 0 m/s, the core part of the temperature and velocity of the water jet is shifted along the direction of the wind speed due to the transverse wind blowing.

The temperature and velocity attenuation of the water jet are negatively correlated with the air entrainment along the way or the jet distance [33]. Figure 6 shows the velocity vector diagram at the nozzle outlet; the color difference represents the difference in speed. It can be seen from the figure that the high-speed water jet and the ambient air converge at the nozzle outlet, and the velocity value and direction of the two fluids are different. Under the entrainment of the longitudinal vortex [33], air approaches the main body of the water jet and turbulently moves forward with it, forming a gas–liquid two-phase mixed layer at the jet boundary. The violent momentum exchange and heat transfer occur between the entrained air and the water column, so the entrainment degree of the water jet is an essential factor affecting the jet's temperature and velocity.

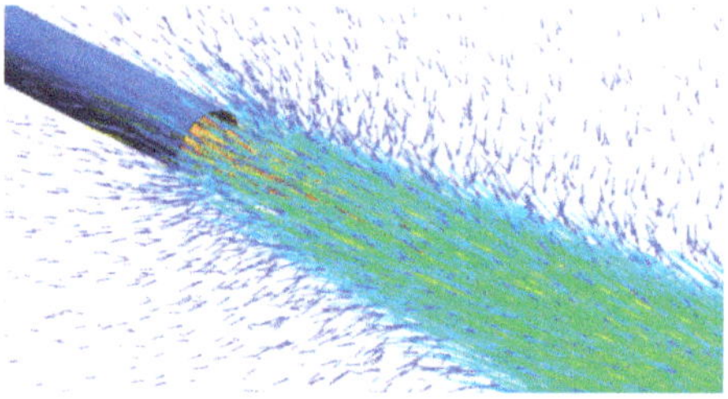

Figure 6. Enlarged view of the velocity vector at the nozzle outlet.

The velocity of the jet along the axis decays, and most of the energy is transferred to the surrounding fluid. The velocity of the jet body decreases, but more and more surrounding fluid moves forward together with the jet body under the entrainment of the longitudinal vortex, so that the flow rate of the jet fluid increases gradually along the path.

The mass entrainment rate J is defined to represent the entrainment degree of the jet to the air [34,35]; that is:

$$J = \frac{Q_x - Q_0}{Q_0} \tag{5}$$

where Q_x is the mass flow rate (kg/s) in the range where the minimal axial jet velocity is $0.5U_m$ on the radial cross-section at the distance of x away from the nozzle outlet, U_m is the maximum axial velocity on the radial cross-section (m/s), and Q_0 is the mass flow rate at the nozzle outlet (kg/s).

Figure 7 shows the mass entrainment rate and the axial distance when the outlet pressures are 4, 3, and 2 MPa. It can be seen from the figure that the mass entrainment rates under the three pressures increase with the increase in the axial distance and, especially when the distance $x > 2.5$ m, the mass entrainment rates increase rapidly. The above phenomenon indicates that, as the distance increases, more and more air is gradually entrained at the jet edge and approaches the center of the water jet. The edge of the water jet is "torn" by the air and gradually separates from the main body of the water column, forming a boundary layer with a lower velocity. Therefore, the jet's temperature and velocity decrease progressively along the axis, and its diffusion width gradually increases.

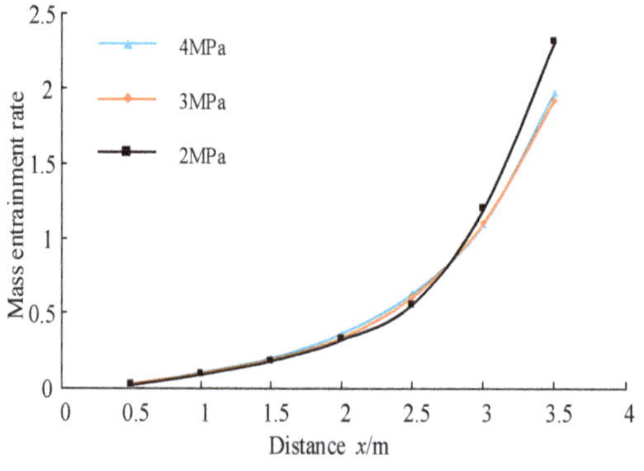

Figure 7. Mass entrainment rates under different outlet pressures.

4.2. Experimental Validation

The diffusion state of the water jet in the experiment is shown in Figure 8. The water column was photographed in sections with a high-speed camera to observe the internal flow of the water jet. The photographed results are shown in Figure 9 (test condition: the frame rate of the high-speed camera is 100 frames/second; in Figure 9a–c the nozzle diameter is 1.2 mm, and the outlet pressure is 1.5 MPa; Figure 9d shows the comparison of experimental results under different distance away from the nozzle, outlet pressures and nozzle diameters).

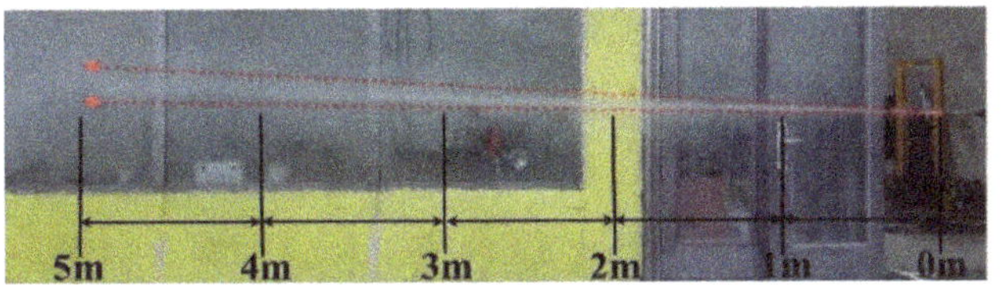

Figure 8. The diffusion state of the water jet.

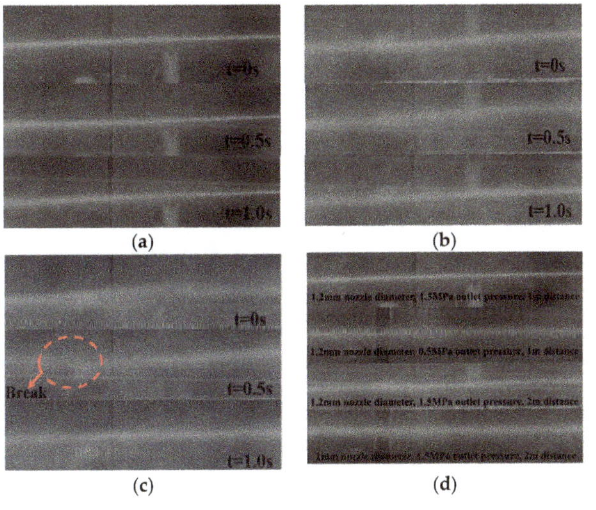

Figure 9. The state of the water jet (high-speed photography): (**a**) 1 m; (**b**) 2 m; (**c**) 3 m; (**d**) Comparison of experimental results under different distance away from the nozzle, outlet pressures and nozzle diameters.

Time represents the acquisition time of a high-speed camera, which is in a steady state. Steady state refers to the obvious law of the water jet at this distance under the condition of a specific nozzle outlet, as can be obtained from the following results:

(1) The distance affects the shape of the water jet. The water jet within 1 m is compact in Figure 9a, and the water column is straight, transparent, and dense. The cross-section width of the water column is clear, and the air is less mixed. According to the experiment results in Figure 9d, the compact segment's length is related to the nozzle diameter and outlet pressure. To be specific, by comparing the water column states in the first two pictures, we can see that the compactness of the water column of 1.5 MPa is better, and the cross-section width of the water column is clearer than that at 0.5 MPa. Therefore, the length of the compact segment grows larger when the outlet pressure rises. Similarly, by comparing the water column states in the last two pictures, it can be seen that the water column of 2 mm nozzle diameter at 2 m is more compact than that of 1.2 mm nozzle diameter—that is, the compact segment extends as the nozzle diameter increases;

(2) After the dense water column enters the air, it has strong friction with the air, the section widens, and the speed slows down. The water column is gradually doped with air, and the jet edge begins to diverge into water mist. The boundary layer of the water column gradually blurs, as shown in Figure 9b;

(3) With the increase in distance, the jet entrains more air, causing the randomness of movement to become greater, and the water column diverges accordingly. For example, "break" occurs after the jet distance reaches 3 m in the condition of 1.2 mm nozzle diameter and 1.5 MPa outlet pressure, as shown in Figure 9c. With the increase in water jet injection distance, air entrainment intensifies, and the water column gradually absorbs air. Due to the doping of air, the motion of the water jet is slow and random, and the velocity is reduced to a slower speed. With the increase in jet length, the jet absorbs more and more air and is isolated into multiple water blocks by the air, resulting in a "break";

(4) Figure 9d shows that we have carried out experiments under different nozzle outlet conditions, and the test results show that the length of the water jet compact section is related to the nozzle diameter and the outlet pressure. Because the jet direction is from the right side to the left side of the picture, the water jet's gravity effect is more obvious after the speed is reduced, resulting in the water jet state showing a low left and high right state.

The jet temperature was measured at different distances. Table 3 shows the experimental conditions of each group. Each group was measured three times, and the average value was taken. The inlet and boundary conditions of the simulation model were set according to the experimental conditions, and the corresponding numerical calculation results were obtained. Since the temperature measured in the test was the average temperature of the water jet, it was necessary to integrate the radial temperature of the numerical results in order to obtain the average numerical calculation temperature T_{ave}.

$$T_{ave} = \frac{\sum_{1}^{n} T_i \cdot \Delta h}{H} \tag{6}$$

where T_i is the temperature of the grid nodes, n is the number of grid nodes, Δh is the grid height, and H is the diffusion width of the water jet.

Table 3. The validation experiment parameters of water jet temperature.

Test Serial Number	Ambient Temperature (°C)	Average Wind Speed (m/s)	The Initial Temperature of Hot Water (°C)	Outlet Pressure (MPa)
1	2	3.4	80	2.2
2	0.2	3.9	65	3.0
3	1.5	4.7	55	4.0

Figure 10 depicts a comparison between numerical results with experimental water jet temperature readings. It can be seen that the two are in excellent accordance, showing that the numerical water jet calculation model is reasonable and that the attenuation law of the water jet temperature may be derived from it. The difference is concentrated in 0–2 m and 4–5 m, because the entrainment rate of the jet increases with the increase in the axial distance—especially when the distance $x > 2.5$ m—and the mass entrainment rate increases rapidly.

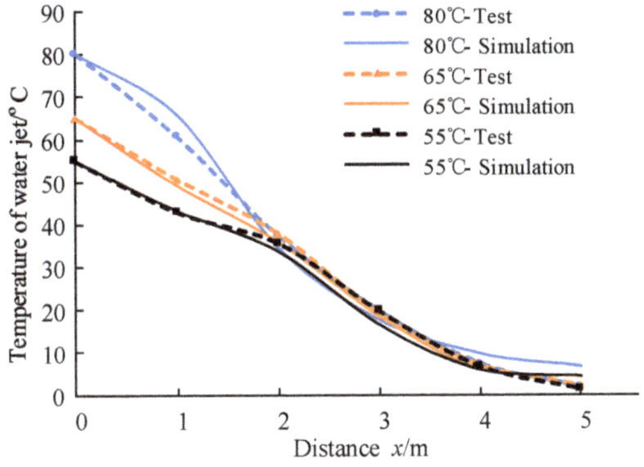

Figure 10. Comparison between experimental and simulation values of water jet temperature.

At the same time, in order to verify the rationality of the numerical calculation of water jet velocity, the simulation calculation of the axial velocity of a 1.5 mm nozzle at outlet pressures of 4, 3, and 2 MPa was carried out. The comparison between the calculated value of the normalized axial velocity U_m/U_0 at different distances and the experimental value is shown in Figure 11, where U_m is the maximum axial velocity on the radial cross-section (m/s), and U_0 is the velocity at the nozzle outlet (m/s).

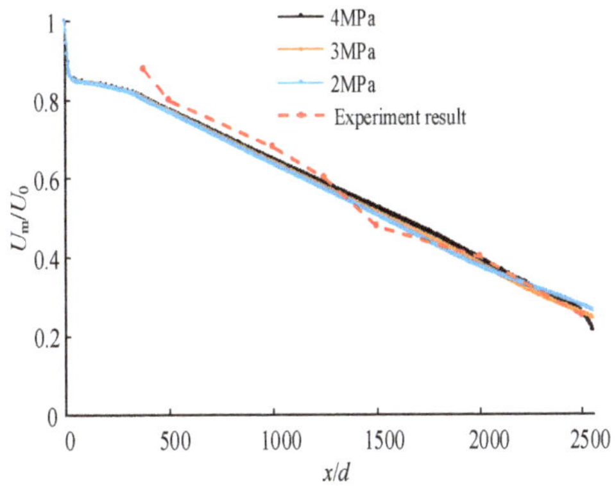

Figure 11. Comparison between numerical calculation results and experimental value (References [36,37]) of axis velocity.

The following can be seen from Figure 11:

(1) Under the three outlet pressures, although the velocity at the nozzle outlet is different, the attenuation law of the normalized axial velocity along the axial direction is virtually the same;

(2) The fluid velocity at the nozzle inlet is low. The velocity increases sharply after the conical contraction section, and reaches its maximum at the nozzle outlet. The jet velocity gradually decreases after leaving the nozzle, and the velocity decays rapidly within a short distance from the nozzle outlet, and then approximately declines linearly;

(3) Taking the numerical result of 4 MPa as an example, when $x = 1000d$—that is, 1.5 m away from the nozzle outlet—the axial velocity U_m = 57.11 m/s, which decreases to 64.73% of the outlet velocity (U_0 = 88.23 m/s). When $x = 2500d$—that is, 3.75 m away from the nozzle outlet—the axial velocity U_m = 22.25 m/s, which decreases to 25.22% of the outlet velocity U_0. The numerical simulation results agree well with the experimental results in the literature [36,37]; that is, the axial velocity of the water

jet shows linear attenuation, and the axial velocity is ~$0.7U_0$ at $1000d$ and ~$0.25U_0$ at $2500d$.

In summary, the relative error between the numerically calculated results of water jet temperature and axial velocity and the experimental result is small. Therefore, the simulation calculation of the water jet in this paper can reflect the characteristics of the water jet flow field well, and the reliability of the numerical calculation model is validated.

According to the optimization method given in the literature [27], the optimal deicing effect parameters in a specific environment can be obtained. For the 1.2 mm diameter nozzle, the ambient temperature is −3 °C, the wind speed is 3 m/s, the icing weight is 2 kg/piece, and the distance is 2 m. The optimal initial hot water temperature and jet outlet pressure values are 86.07 °C (35–100 °C) and 3.49 MPa (1–7 MPa), respectively. The total deicing efficiency is optimal when the hot water outlet temperature is around 86 °C and the jet pressure is around 3.5 MPa.

5. Analysis of Factors Affecting the Spatial Distribution of Water Jet Temperature

5.1. The Effect of the Initial Temperature of the Hot Water

In order to ascertain the relationship between the water jet temperature at different distances and the initial temperature, the test conditions were set as 1.2 mm, −1 °C, 3 m/s, and 3 MPa (from left to right are the nozzle diameter, ambient temperature, wind speed, outlet pressure, and the same below), and the results are shown in Figure 12.

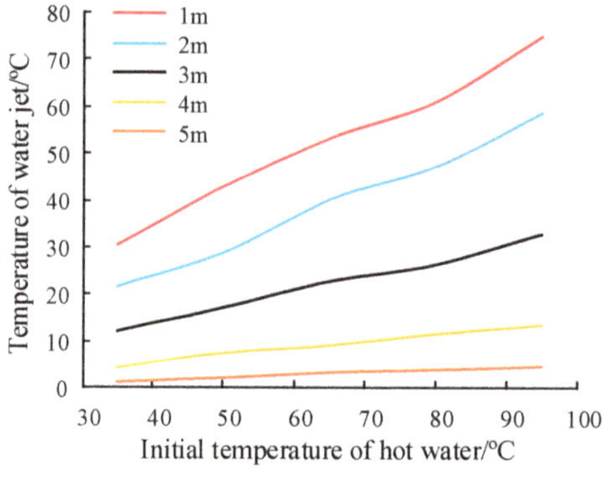

Figure 12. Relationship between water jet temperature and the initial temperature of the hot water.

The following can be concluded from the above Figure:

(1) The temperature of the water jet is positively correlated with the hot water's initial temperature. Compared with 35 °C, the water temperatures of 50–95 °C increased the jet temperature by 33.67%, 86.73%, 119.48%, and 172.82%, showing an approximately linear increase;

(2) The rate of water jet temperature increase with the initial temperature of the hot water is negatively correlated with distance—that is, the longer the distance, the lower the growth rate. As shown in Figure 12, when the distance increases from 1 m to 5 m, the increase rate of the jet temperature decreases from 0.72 to 0.06. With the increase in the jet distance, the difference between the initial temperature of the hot water and the temperature when it reaches the ice layer becomes larger—that is, the effect of the former weakens.

5.2. The Effect of Jet Distance

In order to study the change rule of water jet temperature in the axial direction, the test conditions were set as 1.2 mm, −1 °C, 3 m/s, and 4 MPa in order to ascertain the temperature of the water jet at different distances.

As shown in Figure 13, the temperature of the water jet is negatively correlated with the distance when the environmental parameters remain constant.

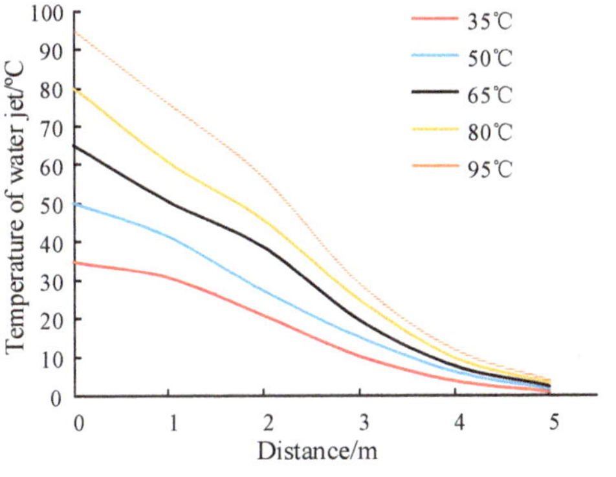

Figure 13. Relationship between water jet temperature and distance.

5.3. The Effect of Outlet Pressure

The outlet pressure was set as 2 MPa, 3 MPa, and 4 MPa, the ambient temperature was −1 °C, and the wind speed was 3 m/s. The relationship between the temperature at 2 m and the outlet pressure at different initial temperatures of hot water was determined, as shown in Figure 14.

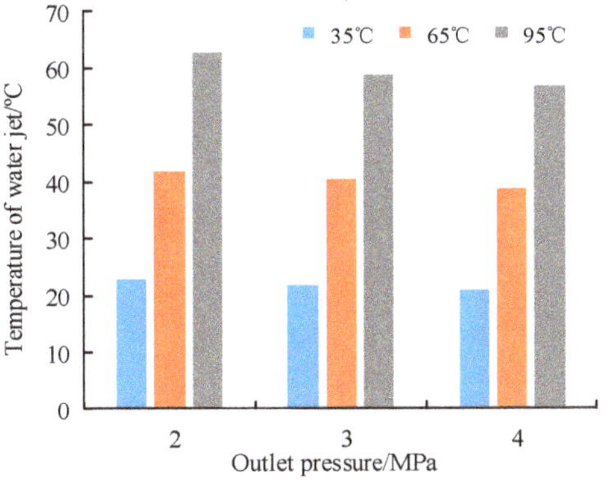

Figure 14. Relationship between the temperature of the water jet and the pressure at the outlet.

As can be seen from Figure 14, the water jet temperature decreases slightly with outlet pressure. Compared with 2 MPa, the temperatures at 3 MPa and 4 MPa were reduced by 3.66% and 7.49%, respectively.

The temperature change of the water jet in the air is a convective heat transfer problem. According to the literature [21,38], this kind of problem can be qualitatively analyzed based on the following equations:

$$q'' = h(T_s - T_\infty) \quad h = kN_u/L \quad \mathrm{Re} = \frac{Lu_\infty}{\nu} \tag{7}$$

where q'' is the convective heat flux, representing the convective heat transfer in units of time; h is the convective heat transfer coefficient; T_s is the water jet temperature; T_∞ is the air temperature; k is the thermal conductivity of air; L is the heat transfer characteristic length, here referring to jet distance; the Nusselt number Nu is positively correlated with the Reynolds number Re; u_∞ is the relative velocity between the water jet and the air, which is mainly determined by outlet pressure and wind speed; and v is the kinematic viscosity of air.

When other factors such as ambient temperature and wind speed remain constant, u_∞ increases as the outlet pressure increases. This causes the Reynolds number Re, the convective heat transfer coefficient h, and convective heat flux q'' to increase accordingly at the same distance from the nozzle outlet. The increase in q'' means more heat is transferred from the water jet to the ambient air within the same time. Therefore, the water jet temperature decreases with the rise in the outlet pressure.

5.4. The Effect of Ambient Temperature

Taking the conditions of −1 °C, 0 m/s, and 3 MPa as the control group, the simulation conditions of the other four groups were −7, −6, −4, and −2 °C, respectively. The water jet temperature at different ambient temperatures is shown in Figure 15.

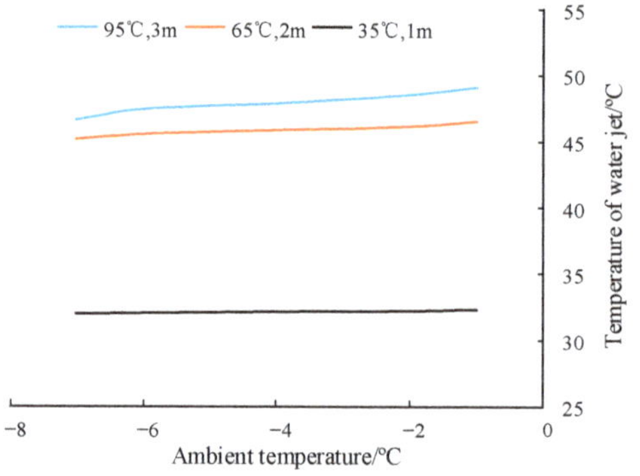

Figure 15. Relationship between water jet temperature and ambient temperature.

When the initial temperature of hot water and the ambient wind speed are constant, the temperature of the water jet decreases with the decrease in the ambient temperature, which is generally linear. Each time the ambient temperature decreases by 1 °C under this condition, the water temperature decreases by approximately 0.375 °C.

It can be speculated from Equation (4) that as the ambient temperature T_∞ drops, the temperature difference T_s–T_∞ between the water jet and the air increases. At the same time, the effect of the ambient temperature change (from −7 °C to −1 °C) on the convective heat transfer coefficient h can be ignored. Therefore, the convective heat flux q'' increases and, thus, the water jet temperature decreases with the decrease in the ambient temperature.

At the same time, by comparing the slopes of the curves at different distances, it can be seen that the further the distance, the faster the water temperature will rise with the increase in ambient temperature. The ambient temperature change at close range has almost no effect on the water jet temperature. The air content in water jets may cause this variation trend to rise with the increasing distance.

5.5. The Effect of Environmental Wind Speed

To study the degree of influence of wind speed on the attenuation of the water jet temperature, the conditions of −1 °C, 0 m/s, and 3 MPa were taken as the control group, and the other two groups were −1 °C, 1.5 m/s, 3 MPa and −1 °C, 3 m/s, 3 MPa. The relationship between water jet temperature and wind speed was calculated, as shown in Figure 16.

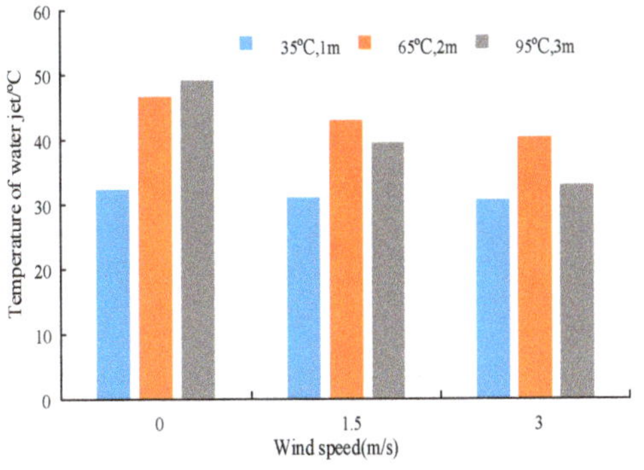

Figure 16. Relationship between water jet temperature and wind speed.

As shown in Figure 16, when the initial temperature of the hot water and the ambient temperature are constant, the temperature of the water jet decreases as the wind speed increases. Compared with the windless condition, the water jet temperatures of 1.5 m/s and 3 m/s reduced by 7.59% and 13.46%, respectively.

The effect of wind speed on the water jet temperature is greater than that of the ambient temperature. For example, when the initial temperature of the hot water is 95 °C, compared with the calculated result of the control group (−1 °C, 0 m/s), as the ambient temperature decreases by 3 °C, the temperature of water jet at 3 m drops from 49.15 °C to 47.9 °C, which is a decrease of 1.25 °C. However, when the ambient wind speed increases by 3 m/s, the temperature of the water jet at 3 m decreases by 16.23 °C.

When the wind speed increases, the relative velocity u_∞ between the water jet and the air increases, which eventually causes the heat loss of the water jet to increase within the same time. Moreover, according to our field tests on water jets in windy conditions, it appears that the compactness of the water column becomes worse with increasing wind speed. The mixing of air into a water jet would enhance the convective heat transfer and, thus, accelerate the attenuation of the water jet temperature. Therefore, the temperature of the water jet decreases as the wind speed increases, and the increase in wind speed has a more significant impact on the water jet temperature than the decrease in ambient temperature.

5.6. The Effect of Nozzle Diameter

Taking three nozzle diameters of 1.2 mm, 1.5 mm, and 2 mm as examples, the relationship between water jet temperature and nozzle diameter at 2 m was calculated under the boundary conditions of 1.5 MPa outlet pressure, with environmental parameters of −1 °C and 0 m/s, as shown in Figure 17.

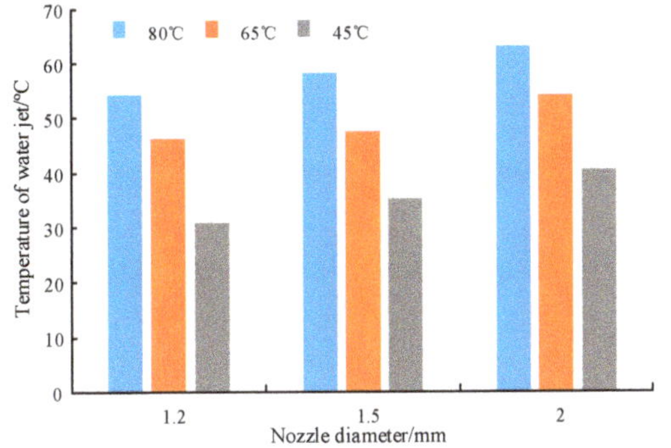

Figure 17. Relationship between water jet temperature and nozzle diameter.

Increasing nozzle diameter can improve water jet temperature when the other parameters are the same. Since air damages the water column structure from the outer layer to the inner core, the thicker water column has a greater ability to withstand damage to its core, and a better ability to retain heat. Therefore, when the initial temperature of the hot water is constant, the larger the nozzle diameter, the smaller the temperature drop would be at the same distance.

For example, when the nozzle diameter increases from 1.2 mm to 2 mm, the water jet temperatures at 2 m distance of 80 °C, 65 °C, and 45 °C initial temperature increase from 54.32 °C, 46.41 °C, and 30.87 °C to 63.15 °C, 54.02 °C, and 40.58 °C respectively, with increases of 16.27%, 16.40%, and 31.45%. The lower the initial temperature of the hot water, the more the nozzle diameter will increase the temperature of the water jet.

6. Conclusions

This paper studied the characteristics of a hot water jet in the air. The numerical calculation model of the water jet flow field was established and verified. The effects of the initial temperature of the hot water, jet distance, outlet pressure, nozzle diameter, ambient temperature, and wind speed on water jet temperature were studied. The main conclusions are as follows:

(1) The water column in the initial segment of the water jet is compact. The increase in distance will cause the water jet to diverge gradually, and the boundary layer will become blurred. After reaching a certain distance, the water jet "breaks";

(2) The numerical results of the hot water jet calculations are consistent with the results of the field studies. The mass entrainment of the water jet increases with the increasing distance. Under the three outlet pressures, the attenuation law of normalized axial velocity of the water jet along the axial direction is the same, and decreases approximately linearly. When the distance $x = 1000d$, $U_m = 0.63U_0$, and when the distance $x = 2500d$, $U_m = 0.25U_0$;

(3) The water jet temperature is negatively correlated with distance and outlet pressure, and the distance has a greater impact. The water jet temperature is positively correlated with the initial temperature of the hot water, and the increase in distance will weaken the influence of the initial temperature of the hot water. When the ambient temperature and wind speed change the same value, the wind speed is greater than the ambient temperature. Increasing the nozzle diameter can increase the flow rate, effectively increasing the water jet temperature. According to the effective laws of the above factors on the spatial distribution of water jet temperature, the optimal parameter for ice melting can be obtained theoretically, which will be discussed in future work.

The convective heat transfer coefficient between the water jet and ice is more significant than that between hot air and ice at the same flow velocity; therefore, hot water deicing has great advantages over laser deicing. In addition, based on the results of the water jet research in this paper, the efficiency of hot water deicing can be further optimized. Furthermore, the water jet properties may also have other applications, such as the generation of a photoacoustic effect [39–41], water jet cutting [42], trenching the seafloor using specifically designed water jetting tools [43], deburring technology, or cavitation [44,45].

Author Contributions: Conceptualization, Q.Y. and Z.Z.; methodology, S.Y. and G.P.; software, S.Y.; validation, S.Y., H.Z. and X.M.; formal analysis, G.P.; investigation, X.M.; resources, H.H.; data curation, S.Y.; writing—original draft preparation, S.Y.; writing—review and editing, G.P.; project administration, Q.Y. All authors have read and agreed to the published version of the manuscript.

Funding: This research was funded by the Guizhou Science and Technology Plan Project, grant number [2020]4Y051, and by the Science and Technology Project of Guizhou Power Grid Corporation, grant number GZKJXM20200656.

Institutional Review Board Statement: Not applicable.

Informed Consent Statement: Not applicable.

Conflicts of Interest: The authors declare no conflict of interest.

Abbreviations

EMUs	Electric multiple units
VOF	Volume of fluid

Appendix A

(1) The governing equations of the VOF model

The governing equations of the VOF model mainly include the volume fraction equation, momentum equation, and energy equation.

(a) Volume fraction equation

For the q-phase fluid, the continuity equation containing its volume fraction is as follows:

$$\frac{1}{\rho_q}\left[\frac{\partial}{\partial t}(\alpha_q\rho_q) + \nabla\cdot\left(\alpha_q\rho_q\vec{v}_q\right) = S_{\alpha_q} + \sum_{p=1}^{n}\left(\dot{m}_{pq} - \dot{m}_{qp}\right)\right] \tag{A1}$$

where α_q is the volume fraction of the q-phase fluid, $\vec{v}_q$ is the fluid velocity (m/s), $\dot{m}_{qp}$ represents the mass transfer from the p-phase fluid to q-phase fluid, $\dot{m}_{qp}$ is the mass transfer from the q-phase fluid to p-phase fluid, and the default source term S_{α_q} is zero. The volume fraction of each phase can be calculated from the following formula:

$$\sum_{q=1}^{n}\alpha_q = 1 \tag{A2}$$

The fluid properties in each unit (such as density, viscosity, and other parameters) are jointly determined by each phase fluid, and the weight is the volume fraction of each phase. For example, for systems containing p and q two-phase fluids, the density of each calculation unit is calculated as follows:

$$\rho = \alpha_p\rho_p + (1-\alpha_p)\rho_q \tag{A3}$$

(b) Momentum equation

Since the fluid properties of each computational unit take the weighted average of each phase, the momentum equation in the computational domain applies to each phase. The velocity field shared by all phases is obtained by solving the same momentum equation, as shown in Equation (6). The volume fraction of each phase jointly determines the density and viscosity in the momentum equation.

$$\frac{\partial}{\partial t}\left(\rho\vec{v}\right) + \nabla\cdot\left(\rho\vec{v}\vec{v}\right) = -\nabla p + \nabla\cdot\left[\mu\left(\nabla\vec{v} + \nabla\vec{v}^{T}\right)\right] + \rho\vec{g} + \vec{F} \tag{A4}$$

where μ is the viscosity (kg/m·s), $\vec{v}$ is the fluid velocity (m/s), and $\vec{F}$ is the force (N).

(c) Energy equation

The energy equation between each phase is as follows:

$$\frac{\partial}{\partial t}(\rho E) + \nabla\cdot\left(\vec{v}(\rho E + p)\right) = \nabla\cdot\left(k_{eff}\nabla T\right) + S_h \tag{A5}$$

The VOF model takes energy (E) and temperature (T) as the mass average:

$$E = \frac{\sum_{q=1}^{n} \alpha_q \rho_q E_q}{\sum_{q=1}^{n} \alpha_q \rho_q} \tag{A6}$$

where E is energy (J/kg), k_{eff} is the effective thermal conductivity (W/m·K), and S_h is the source term, including the volume heat source from radiation;

(2) Experimental equipment

In this test, the rated power of the high-pressure source is 2.2 kW, the working pressure of the pump head is adjustable, and the maximum flow rate is 14 L/min. The main working element of the high-pressure source is a high-pressure plunger pump. The plunger's external pulling and internal pushing make the pump pressure different from the inlet and outlet pressure, thereby sucking or discharging liquid.

The high-pressure water pipe has a rated pressure of 28 MPa. To prevent excessive heat loss when hot water circulates within it, the high-pressure water pipe is wrapped with an NBR (nitrile rubber) pipe with an inner diameter of 16 mm and a thickness of 5 mm. After measurement, the temperature difference between the inlet and outlet of the hot water after the rubber tube is within 5 °C, and the heat preservation effect is good.

The nozzle is a key element that converts the pressure energy of the water flow in the high-pressure water pipe into kinetic energy. In the experiment, two stainless steel cone-shaped nozzles of 1.2 mm and 2 mm in diameter are used, as shown in Figure 1. There is pressure loss during the transmission of water flow in the pipeline system, so the outlet pressure of the nozzle affects the spray state and spatial temperature distribution of the water jet; thus, a stainless steel pressure gauge is installed between the end of the nozzle and the high-pressure water pipe in order to measure the outlet pressure; its measurement range is 0–10 MPa, and its resistant medium temperature is −40 °C~260 °C.

References

1. Yin, F.; Farzaneh, M.; Jiang, X. Electrical characteristics of an energized conductor under various weather conditions. *High Volt.* **2017**, *2*, 102–109. [CrossRef]
2. Ale-Emran, S.M.; Farzaneh, M. Flashover performance of ice-covered post insulators with booster sheds using experiments and partial arc modeling. *IEEE Trans. Dielectr. Electr. Insul.* **2016**, *23*, 979–986. [CrossRef]
3. Hu, Q.; Wang, S.; Shu, L.; Jiang, X.; Liang, J.; Qiu, G. Comparison of AC icing flashover performances of 220 kV composite insulators with different shed configurations. *IEEE Trans. Dielectr. Electr. Insul.* **2016**, *23*, 995–1004. [CrossRef]
4. Farzaneh, M.; Chisholm, W.A. 50 years in icing performance of outdoor insulators. *IEEE Electr. Insul. Mag.* **2014**, *30*, 14–24. [CrossRef]
5. Jiang, X.; Fan, C.; Xie, Y. New method of preventing ice disaster in power grid using expanded conductors in heavy icing area. *IET Gener. Transm. Distrib.* **2019**, *13*, 536–542. [CrossRef]
6. Jiang, X.; Xiang, Z.; Zhang, Z.; Hu, J.; Hu, Q.; Shu, L. Predictive Model for Equivalent Ice Thickness Load on Overhead Transmission Lines Based on Measured Insulator String Deviations. *IEEE Trans. Power Deliv.* **2014**, *29*, 1659–1665. [CrossRef]
7. Hu, Q.; Yuan, W.; Shu, L.; Jiang, X.; Wang, S. Effects of electric field distribution on icing and flashover performance of 220 kV composite insulators. *IEEE Trans. Dielectr. Electr. Insul.* **2014**, *21*, 2181–2189. [CrossRef]
8. Polhman, J.C.; Landers, P. Present state-of-the-art of transmission line icing. *IEEE Trans. Power Appar. Syst.* **1982**, *101*, 2443–2450. [CrossRef]
9. Losowski, E.P.; Gayet, J.F. Atmospheric icing: A review. *Process. IWAIS* **1988**, *3*, 1–6.
10. Laforte, J.; Allaire, M.; Laflamme, J. State-of-the-art on power line de-icing. *Atmos. Res.* **1998**, *46*, 143–158. [CrossRef]
11. Chang, G.; Su, S.; Li, M.; Chao, D. Novel Deicing Approach of Overhead Bundled Conductors of EHV Transmission Systems. *IEEE Trans. Power Deliv.* **2009**, *24*, 1745–1747. [CrossRef]
12. Farzaneh, M.; Jakl, F.; Arabani, M.P.; Eliasson, A.J.; Fikke, S.M.; Gallego, A.; Haldar, A.; Isozaki, M.; Lake, R.; Leblond, L. *Systems for Prediction and Monitoring of Ice Shedding, Anti-Icing and De-Icing for Power Line Conductors and Ground Wires*; CIGRE: Paris, France, 2010; p. 438.
13. Yan, B.; Chen, K.; Guo, Y.; Liang, M.; Yuan, Q. Numerical Simulation Study on Jump Height of Iced Transmission Lines after Ice Shedding. *IEEE Trans. Power Deliv.* **2012**, *28*, 216–225. [CrossRef]
14. Chen, Y.; Wang, Q.; Gu, B.; Cai, Y.; Wen, F. Research on the operating process of DC De-icer and its De-icing efficiency in China Southern Power Grid. In Proceedings of the 2012 10th International Power & Energy Conference (IPEC), Ho Chi Minh City, Vietnam, 12–14 December 2012; pp. 583–588.

15. Wei, S.; Wang, Y.; Yang, Y.; Yin, F.; Cao, W.; Tang, Y. Applying Q-learning Algorithm to Study Line-Grasping Control Policy for Transmission Line Deicing Robot. In Proceedings of the 2010 International Conference on Intelligent System Design and Engineering Application, Changsha, China, 13–14 October 2010; pp. 382–387.
16. Zhu, W. Research on De-Icing of Power Line Using CO_2 Laser. Master's Thesis, School of Optoelectronic Science and Engineering, Huazhong University of Science and Technology, Wuhan, China, 2007.
17. Qi, L.; Zhu, X.; Guo, F.; Gu, S. Deicing with Nd:YAG and CO_2 lasers. *Opt. Eng.* **2010**, *49*, 114301. [CrossRef]
18. Xie, T.; Dong, J.; Chen, H.; Jiang, Y.; Yao, Y. Experiment investigation on de-icing characteristics and energy efficiency using infrared ray as heat source. *Energy* **2016**, *116*, 998–1005. [CrossRef]
19. Li, B.; He, L.; Liu, Y.; Zhang, G. Influences of key factors in hot-air de-icing for live substation equipment. *Cold Reg. Sci. Tech.* **2019**, *160*, 89–96. [CrossRef]
20. Xie, T.; Dong, J.; Chen, H.; Jiang, Y.; Yao, Y. Experimental investigation of de-icing characteristics using hot air as heat source. *Appl. Therm. Eng.* **2016**, *107*, 681–688. [CrossRef]
21. Incropera, F.P.; Dewitt, D.P. *Fundamentals of Heat and Mass Transfer*; John Wiley & Sons Press: New York, NY, USA, 2007.
22. Takahashi, K.; Usami, N.; Shibata, T.; Goto, K.; Uehara, T.; Kondo, H.; Ishikawa, R.; Saeki, H. Experimental study on a deicing method using a water jet. In Proceedings of the Oceans '04 MTS/IEEE Techno-Ocean '04 (IEEE Cat. No.04CH37600), Kobe, Japan, 9–12 November 2004; pp. 1668–1673.
23. Guo, Z.; Long, X.; Liu, Q.; Zhang, M. Experimental study on water-jet de-icing. In Proceedings of the 2014 ISFMFE—6th International Symposium on Fluid Machinery and Fluid Engineering, Wuhan, China, 22–22 October 2014; pp. 1–4.
24. Zhu, Z.; Zhang, X.; Wang, Q.; Chu, W. Research and experiment of thermal water de-icing device. *Trans. Can. Soc. Mech. Eng.* **2015**, *39*, 783–788. [CrossRef]
25. Zhou, J.Z.; Xu, X.P.; Chu, W.J.; Dai, J.Y.; Chen, Y.H.; Zhu, Z.C.; Lai, S.W. R&D of a New De-Icing Device Based on Synthetically De-Icing Techniques. *Appl. Mech. Mater.* **2013**, *372*, 381–386. [CrossRef]
26. Ji, C. Experimental Research and Simulation Analysis for EMU de-icing method. Master's Thesis, Mechanical Engineering Tianjin University, Tianjin, China, 2012.
27. Zhang, Z.; Yang, S.; Jiang, X.; Ma, X.; Huang, H.; Pang, G.; Ji, Y.; Dong, K. Hot Water Deicing Method for Insulators Part 2: Analysis of Ice Melting Process, Deicing Efficiency and Safety Distance. *IEEE Access* **2020**, *8*, 130729–130739.
28. Wen, J.; Qi, Z.; Behbahani, S.S.; Pei, X.; Iseley, T. Research on the structures and hydraulic performances of the typical direct jet nozzles for water jet technology. *J. Braz. Soc. Mech. Sci. Eng.* **2019**, *41*, 570. [CrossRef]
29. Zhou, J.Z.; Xu, X.P.; Chu, W.J.; Zhu, Z.C.; Chen, Y.H.; Lai, S.W. Analysis and Simulation of the Fluid Field in Thermal Water-Jet Nozzle Based on ANSYS FLUENT & ICEM CFD. *Appl. Mech. Mater.* **2013**, *423–426*, 1677–1684. [CrossRef]
30. Guha, A.; Barron, R.M.; Balachandar, R. Numerical simulation of high-speed turbulent water jets in air. *J. Hydraul. Res.* **2010**, *48*, 119–124. [CrossRef]
31. Srinivasan, V.; Salazar, A.J.; Saito, K. Modeling the disintegration of modulated liquid jets using volume-of-fluid (VOF) methodology. *Appl. Math. Model.* **2011**, *35*, 3710–3730. [CrossRef]
32. Tikhomirov, R.A.; Petukhov, E.N.; Babanin, V.F. *High-Pressure Jet Cutting*; ASME Press: New York, NY, USA, 1992.
33. Turner, J.S. Turbulent entrainment: The development of the entrainment assumption, and its application to geophysical flows. *J. Fluid Mech.* **1986**, *173*, 431–471. [CrossRef]
34. Yu, Y.; Li, X.; Meng, H.; Wang, Y.; Song, M.; Wu, J. Jet flow and entrainment characteristics of nozzles with different shapes. *Chin. J. Process Eng.* **2014**, *14*, 549–555.
35. Li, D.; Fan, J.; Cen, K. Direct numerical simulation of the entrainment coefficient and turbulence properties for compressible spatially evolving axisymmetric jets. *Fuel* **2012**, *102*, 470–477. [CrossRef]
36. Rajaratnam, N.; Rizvi, S.A.H.; Steffler, P.M.; Smy, P.R. An experimental study of very high velocity circular water jets in air. *J. Hydraul. Res.* **1994**, *32*, 461–470. [CrossRef]
37. Guha, A.; Barron, R.M.; Balachandar, R. An experimental and numerical study of water jet cleaning process. *J. Mater. Process. Technol.* **2011**, *211*, 610–618. [CrossRef]
38. Holman, J.P. *Heat Transfer*; China Machine Press: Beijing, China, 2011.
39. Pang, G.A.; Laufer, J.; Niessner, R. Photoacoustic Signal Generation in Gold Nanospheres in Aqueous Solution: Signal Generation Enhancement and Particle Diameter Effects. *J. Phys. Chem. C* **2016**, *6*, 09374. [CrossRef]
40. Gandolfi, M.; Banfi, F.; Glorieux, C. Optical wavelength dependence of photoacoustic signal of gold nanofluid. *Photoacoustics* **2020**, *20*, 100199. [CrossRef]
41. Chen, Y.S.; Frey, W.; Kim, S. Silica-Coated Gold Nanorods as Photoacoustic Signal Nanoamplifiers. *Nano Lett.* **2011**, *11*, 348–354. [CrossRef] [PubMed]
42. Yuyong, L.; Puhua, T.; Daijun, J.; Kefu, L. Artificial neural network model of abrasive water jet cutting stainless steel process. In Proceedings of the 2010 International Conference on Mechanic Automation and Control Engineering, Wuhan, China, 26–28 June 2010; pp. 3507–3511. [CrossRef]
43. Li, J.H.; Kim, J.T.; Lee, M.J.; Jee, S.C.; Kang, H.J.; Kim, M.K.; Kwak, H.W.; Kim, S.B.; Oh, T.W. Water jetting arm optimal design consideration for a ROV trencher. In Proceedings of the OCEANS 2015—Genova, Genova, Italy, 18–21 May 2015; pp. 1–5. [CrossRef]

44. Wang, G.; Senocak, I.; Shyy, W.; Ikohagi, T.; Cao, S. Dynamics of attached turbulent cavitating flows. *Prog. Aerosp. Sci.* **2001**, *37*, 551–581. [CrossRef]
45. Sun, Z.; Kang, X.; Wang, X. Experimental system of cavitation erosion with water-jet. *Mater. Des.* **2004**, *26*, 59–63. [CrossRef]

energies

MDPI

Review

Dielectric Strength of Polymeric Solid–Solid Interfaces under Dry-Mate and Wet-Mate Conditions

Emre Kantar [1,2]

[1] Electric Power Technology, SINTEF Energy Research, Sem Sælands vei 11, 7034 Trondheim, Norway; emre.kantar@sintef.no
[2] Electric Power Engineering, Norwegian University of Science and Technology, Høgskoleringen 1, 7491 Trondheim, Norway

Abstract: One of the most important causes of insulation system failure is the breakdown of the interface between two solid dielectrics; understanding the mechanisms governing this breakdown phenomenon is therefore critical. To that end, investigating and reviewing the practical limitations of the electrical breakdown strength of solid–solid interfaces present in insulating components is the primary objective of this work. The published literature from experimental and theoretical studies carried out in order to scrutinize the effects of the presence of solid–solid interfaces is investigated and discussed, considering macro, micro, and nano-scale characteristics. The reviewed literature suggests that solid–solid interfaces in accessories have non-uniform distributions of electrical fields within them in comparison to cables, where the distribution is mostly radial and symmetrical. Many agree that the elastic modulus (elasticity), radial/tangential pressure, surface smoothness/roughness, and dielectric strength of the ambient environment are the main parameters determining the tangential AC breakdown strength of solid–solid interfaces.

Keywords: cable insulation; contact surface; dielectric breakdown; electrical breakdown; polymeric insulation; solid–solid interface; surface roughness

Citation: Kantar, E. Dielectric Strength of Polymeric Solid–Solid Interfaces under Dry-Mate and Wet-Mate Conditions. *Energies* **2021**, *14*, 8067. https://doi.org/10.3390/en14238067

Academic Editors: Zhijin Zhang and Hualong Zheng

Received: 15 October 2021
Accepted: 26 November 2021
Published: 2 December 2021

1. Introduction

Various insulating and conductive materials are used in electrical insulation systems. The alternating current (AC) breakdown strength (BDS) of an insulation system can be as high as the dielectric strength of the weakest link in the system. One of the most critical failure mechanisms within the insulation system is the electrical breakdown of the interface between two solid dielectrics. Therefore, it is essential to discover the breakdown mechanisms occurring at solid–solid interfaces. An electrical breakdown at the interface between two solid dielectrics has been reported to be the leading cause of insulation system failure [1]. Hence, studying the mechanisms governing this breakdown phenomenon is essential for improving insulation systems.

Surface irregularities cause discrete contact points that form when two rough, nominally flat surfaces are brought into contact. Numerous cavities arise between the contact spots. The factors influencing the interfacial dielectric strength are then related to the insulation properties of cavities (dimension, shape, and dielectric medium inside) and the contact area (treeing/tracking resistance, etc.). Therefore, this review concentrates on the dielectric properties of both cavities and contact spots in the macro, micro, and nano-scale to offer a holistic overview of the problem.

In this paper, a brief introduction to state-of-the-art connectors and cable apparatuses containing interfaces is provided. Particular attention is paid to the solid–solid interfaces, considering the causes of interfacial failure and important factors that influence the dielectric strength of an interface. Second, publications relevant to the scope of this review paper are summarized into two main categories: empirical studies on solid–solid interfaces and theoretical studies modeling the rough contact surfaces at solid–solid interfaces. The scope of the literature survey is illustrated in Figure 1.

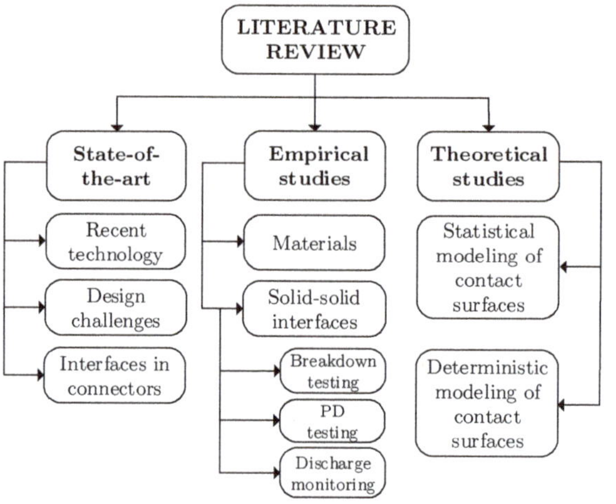

Figure 1. Categories covered in the literature review.

2. Interfaces in State-of-the-Art Cable Connectors

In this section, an overview of modern connectors and the limiting factors involved in the design phase and in practice are given. In addition, a more general overview of solid–solid interfaces in polymer-insulated power cables is provided.

2.1. Interfaces in Subsea Cable Connections

Subsea cable connectors are a pertinent example of modern, sophisticated connector solutions available for cable connectors. They are categorized as "wet-mate" connectors, "dry-mate" connectors, and penetrators. A modern wet-mate connector is composed of a plug and a receptacle, as presented in Figure 2 [2]. In subsea applications, the retrieval of pumps or transformers for repairs on the surface is of paramount importance, and wet-mate connectors significantly facilitate performing this task [3–5]. Wet-mate connectors can be connected/disconnected underwater, allowing the equipment to be disconnected before retrieval to the surface and to be connected after being installed in the subsea grid [3–7]. On the other hand, dry-mate connectors require equipment to be assembled on a vessel along with the cable before being lowered to the seabed. Penetrators are essentially cable terminations, allowing high-voltage cables to be run through equipment enclosures [3]. High differential pressures are often tolerated by penetrators, and hence they allow equipment requiring a 1-atm environment to be connected [3].

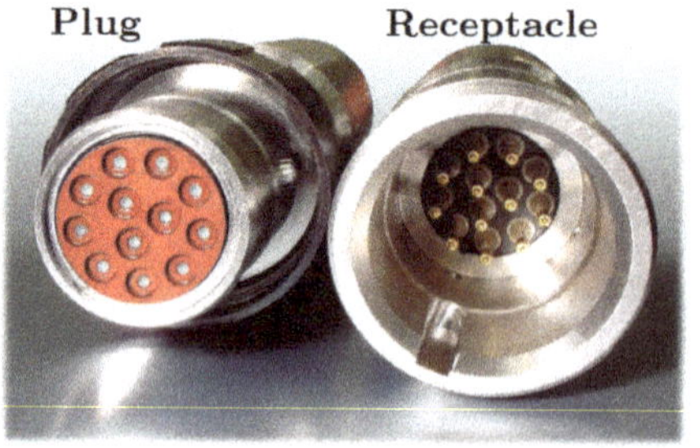

Figure 2. Plug and receptacle of a subsea connector [2].

Wet-mate subsea connectors have been in operation in the oil and gas industry for a long time because the plugging can take place in water/underwater effortlessly [3–5,7,8]. As the transition to "green energy" is gaining pace recently thanks to the increased awareness of climate change, wet-mate connectors are becoming increasingly popular in renewable energy applications such as offshore wind farms (both floating and fixed-bottom), tidal energy systems, and floating-type photovoltaic power stations (solar farms). However, recent and future subsea extensions require that the wet-mate connector technology be improved in a cost-effective manner so that it can provide higher power ratings with reduced losses and operate at higher voltages and higher temperatures, in deeper waters, and with longer tiebacks [3–5].

Currently, wet-mate cable connectors up to 45 kV (dry-mate connectors/penetrators up to 145 kV) are commercially available and applicable for deepwater and dynamic applications, including power umbilicals. Wet-mate connectors are crucial to future subsea substations or mid-point compensation for long high-voltage AC cables. The inter-array voltage level of 33 kV has been upgraded to 66 kV in modern offshore wind parks. The average capacity of an individual wind turbine is likely to increase to above 14 MW, which means the average total capacity of modern wind parks will also rise along with it. Within a few years, inter-array connections in offshore systems will likely require even higher voltage levels above 100 kV. This calls for higher voltage ratings, especially for wet-mate connectors used as vital components in future subsea inter-array grids (e.g., in junction boxes or subsea substations). Thus, further steps must be taken to achieve higher voltage levels (>220 kV for dry-mate, >100 kV for wet-mate) needed for AC longer step-out offshore electrification as a cost-efficient alternative to high-voltage DC.

A connector includes two different insulation systems within a controlled environment: one oil chamber is placed in the other, separated by a diaphragm, as illustrated in Figure 3. The main potentially weak parts in subsea connectors are the interfaces between the solid–solid and solid–liquid dielectric materials. Leading causes of failure are the presence of imperfections, defects, impurities at the interfaces, and water intrusion [1]. They are likely to result in locally high field stresses that, in turn, initiate surface discharges through the guide pin, possibly leading to a premature electrical breakdown.

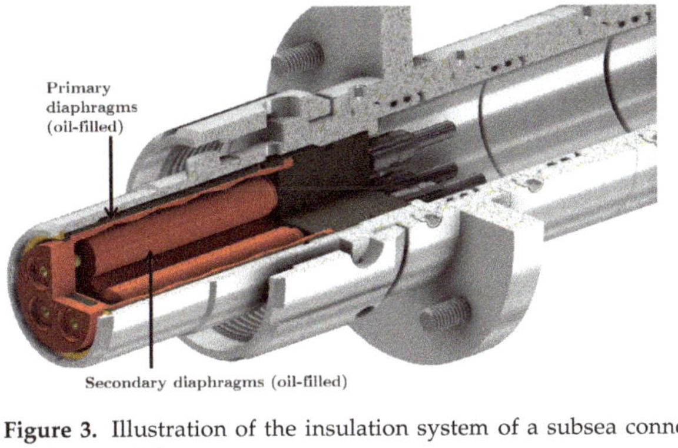

Figure 3. Illustration of the insulation system of a subsea connector with two nested, oil-filled diaphragms [2].

Figure 4a illustrates an engaged subsea connector in operation. The main causes of failure are illustrated in Figure 4b–e at the plug component. Water ingress adversely influences the dielectric performance of insulating oil [6]. Several options exist, allowing water to penetrate the connector's insulation; diffusion through the primary diaphragm is the most common mechanism. In fact, diffusion through the diaphragms inevitably occurs to some extent. Although diffusion is a gradual process, it leads to moderately elevated relative humidity (RH) inside the connector. Increased RH can easily be measured using pertinent readily available humidity sensors [6]. Additionally, dirt or water residue on the guide pin or diaphragms as well as mechanical damage may lead to water ingress (Figure 4b,c) [6]. Last but not least, while wet-mating, water may enter the oil. A thin conductive layer may also form on the guide pin as a result of inadequate contact pressure between the seal and the guide pin (Figure 4d) as well as the deformation/damage/aging/fatigue of the sealing material. The presence of conductive material will lead to locally high field stresses. Surface discharges through the guide pin may ensue, possibly leading to a complete flashover [6].

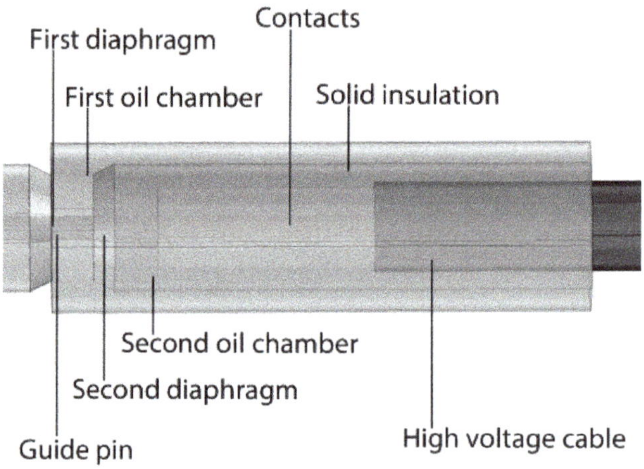

(**a**) Close-up of the insulation system of a subsea connector.

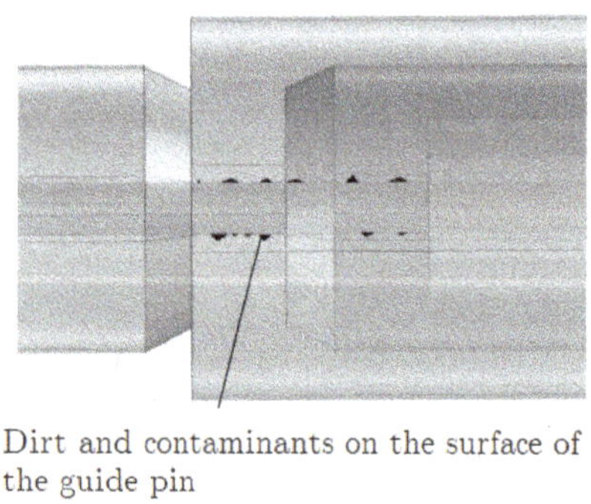

(**b**) Dirt formation around the pin.

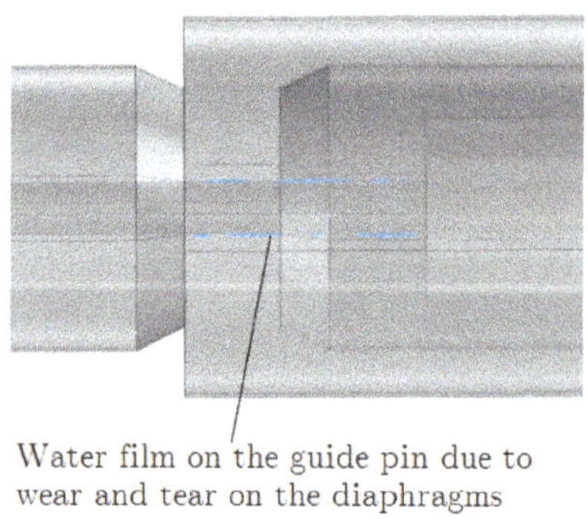

(**c**) Water film on the pin.

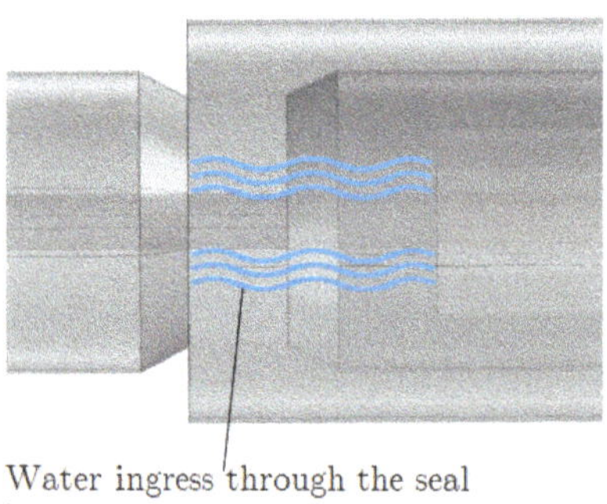

(**d**) Water intrusion through the pin.

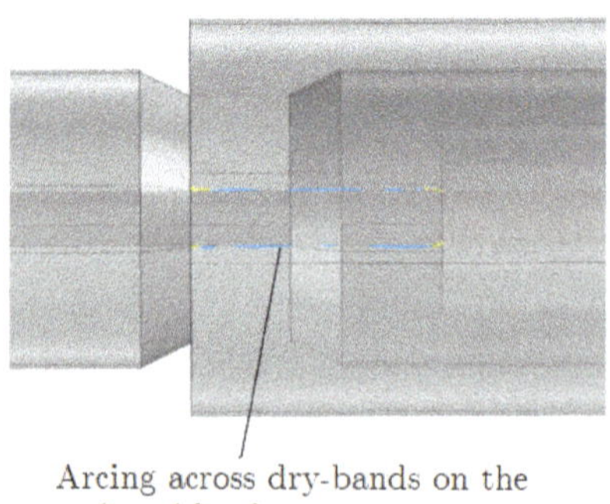

(**e**) Discharges at dry-bands on the pin.

Figure 4. Illustration of a subsea connector and possible failure mechanisms. Reproduced from [6], NTNU: 2014.

Analyzing a simplified geometry of a subsea connector using finite element methods (FEM) can help reveal problematic areas, as presented in Figure 5a. FEM analysis indicates that the edges where the guide pin engages with the first diaphragm have the greatest local electric field strength. The locally high electric stress holds a high potential for failure based on the voltage rating and material properties of the connector. Electric field grading techniques seem necessary to distribute the electric field more uniformly inside the connector. The tangential component of the electric field (along the pin), depicted in Figure 5b, suggests that the primary diaphragm and contact areas are likely to experience surface discharges. The reason for this is that, in the case that several of the mechanisms become present simultaneously, surface tracking or arcing along the dry bands between impurities, e.g., water droplets, may ensue, as illustrated in Figure 4e. The causes of failure presented here are not inherent only to subsea connectors. In the next section, we focus on solid–solid interfaces.

(**a**) (**b**)

Figure 5. Illustration of the tangential electric field distribution in a subsea connector: (**a**) Contour plot. (**b**) Surface plot. Reproduced from [6], NTNU: 2014.

2.2. *Interfaces in Polymer-Insulated Cable Connections*

The simple structure of the insulation in cross-linked polyethylene (XLPE) cables led to the early development of easy-fit prefabricated joints and terminations, whose field control elements are prefabricated and tested in the factory [8]. In the past, such accessories were initially developed for medium-voltage (MV) applications, which were then upgraded for the high-voltage (HV) and extra-high-voltage fields (EHV). However, a large number of alternative solutions are currently competing with these prefabricated elements.

Recent developments have progressed to prefabricated and routine-tested slip-on units, even for straight joints and polymer-insulated cables [8]. Field control components are already incorporated in these joints [9]. Accessories with slip-on stress cones for HV and EHV cables usually utilize field control deflectors, as shown in Figure 6a [8,10]. Properly contoured deflectors made from an elastic conductive material are positioned into a similar elastic insulator permanently, then pressed in one piece onto the suitably prepared polymer-insulated cable precisely, such as EPR (ethylene propylene rubber), PE (polyethylene), LDPE (low-density polyethylene), or XLPE.

Figure 6a shows the cross-section of a slip-on joint, consisting of two opposing control deflectors and a field smoothing sheet for the conductor connection [8]. A conductive coating for the surface of the joint is needed to provide the outer screening. Lastly, a metal housing (durable against corrosion) is used to avoid ingress of moisture and mechanical damage [8].

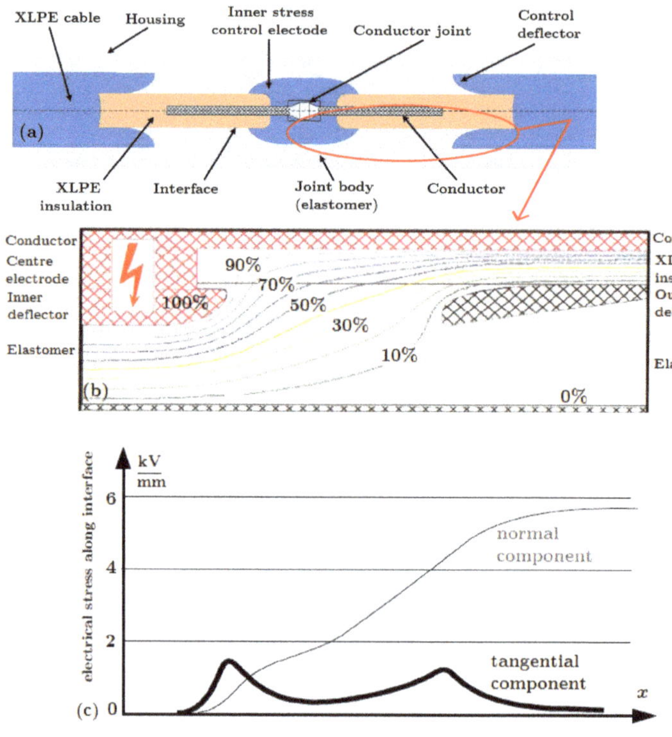

Figure 6. Illustration of a cable joint highlighting the parts where the solid–solid interfaces exist and why the tangential electric field is of concern: (**a**) prefabricated EHV silicone joint for 400 kV XLPE-insulated cables. Reproduced from [11], NTNU: 2016. (**b**) Calculated potential distribution in a prefabricated slip-on joint. Reproduced from [1], CIGRE: 2000. (**c**) Calculated field patterns in a prefabricated slip-on joint. Reproduced from [1], CIGRE: 2000.

2.3. Factors Affecting the Interfacial Breakdown Strength

Solid–solid interfaces in cable joints usually arise between a soft material (elastomer/polymer) and a hard (polymer) material such as XLPE–EPDM (EPDM: ethylene propylene diene monomer), XLPE–SiR (SiR: silicone rubber), XLPE–EPR, and XLPE–PEEK (PEEK: polyether ether ketone), or between the same materials. With soft materials, improved contact and sealing is possible even at low and moderate contact pressures.

Despite the presence of deflectors with identical structures, different field conditions arise in joints to those in sealing ends. In particular, the tangential component of the electrical field that is locally enhanced at the interface between the cable dielectric and joint insulation becomes more significant relative to the maximum field strength within the body of the joint [8]. Accurate field calculations, as illustrated in Figure 6, are essential to avoid intolerably high stresses and to optimize the shape of the joint. Figure 6b,c depicts the results of field calculations in the form of the potential distribution and the field distribution of the normal and tangential components in a 400 kV slip-on joint, respectively [8].

The electrical performance of solid–solid interfaces are dependent on the following:

- Surface roughness;
- Contact force;
- Mechanical and electrical characteristics of the insulation materials, such as elasticity and tracking resistance;
- Surrounding/insulating dielectric medium; and

- Care exercised and conditions during assembly.

In the next section, the individual effect of each above-listed parameter is present.

2.4. Contact Surfaces at Solid–Solid Interfaces

Prefabricated and pretested/qualified cable accessories do not necessarily guarantee ideal assembly conditions because the site conditions in which they are usually assembled may be suboptimal and hard-to-control [8]. As long as the fitting is not performed in a laboratory or cleanroom environment, the interfaces will be vulnerable during installation. Consequently, cavities, protrusions, and impurities are likely to develop at solid–solid interfaces [1]. Figure 6c illustrates the locally enhanced electric field stresses originating from imperfections at an interface (complementary to Figure 5). To be more specific, rough surfaces lead to various cavities at the interfaces, whereas contact force affects the size and deformation of the cavities and contact areas, as shown in Figure 7. Mechanical and electrical characteristics of the insulation materials, such as elasticity and interfacial tracking resistance, strongly affect the interfacial BDS. The type and quantity of lubricant/grease used during assembly, water penetration to the interface, or assembly at dry and optimal conditions change the insulating dielectric medium filling the cavities. Lastly, poor workmanship, wear and tear of materials, contaminants, and impurities cause a substantial reduction in the BDS [8].

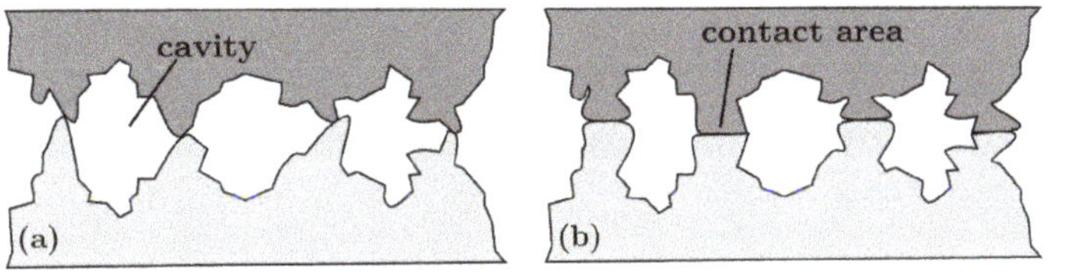

Figure 7. Illustration of surface asperities leading to cavities and contact spots at solid–solid interfaces at (**a**) no-load, (**b**) increased contact area and reduced cavity size under load. Reproduced from [12], Macedonian Journal of Chemistry and Chemical Engineering: 2018.

As previously stated, the presence of the solid–solid interfaces (i.e., imperfect contact) increases the risk of locally high electric field stresses, leading to partial discharges (PD) and eventually a premature flashover [1,8,13–19].

A major failure process for power cable connectors is the breakdown of the interfacial layer between two solid insulating materials, as reported in [1,8,13,15,17,18]. When dimensioning the thickness of the insulation walls of the cable and joint body, Peschke and Olshausen [8] recommend restricting the operational stress on the outer conductive layer to around 6–7 kV/mm even if the installation is performed with due care, and the mechanical and electrical design of the apparatus is optimal [8].

Cavities on a dielectric surface differ in size and distribution based on the surface roughness, contact force, and mechanical properties of the material, as well as the care taken during manufacturing and installation [1,8]. Surface irregularities cause discrete contact points that form when two rough, nominally flat surfaces are brought into contact. Figure 8a,b illustrates the formation of numerous cavities between contact spots. Figure 8b elucidates the significantly low ratio of "actual contact area" to "nominal contact area". An interfacial cavity along the tangential axis (x- or y-axis) is generally considerably larger, as illustrated in Figure 8c [20].

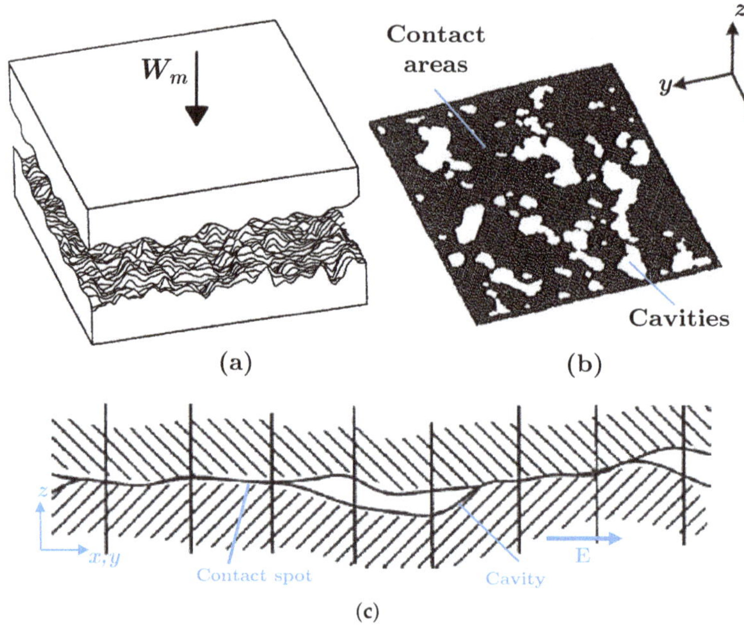

Figure 8. (**a**) Demonstration of two rough surfaces in contact. (**b**) Contact area and cavities at the interface in 3D. (**c**) A two-dimensional illustration of cavities at the interface. Reproduced from [21], ASME Journal of Tribology: 1996.

In light of the overview of the important parameters discussed in this section, the primary parameters influencing the distribution and size of microcavities and hence the electrical breakdown strength of solid–solid interfaces are summarized in Figure 9a with their individual effects on the interfacial dielectric strength illustrated in Figure 9b.

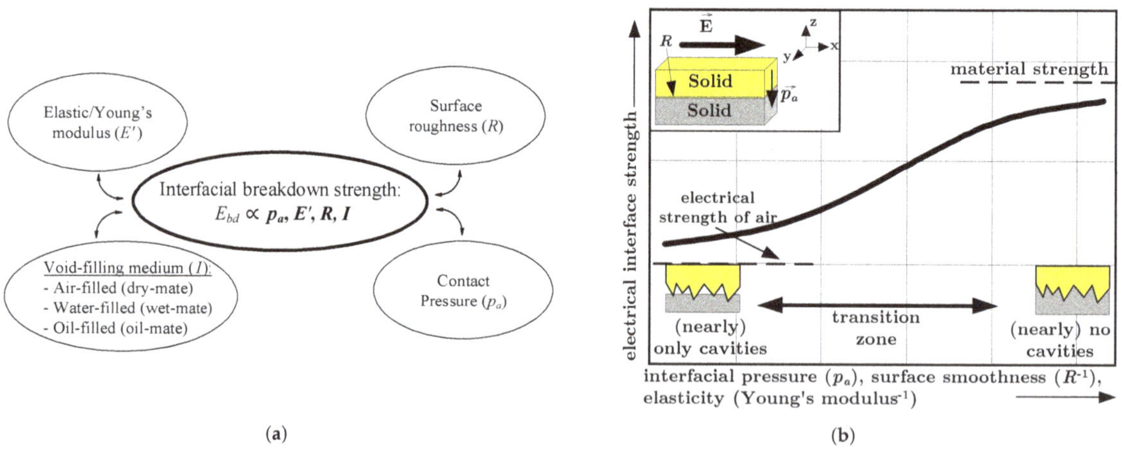

Figure 9. Parameters influencing the breakdown voltage (dielectric strength) of solid–solid interfaces. (**a**) Balloon chart summarizing the parameters. (**b**) Electrical interface strength vs. pressure and roughness. Reproduced from [22], CIGRE: 2002.

3. Empirical Studies on Dielectric Strength of Polymers and Solid–Solid Interfaces

Experimental work regarding the electrical properties of solid–solid interfaces in insulating materials, HV apparatuses, and cable accessories has been extensively studied in the literature. The papers referred to in this section have predominately focused on the BDS and PD inception field strength (PDIV) of solid–solid interfaces by focusing on interfacial discharge and breakdown mechanisms induced by enclosed cavities (either spherical or elongated in the field direction) and interfacial tracking resistance (PD resistance) of the polymers. Besides this, a few papers have correlated the intensity of discharge images with the interfacial BDS values. In the experiments, cable joints as a whole, interfaces assembled between pieces of polymers cut from commercial cables, or polymers cast in laboratories have reportedly been used. Casting custom-made polymers in laboratories using molds in desired shapes and sizes has been popular among researchers because it allows metal electrodes or air-filled cavities to be embedded in diverse shapes and sizes in the specimens. Below, a summary of findings from the selected publications is provided.

3.1. Studies on the Insulation Properties of Polymers

Albayrak et al. [23], Roy et al. [24], and Ding and Varlow [25] observed improved electrical insulation properties when the elastic modulus was increased by putting in micro and nano-scaled zinc oxide, nano-scaled zirconia particles, and silica nanoparticles. In these studies, the dielectric strengths of different dielectric materials were tested by changing the chemical and material properties of the bulk insulation material.

In addition, Du et al. [26,27] studied the interface charge behavior of multi-layer insulations extensively. These studies were performed under direct current (DC) voltage, in which the influence of fillers in bulk materials of LDPE and EPDM on the overall breakdown strength of interfaces was investigated. It was concluded that suppressed/reduced charge accumulation by 50% at the interface increases the BDS and PDIE. Besides, in [28], the correlation between space charge behavior and the mechanical stress was studied using polypropylene (PP) blended with polyolefin elastomer (POE), i.e., PP/POE blend, under DC excitation. After mechanical stretching, the microstructure of the PP/POE blend altered, causing a greater accumulation of space charges in stretched specimens. All in all, these results provide a useful insight into the electrical performance of polymer interfaces.

Tracking failure in HV cable insulations has been the subject of various studies since a solid insulation under high electrical stress is likely to undergo breakdown due to tracking [29]. Several experimental configurations of needle-plane electrodes have extensively been used in the literature to test insulation materials for tracking resistance by inducing a high non-homogeneous field to stimulate the initiation of surface or interface tracking [29–32]. Using empirical data, Fothergill [31] developed an analytical expression to estimate the interfacial tracking resistance of polymers. There, the interfacial tracking resistance is linearly correlated to the fourth root of the elastic modulus. Chen et al. [30] used Fothergill's model to investigate the interfacial tracking behavior in XLPE cable insulation samples. Both Fothergill [31] and Chen et al. [30] used needle-plane electrode configurations to generate a strong electric field owing to the non-homogeneous field generated. Mason [33] investigated the PD resistance of XLPE samples using nine different combinations of needle, plane, and rod electrodes.

Eichhorn [32] published a review paper on interfacial tracking in solid dielectrics in 1977. In his conclusion, the most commonly investigated interfacial tracking phenomena were those that result from the degradation of organic materials and most dielectrics by internal electrical discharges. Even though the presence of voids and contaminants in electrode–insulation interfaces that contain defects is undesirable, the damage which is caused by moderate AC voltages is of greater commercial importance [32]. There may be very high, localized stress gradients in this case, which would facilitate tree initiation and growth and eventually would lead to tree breakdown. To estimate these localized stress gradients, Eichhorn [32] provided a thorough review of the mathematical formulas for enhanced field stress at the tip of sharp metallic electrodes.

Finally, Gubanski et al. [29,34–43] contributed to the literature with numerous extensive and thorough studies on the interfacial tracking resistances of polymeric materials, such as SiR, PE, XLPE, and LDPE under AC excitation. The essence of the findings from [29,34–43] is that charges injected from the needle substantially reduce the electric stress at the needle tip. The results indicate that the maximum electric field emerges in the bulk material very close to the needle tip, and an electrical tree might form as a result of induced changes in the material.

3.2. Studies on the Dielectric Strength of Solid–Solid Interfaces

3.2.1. Studies with a Focus on Electrical Breakdown

In the 1990s, Fournier et al. [14–16,44–46] studied solid–solid interfaces thoroughly using needle-plate electrodes under AC or DC excitation across the interfaces formed between XLPE–XLPE, EPDM–EPDM, and EPDM–XLPE samples. In [14,15,45], dry interfaces and greased/lubricated interfaces were examined. In both cases, the interfacial breakdown strength was reported to increase by a factor of 2.7 and 1.5, respectively, when the applied pressure was increased from 0 kPa to 80 kPa. It was also shown that lubricated interfaces had a six times higher breakdown strength than that of an interface without grease. Moreover, Fournier [16] studied the influence of surface roughness at dry and greased (similar to oil-mate) EPDM–XLPE and EPDM–EPDM interfaces. After sanding and greasing the XLPE surface, EPDM–XLPE interfaces had dielectric strengths that were three to four times greater than those of unsanded and ungreased interfaces. The dielectric strength of unsanded EPDM–XLPE interfaces improved slightly after greasing the interface. Moreover, EPDM–EPDM interfaces demonstrated higher breakdown strengths in comparison to EPDM–XLPE interfaces, whose strengths depended on the chosen grease type. As EPDM is softer than XLPE, the interfaces between softer materials achieve higher BDS than those between hard materials. Lastly, Dang and Fournier [14,44] found that the interfacial breakdown voltage increased at elevated interface pressures; however, aged cable accessories could cause a reduction in the interface pressure, leading to a reduced interface dielectric strength. These findings fully agree with the trends Kantar et al. observed in [47,48]. They tested the breakdown strength of dry-mate and oil-mate XLPE–XLPE and SiR–SiR interfaces. The breakdown strength of oil-mate interfaces were significantly higher than those for dry-mate samples: 2.8-fold and 1.6-fold improvements were detected in XLPE–XLPE and SiR–SiR interfaces, respectively.

Kunze et al. [1,49] studied the design of interfaces in HV cable accessories by varying the surface roughness and contact pressure. XLPE–SiR interfaces under AC and impulse voltages were tested, and the experimental results indicated that surface roughness and radial pressure significantly influence the longitudinal electrical strength of interfaces. To be more specific, the increase in the surface roughness (mean height of surface asperities varied from 5 μm to 50 μm) of the XLPE samples reduced the interfacial BDS by 50%.

Takahashi et al. [13] studied the interfacial breakdown strength and PD patterns of interfaces between the SiR and epoxy utilizing two different samples with orthogonal orientations that allow for the electric field to be applied tangentially or perpendicularly to the interface. The effect of delaminations between interfaces filled with air on the interfacial breakdown voltage was also investigated. They concluded that the tangential component of the electrical field governs the interfacial breakdown, where AC surface breakdown voltage values rose by a factor of 1.5 as the thickness of the air delamination was decreased from 1 mm to 0.01 mm.

Du et al. [17,50] studied the impact of contact pressure using XLPE and SiR samples under AC voltage using needle-plane electrodes. A 1.7-fold rise in the initial discharge voltage was observed as the contact pressure was raised from 20 kPa to 300 kPa. They concluded that in order to increase the lifetime of power cables, it is necessary to avoid the loss of interfacial pressure between solid materials; in this case, XLPE–SiR. In [18], the effect of surface roughness on the tracking mechanisms at XLPE–SiR interfaces under AC voltage was studied. The results concluded that as the surface roughness was decreased (surfaces sanded with #100 to #1000 in a sequence), rising trends in the PDIV, breakdown voltage, and time to breakdown were observed by factors of 1.8, 1.4, and 2.3, respectively, whereas the intensity of the emitted discharge light decreased. In a similar research paper, Chen et al. [51] investigated the tracking failure of XLPE–SiR interfaces under AC and impulse voltages and concluded that at AC excitation, there is a longer period of overvoltage in each cycle, resulting in an immediate interface tracking failure.

Hasheminezhad [11,52–56] studied the interfacial breakdown strength of solid–solid interfaces between 2007 and 2011 within the scope of his Ph.D. work. During this time, he investigated the BDS and PD inception field strength of XLPE–XLPE interfaces under a homogeneous AC field by varying the contact pressure (p_a) and surface roughness. The core results from his thesis are summarized in Figure 10, and they agree with the reported results in the literature above and with the results to be shown in the following work performed by Kantar and his peers.

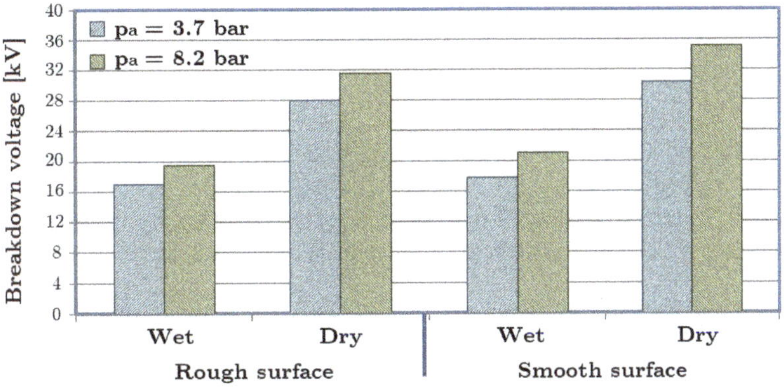

Figure 10. Overview of the resulting breakdown strength values of dry and wet-mate interfaces using XLPE-insulated cable samples in Hasheminezhad's Ph.D. thesis [11]. Reproduced from [11], NTNU:2016.

Kantar et al. [47,48,57–63] continued Hasheminezhad's work. They studied the effect of mechanical pressure, elasticity, and surface roughness on the longitudinal electrical AC breakdown strength using different insulation materials (i.e., solid–solid interfaces between rectangular-shaped lab-made samples) and various surface roughnesses under dry-mate and wet-mate conditions. The surface of the materials was prepared using sandpapers with different grit sizes attached to a rotating sanding machine at a predetermined pressure to examine the relation between the surface roughness and the longitudinal AC BDS at solid–solid interfaces. Interfaces were named after the sandpaper grit that they were sanded with: #180 (roughest), #500, #1000, and #2400 (smoothest). Different mechanical loads to apply various contact pressures were used to test the interfaces between the same materials: XLPE, epoxy, PEEK, and SiR (XLPE–XLPE, SiR–SiR, etc.). Details of the test setup and experimental design can be found in [48,60,61].

Dry-assembled vs. wet-assembled interfaces under various contact pressure [47,48]

The results of the dry-mate and wet-mate interfaces formed using the XLPE #500 and SiR #500 samples are summarized in Figure 11. The 63.2% BDS values (Weibull distribution used) in the case of dry-mate XLPE–XLPE #500 were higher than those in the case of wet-mate XLPE–XLPE #500 by a factor ranging from 2.6 to 2.9 as the contact pressure was increased from 0.5 to 1.16 MPa. Increasing the contact pressure from 0.5 to 1.16 MPa resulted in an increase by a factor of 1.2 in the 63.2% BDS for the dry-mate XLPE–XLPE #500. On the other hand, in the case of wet-mate XLPE–XLPE #500, from 0.5 to 1.16 MPa, the 63.2% BDS increased by a factor of 1.7. These findings indicate that the increase in contact pressure is likely to squeeze some water droplets out of the interface. Thus, air-filled and water-filled cavities are likely to coexist at higher contact pressures, which in turn increases the interfacial BDS significantly.

In Figure 11, the BDS of air is also shown for reference. It was measured using the same test setup in the air with a 4 mm distance between the electrodes. In accordance with the field simulations shown in [47], Figure 11 indicates that having water at the interface adversely impacts the BDS in the AC breakdown experiments. Particularly at low interfacial pressures, the dielectric strength of an interface is comparable with that of air.

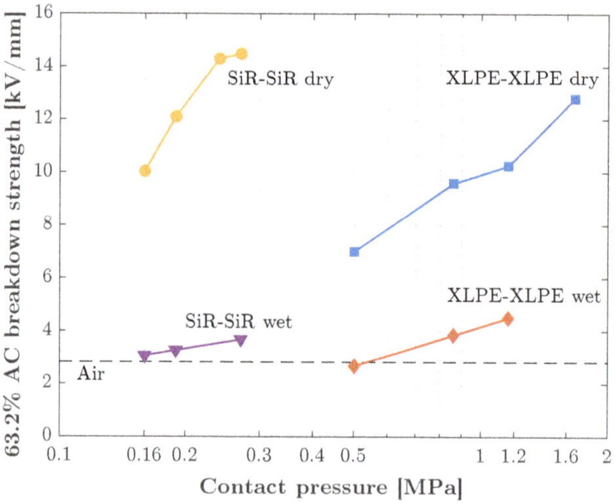

Figure 11. Longitudinal AC BDS values of dry-mate and wet-mate XLPE–XLPE #500 and SiR–SiR #500 interfaces vs. the contact pressure (the BDS of air (~2.8 kV/mm (RMS)) is measured using the same experimental setup in ambient air at the laboratory with an electrode distance of 4 mm, as performed in the interface breakdown experiments).

Similarly, the 63.2% BDS values in the case of dry-mate SiR–SiR #500 were higher than those for the wet-mate SiR–SiR #500 interfaces by a factor ranging from 3.3 to 3.9, following the contact pressure change from 0.16 to 0.27 MPa. In the case of dry-mate SiR–SiR #500, the 63.2% BDS increased by a factor of 1.4 as the pressure increased from 0.16 to 0.27 MPa. Similar to wet-mate XLPE–XLPE, the BDS of the wet-mate SiR–SiR is comparable to that of air at 0.16 MPa. Hence, it can be deduced that water intrusion is a serious concern in the design of any HV equipment incorporating solid–solid interfaces.

The increased pressure from 0.16 to 0.27 MPa led to higher BDS for the wet-mate SiR–SiR #500 by a factor of 1.2. It can be argued that the SiR samples are more hydrophilic than the XLPE samples [64], which might have resulted in the removal of fewer water droplets from the interface.

Effect of surface roughness under various contact pressures [60,62]

In order to study the correlation between surface roughness and electrical interface breakdown, XLPE samples of four different surface roughnesses were used in the experiments to form the dry-mate polymer interfaces in [60,62]. At 0.5, 0.86, and 1.16 MPa contact pressures, the effect of surface roughness on the interfacial BDS was explored, and the main results are shown in Table 1. The surface roughness of each interface was determined using a 3D optical profilometer that returned the mean asperity height R_a as the first-hand quantitative parameter to compare the surface roughnesses resulting from the roughest to the smoothest sandpaper.

Table 1. Overview of the experimental results on the effect of surface roughness. '#' stands for the sandpaper grit no. used for each interface while R_a is the mean arithmetic height of the asperities at each interface.

XLPE–XLPE Interface	63.2% BDS [kV/mm] at p_a		
	0.5 MPa	0.86 MPa	1.16 MPa
#180 ($R_a = 8.9$ µm)	5.92	7.13	8.67
#500 ($R_a = 7.8$ µm)	6.99	9.61	10.26
#1000 ($R_a = 1.7$ µm)	7.56	10.13	11.62
#2400 ($R_a = 0.3$ µm)	10.98	14.69	18.70

The results suggest that increased surface roughness gives rise to a lower BDS, whereas a higher contact pressure yields augmented BDS. In fact, the 63.2% BDS was nearly twice as high in the case of #2400 as it was for #180 at each contact pressure. A 30-fold reduction in the mean asperity height R_a from #180 to #2400 yielded a 1.85-fold rise in the BDS at $p_a = 0.5$ MPa and a 2.15-fold increase at $p_a = 1.16$ MPa. At all pressure levels, the BDS increased as the interfaces became smoother—i.e., from #1000 to #2400—where the highest increase (1.6-fold) was detected at 1.16 MPa. To sum up, the surface roughness of the interfaces has a significant impact on the overall BDS in dry-mate conditions.

From 0.5 to 1.16 MPa, the 63.2% BDS increased by factors of 1.4 and 1.7 in the cases of #180 ($R_a = 8.9$ µm) and #2400 ($R_a = 0.30$ µm), respectively. Thus, the smoothest interface shows the strongest dependency on the contact pressure.

Effect of elasticity under various contact pressures [61,63]

In [61], the tangential AC breakdown strength at solid–solid interfaces as a function of the elastic modulus was studied. Interfaces between identical materials of SiR, XLPE, EPOXY, and PEEK were analyzed at several interfacial contact pressures, and all sample surfaces were prepared using the same sandpaper with grit #500. Preliminary tests were conducted to determine the applied pressure levels, in which samples and interfaces were examined for deformation. The SiR–SiR interface, for example, could not be tested at contact pressures above 0.27 MPa due to the deformation of the samples.

Table 2 presents the 63.2% values for each interface. The results suggest that a higher elastic modulus gives rise to a reduction in the BDS. It is also evident that contact pressure has a significant effect such that an increase of around three times the contact pressure resulted in a 1.4-fold increase in interfacial BDS at SiR–SiR (softest interface–lowest elastic modulus). On the other hand, the BDS for the highest modulus (PEEK–PEEK) was higher by a factor of 2.4. These results signify a strong influence of elasticity on the BDS of solid–solid interfaces. BDS values for SiR and XLPE, both with relatively low moduli, were thus found to be higher even at relatively low contact pressures.

Table 2. Overview of the experimental results on the effect of elasticity.

Contact Pressure	SiR–SiR		XLPE–XLPE		EPOXY–EPOXY		PEEK–PEEK	
	p_a [MPa]	63.2% BDS [kV/mm]	p_a [MPa]	63.2% BDS [kV/mm]	p_a [MPa]	63.2% BDS [kV/mm]	p_a [MPa]	63.2% BDS [kV/mm]
p_{a1}	0.16	10.0	0.5	7.0	1.16	8.9	1.16	6.3
p_{a2}	0.19	12.1	0.86	9.6	1.67	10.0	1.67	8.1
p_{a3}	0.24	14.3	1.16	10.3	2.25	12.6	2.25	11.1
p_{a4}	0.27	14.5	1.67	12.8	3.34	15.6	3.34	15.1

3.2.2. Studies with a Focus on Partial Discharges

Illias et al. [65–67] performed thorough studies on the measurement and modeling of partial discharges in solid dielectric materials and at polymer interfaces. They extensively used phase-resolved partial discharge analysis (PRPDA) and pulse sequential analysis (PSA) methods to display both experimental and simulation results. They proposed a 2D finite element analysis (FEA) model to simulate PDs in spherical cavities for different voltage frequencies and amplitudes, material temperatures, and cavity sizes. They concluded in [65,66] that the charge/temperature decay time constant, surface conductivity of the cavity, initial electron generation rate, and discharge inception/extinction field in the cavity are the most critical parameters governing the PD activity in cavities. In addition, their findings suggest that as the cavity diameter gets larger, the number of PDs per period diminishes, whereas the total apparent charge per period and the maximum discharge magnitude rise.

Stewart et al. [68–73] examined factors affecting the PD activity in internal cavities and surface properties of the cavities and reported on the characteristics of PD in artificially created cavities. The studied cavity types were enclosed cavities, vented channels, and unvented channels in [68,69], and they reported that changes in gas content and by-products from the PD activity in cavities affect the build-up of space charges on the cavity walls, the generation rate of initiating electrons, and the energy of collisions, thus altering the PD characteristics. These studies are found to be very relevant in determining the effect of gas pressure inside the cavities and the surface roughness on the overall BDS of insulation materials. They also reported in [68] that vented channels were likely to be subjected to decreased degradation due to by-products dispersing and gas refresh through the vent. These findings strongly correlate with the results of the experimental and the theoretical work performed in [74] that suggest that, despite the significant difference in elasticity, at high contact pressures, the presence of enclosed cavities was suspected due to the high BDS achieved in the cases of XLPE–XLPE #500, EPOXY–EPOXY #500, and PEEK–PEEK #500 in the AC breakdown experiments. In that case, cavities possibly experience an increase in gas pressure based on the compression extent in the cavity, and hence the dielectric strength of the cavity is augmented based on the ideal gas law [74].

3.2.3. Studies with a Focus on Interfacial Discharge Monitoring

Gu and He [75] examined the effect of microcavities on the interfacial breakdown between XLPE and SiR using a lab-made cable joint using image processing methods by examining the channel widths of discharge light and carbonization extent. They found that microcavities significantly reduce the interfacial dielectric strength and lead to discharge and tracking failure, while an elongated cavity parallel to the tangential component of the electric field leads to interfacial discharge more easily.

Du et al. [17,50] examined the effects of the interface pressure and the interfacial tracking failure between XLPE and SiR by processing the discharge images. They determined the interfacial tracking failure mechanisms at different contact pressures using the distribution characteristics of discharge light and carbonization patterns. Their results suggest that a higher contact pressure notably inhibits the propagation of discharges, delaying the build-up of carbonized species and the interfacial breakdown. They also claimed that a higher contact pressure resulted in decreased discharge activity and less carbonization formation. In [18], the effect of surface roughness on the discharge processes at XLPE–SiR interfaces at AC voltage was studied. The results indicated that as the surface roughness was decreased (surfaces sanded with #100 to #1000 in a sequence), rising trends in the PDIV, breakdown voltage, and time to breakdown were observed by factors of 1.8, 1.4, and 2.3, respectively, whereas the intensity of the emitted discharge light decreased.

Kantar and Ildstad [76] proposed a novel test method to visualize PD activity in microcavities at solid–solid interfaces. Their primary objective was to examine the initiation, development, and propagation of discharge streamers at a solid–solid interface using a CCD camera. They observed that the discharge streamers were wide and long (continuous,

spots against discharges at an interface is roughly modeled and is estimated employing the tracking resistances of the solid materials forming the interface.

For clarity, a hypothetical case is exemplified as follows. Assume that cavities at two different homogeneous interfaces (interfaces formed between identical pairs of solid dielectrics) are identical (identical size, number, and shape) and are discharged at the same voltage; the interface formed between the materials with a higher "interfacial tracking resistance" is likely to yield a higher breakdown strength, as observed in the AC breakdown experiments [60,61].

In the following sections, further details on the statistical and deterministic roughness models are provided.

4.2. Statistical Interface Contact Models

Zhuravlev's (1940) statistical model of contact between rough elastic solids is an early, novel precedent [88]. A contact model of nominally flat surfaces was later proposed by Greenwood and Williamson (G&W) [89], in which the real area of contact was shown to be proportional to the load applied, as demonstrated by the Gaussian and exponential distributions of the asperity peaks.

Contacts that are either elastic or plastic were analyzed between a rough surface and a smooth surface by G&W [89]. Bhushan [20] retrofitted his approach on top of the G&W model [89] to make it more comprehensive. These two models (actually Bhushan's upgraded model) were utilized to propose a customized contact model in [62].

Kantar et al. [62] proposed a stochastic model for the contact surfaces between solid dielectrics using the above-mentioned tribological principles introduced in [20,21,89–94]. Their proposed model was based on multiple-asperity dry contacts formed at a solid–solid interface, which was employed to calculate the average cavity size at a given interface in a two-dimensional space. The proposed contact model suggest that higher contact pressure, stiffer interfaces, and/or surface smoothness generate smaller average-sized cavities and higher ratios of "real contact area to the nominal contact area", A_{re}/A_a. The model then predicted the AC PD inception field strength of an average-sized cavity using Paschen's law. The model was verified via an experimental study that used XLPE–XLPE interfaces with different surface roughness degrees under various contact pressures. Based on the AC breakdown test results of the XLPE–XLPE interfaces in [62], the cavity discharge inception and the interfacial breakdown phenomenon were found to be closely intertwined, leading to the deduction that the cavity discharge influences the interfacial breakdown phenomenon significantly. This study was extended in [74,95] by using different insulation materials and additional experiments where its performance was compared to a deterministic approach. Details of the extended model are provided in the next section.

Lastly, Zhu et al. [96] utilized the simplified interface contact model proposed in [52,53,58,60]. They studied the correlation between the DC breakdown voltage of XLPE–SiR interfaces and the interfacial morphology using the interface contact model. They combined the analytical model with an image processing algorithm that yielded similar surface simulations as those we obtained by using the deterministic contact model in [74,95], as introduced in the next section. They concluded that although the density of real contact irregularities is high, the real contact area is significantly lower than the nominal contact area. Consequently, there are many interconnected cavities at the interface that are the primary discharge path for an interfacial breakdown, while the contact spots serve as hurdles for the propagating discharge channel. These deductions are found to strongly agree with the main findings reported in [62,74].

4.3. Deterministic Interface Contact Models

With the advent of supercomputers that can perform heavy computations of big data in a matter of hours—if not minutes—deterministic models have increasingly been favored. The outcome is becoming increasingly realistic as computational speed increases. However, simplified models of material topographies are needed, making it necessary to minimize the computing time. In this respect, Almqvist's Ph.D. thesis [82] is found to be of immense benefit, which was built on the numerical model proposed by Tian and Bhushan [21].

Almqvist [82] proposed a contact model based on the theory of minimum potential complementary energy. The model was fed with the surface profile data of the interface of contact between two surfaces. By minimizing an integral energy equation, he determined the displacement of peaks and valleys in the profile as a function of the material properties and the applied contact pressure.

The model simulates the deformation occurring in the surface profile at the interface between the measured surface profile and an ideal plane [82]. The sizes of all cavities at the interface and total area of contact are definite. Hasheminezhad [11] employed Almqvist's deterministic model [82] to estimate the length of the largest cavity (in the direction of the electric field) at the XLPE–XLPE interfaces. His study covered scanned 2D surface profiles of XLPE specimens. Kantar's sequel to Hasheminezhad's work [61,62] unveiled the necessity of a thorough 3D contact model. To that end, in [74], the proposed model in [11] was further extended using the analytical expressions derived in [82] to incorporate 3D surface profiles in addition to the 2D surface profiles. The deterministic interface contact model provides an insight into how the real area of contact at solid–solid interfaces varies as a function of the contact pressure, surface roughness, elasticity, and hardness of the solid material in 3D. The major upgrade to the 2D contact models is that the 3D contact surface simulations can reveal the interconnection between the cavities and visualize how long an air gap can be in all directions, not only in one horizontal direction as in the case of 2D models.

Kantar et al. [74,95] utilized contour and surface plots to present the results of the deterministic model. Contour lines depict the amplitudes of the asperities and the enclosed area of air gaps, as illustrated in Figure 16. Hence, the area between the air gaps represents the contact area at an interface. Color bars were used to represent peaks and the ground level (contact area) with amplitudes in μm. The 3D simulations suggest that the increased contact pressure diminishes the number and length of long air gaps and thus creates more enclosed cavities. The exemplary cases of no-load (Figure 16a) and nominal contact pressure (Figure 16b) for the XLPE–XLPE #180 interface clearly depict this behavior. A selected portion of the original surface profile (yellow) and displacement of surface asperities (red) along with the deformed surface profile (blue) are presented in Figure 17. The surface profiles portray the equivalent rough surfaces while the horizontal axis delineates an ideal smooth plane (no cavities hypothetically). Therefore, the non-flat portions between the neighboring flat areas (contact spots) are the cavities. Differences between the no-load and deformed surface profiles reveal that the increased loading presses the floating edges toward the ideal plane, resulting in new contact points and thus smaller air gaps. Moreover, the discrete distribution of contact pressure along the interface is illustrated in Figure 17 (right-hand side y-axis). As seen, the contact spots enclosing the cavities undergo much higher pressure, in some cases leading to the plastic deformation (equal to the material hardness in the model) of those spots. Otherwise, elastic contact points occur at the contact areas that vary between zero and the material hardness. In [74,95], it was also shown that smoother interfaces lead to more enclosed, smaller cavities. Furthermore, the elasticity of the material proved to affect cavity sizes significantly—interfaces between relatively hard materials are likely to contain longer and larger cavities (channels).

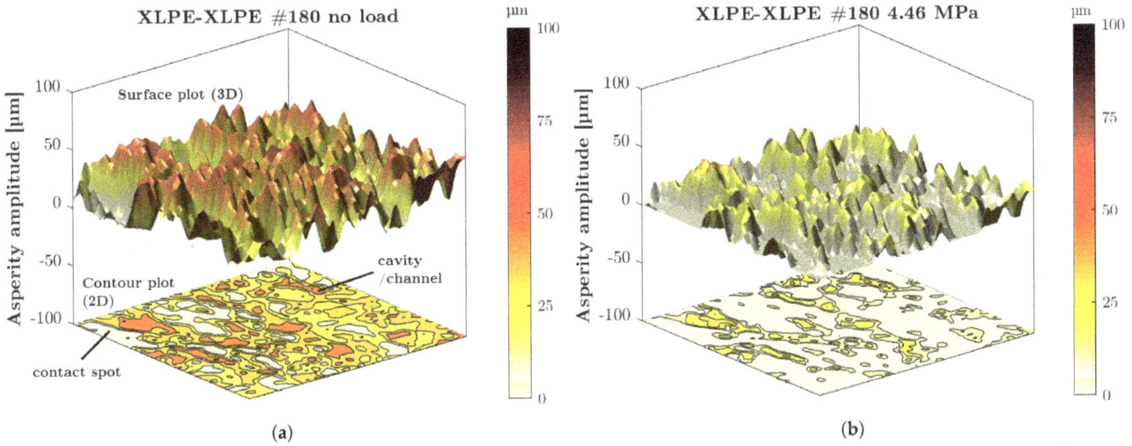

Figure 16. Illustration of a measured 3D surface profile of an XLPE sample (polished by sandpaper grit #180) with a filled-contour plot for the projection of the 3D surface on a 2D surface at (**a**) 0.5 MPa, (**b**) 4.46 MPa.

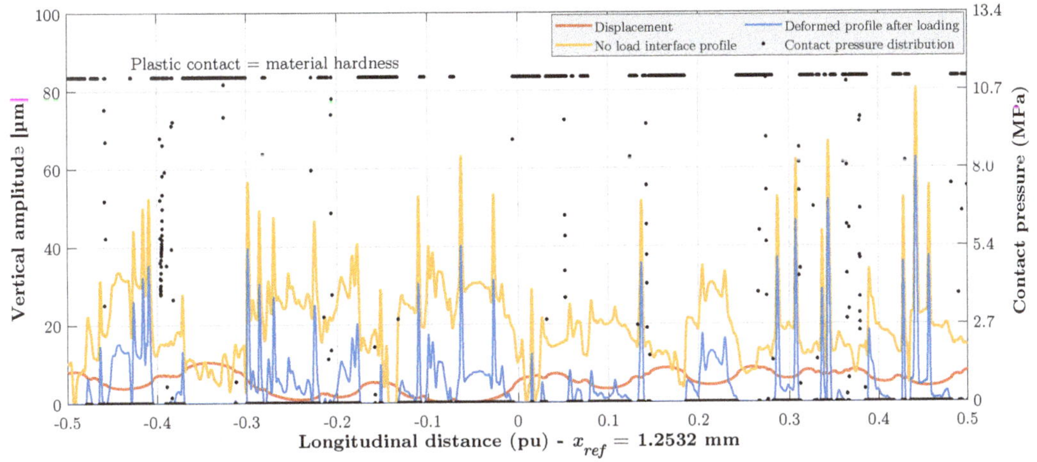

Figure 17. Measured (no-load) 2D surface profile of the XLPE–XLPE #180 interface, displacement of peaks after loading (4.46 MPa), deformed profile, and distribution of contact pressure at contact spots. Nominal longitudinal distance (x_{ref}) is 1.2532 mm.

5. Concluding Remarks

In this paper, a thorough review of the modern and up-to-date literature addressing the electrical performance of cables, connectors, terminations, and other accessories that contain interfaces have been performed. Particular focus has been on the polymeric dielectric solid–solid interfaces, considering the causes of electrical interfacial failure and important factors that influence the dielectric strength of an interface. The following remarks can be made in the light of the thorough literature review:

- The findings from experimental and theoretical studies indicate that different processes control the discharge of air-filled cavities and the breakdown of contact spots. The main conclusion is that the properties of the cavities and contact spots govern the interface breakdown. On the one hand, the size, shape, and insulating medium inside the cavities determine the discharge inception field of the cavities. On the other hand, the tracking resistance of the contact spots between the discharged cavities heavily affects the interfacial breakdown strength. Different tracking resistances of the polymers tend to affect the primary discharge propagation mechanisms of particle bombardment and the light emission from the discharge channel. In contrast, the charge injection, trapping, and de-trapping mechanisms have indirect effects as they result in more intense light emission and local field distortion in the initiation stage.

- The breakdown strengths of the dry-mate interfaces are found to be the highest in the cases where the contact pressure is relatively high, and the interface is as smooth as possible. Consequently, solid–solid interfaces can be made to perform better by introducing a smoother surface and sustaining the interface pressure to be high enough throughout the service life.
- Air-filled (dry-mate) cavities and water-filled (wet-mate) cavities yield breakdown strength values that are significantly lower than those in the case of oil-mate interfaces. Strong local field stresses arising at the edges of the contact area (due to short-circuited/discharged cavities) dramatically reduce the overall breakdown strength when water is present at an interface. Hence, water intrusion is a serious concern in the design of any HV equipment incorporating solid–solid interfaces.
- The surface roughness has a significant influence on the interfacial breakdown strength. A high correlation between the interfacial breakdown strength and the surface roughness is found. The breakdown strength may potentially become twice as high from the roughest to the smoothest surface.
- The elastic modulus stands out as an important material property for solid materials/interfaces because it strongly impacts the interfacial dielectric strength. Specifically, softer materials with low elastic moduli, such as SiR and XLPE, have a much higher breakdown strength when compared with stiffer materials with high elastic moduli.
- Increased contact pressure yields higher BDS values irrespective of the surface roughness and elasticity, where elasticity can be a limiting factor in elastic contacts. In plastic contacts, the real area of contact does not increase even if the contact pressure is further increased.
- Considering practical power cables and connectors, avoiding the loss of interfacial pressure between solid materials and water ingress appears to be of considerable value in practical applications to ensure a high breakdown strength and long service life.

Funding: The project is supported by The Research Council of Norway (Project No. 228344/E30).

Institutional Review Board Statement: Not applicable.

Informed Consent Statement: Not applicable.

Data Availability Statement: Not applicable.

Conflicts of Interest: The author declares no conflict of interest.

References

1. Kunze, D.; Parmigiani, B.; Schroth, R.; Gockenbach, E. *Macroscopic Internal Interfaces in High Voltage Cable Accessories*; CIGRE Session: Paris, France, 2000; pp. 15–203.
2. Siemens. Spectron. Available online: https://assets.siemens-energy.com/siemens/assets/api/uuid:e0e64e11-b801-4629-9f0f-50cd0c544f6f/spectronbrochure.pdf (accessed on 30 November 2021).
3. Midttveit, S.; Monsen, B.; Frydenlund, S.; Stenevik, K. SS on Implications of subsea processing power distribution-subsea power systems—A key enabler for subsea processing. In Proceedings of the Offshore Technology Conference, Houston, TX, USA, 3–6 May 2010.
4. Weiss, P.; Beurthey, S.; Chardard, Y.; Dhedin, J.; Andre, T.; Rabushka, K.; Tourcher, C.; Gauch, F.; Micoli, C. Novel wet-mate connectors for high voltage and power transmissions of ocean renewable energy systems. In Proceedings of the 4th International Conference on Ocean Energy, Dublin, Ireland, 17–19 October 2012.
5. Østergaard, I.; Nysveen, A.; Romanisko, T. MECON: A High Voltage Subsea Connector. In Proceedings of the Offshore Technology Conference, Houston, TX, USA, 3–6 May 1999. [CrossRef]
6. Myklatun, J.T. Condition Monitoring of Subsea Connectors. Master's Thesis, Norwegian University of Science and Technology, Trondheim, Norway, 2014.

7. Nysveen, A. Coupling-and Switch System for Subsea Electrical Power Distribution. U.S. Patent 5834721, 22 May 1998.
8. Peschke, E.; von Olshausen, R. *Cable Systems for High and Extra-High Voltage: Development, Manufacture, Testing, Installation and Operation of Cables and Their Accessories*; Wiley-VCH: Hoboken, NJ, USA, 1999.
9. Argaut, P.; Becker, J.; Dejean, P.M.; Sin, S.; Dorison, E. *Studies and Development in France of 400 kV Cross-Linked Polyethylene Cable Systems*; CIGRE Session: Paris, France, 2000; pp. 15–203.
10. Argaut, P. (Ed.) *Accessories for HV and EHV Extruded Cables*; CIGRE Green Books; Springer: Berlin/Heidelberg, Germany, 2021.
11. Hasheminezhad, S.M. Tangential Electric Breakdown Strength and PD Inception Voltage of Solid-Solid Interface. Ph.D. Thesis, Norwegian University of Science and Technology, Trondheim, Norway, 2016.
12. Bogoeva-Gaceva, G.; Dimeski, D.; Srebrenkoska, V. Friction Mechanism of Polymers and Their Composites. *Maced. J. Chem. Chem. Eng.* **2018**, *37*, 1–11. [CrossRef]
13. Takahashi, T.; Okamoto, T.; Ohki, Y.; Shibata, K. Breakdown strength at the interface between epoxy resin and silicone rubber—A basic study for the development of all solid insulation. *IEEE Trans. Dielectr. Electr. Insul.* **2005**, *12*, 719–724. [CrossRef]
14. Fournier, D.; Dang, C.; Paquin, L. Interfacial breakdown in cable joints. In Proceedings of the 1994 IEEE International Symposium on Electrical Insulation, Pittsburgh, PA, USA, 5–8 June 1994; pp. 450–452. [CrossRef]
15. Fournier, D.; Lamarre, L. Interfacial breakdown phenomena between two EPDM surfaces. In Proceedings of the Sixth International Conference on Dielectric Materials, Measurements and Applications, Manchester, UK, 7–10 September 1992; pp. 330–333.
16. Fournier, D. Effect of the surface roughness on interfacial breakdown between two dielectric surfaces. In Proceedings of the Montreal, Montreal, QC, Canada, 16–19 June 1996; Volume 2, pp. 699–702.
17. Du, B.; Gu, L. Effects of interfacial pressure on tracking failure between XLPE and silicon rubber. *IEEE Trans. Dielectr. Electr. Insul.* **2010**, *17*, 1922–1930. [CrossRef]
18. Du, B.; Zhu, X.; Gu, L.; Liu, H. Effect of surface smoothness on tracking mechanism in XLPE-Si-rubber interfaces. *IEEE Trans. Dielectr. Electr. Insul.* **2011**, *18*, 176–181. [CrossRef]
19. Dissado, L.A.; Fothergill, J.C. *Electrical Degradation and Breakdown in Polymers*; IET: London, UK, 1992; Volume 9.
20. Bhushan, B. Contact mechanics of rough surfaces in tribology: Multiple asperity contact. *Tribol. Lett.* **1998**, *4*, 1–35. [CrossRef]
21. Bhushan, B.; Tian, X. A numerical three-dimensional model for the contact of rough surfaces by variational principle. *ASME J. Tribol* **1996**, *118*, 33–42.
22. CIGRE Technical Brochure 210. *Joint Task Force 21/15: Interface in Accessories for Extruded HV and EHV Cables*; CIGRE: Paris, France, 2002.
23. Albayrak, S.; Becker-Willinger, C.; Aslan, M.; Veith, M. Influence of nano-scaled zirconia particles on the electrical properties of polymer insulating materials. *IEEE Trans. Dielectr. Electr. Insul.* **2012**, *19*, 76–82. [CrossRef]
24. Roy, M.; Nelson, J.K.; MacCrone, R.K.; Schadler, L.S.; Reed, C.W.; Keefe, R. Polymer nanocomposite dielectrics-the role of the interface. *IEEE Trans. Dielectr. Electr. Insul.* **2005**, *12*, 629–643. [CrossRef]
25. Ding, H.Z.; Varlow, B.R. Effect of nano-fillers on electrical treeing in epoxy resin subjected to AC voltage. In Proceedings of the 17th Annual Meeting of the IEEE Lasers and Electro-Optics Society, Boulder, CO, USA, 20 October 2004; pp. 332–335.
26. Li, J.; Du, B.X.; Xu, H. Suppressing interface charge between LDPE and EPDM for HVDC cable accessory insulation. *IEEE Trans. Dielectr. Electr. Insul.* **2017**, *24*, 1331–1339. [CrossRef]
27. Du, B.X.; Li, J. Interface charge behaviors between LDPE and EPDM filled with carbon black nanoparticles. *IEEE Trans. Dielectr. Electr. Insul.* **2016**, *23*, 3696–3703. [CrossRef]
28. Du, B.X.; Xu, H.; Li, J. Effects of mechanical stretching on space charge behaviors of PP/POE blend for HVDC cables. *IEEE Trans. Dielectr. Electr. Insul.* **2017**, *24*, 1438–1445. [CrossRef]
29. Huuva, R.; Englund, V.; Gubanski, S.M.; Hjertberg, T. A versatile method to study electrical treeing in polymeric materials. *IEEE Trans. Dielectr. Electr. Insul.* **2009**, *16*, 171–178. [CrossRef]
30. Chen, X.; Xu, Y.; Cao, X.; Gubanski, S.M. Electrical treeing behavior at high temperature in XLPE cable insulation samples. *IEEE Trans. Dielectr. Electr. Insul.* **2015**, *22*, 2841–2851. [CrossRef]
31. Fothergill, J.C. Filamentary electromechanical breakdown. *IEEE Trans. Electr. Insul.* **1991**, *26*, 1124–1129. [CrossRef]
32. Eichhorn, R.M. Treeing in Solid Extruded Electrical Insulation. *IEEE Trans. Electr. Insul.* **1977**, *EI-12*, 2–18. [CrossRef]
33. Mason, J.H. Assessing the resistance of polymers to electrical treeing. *IEE Proc. A Phys. Sci. Meas. Instrum. Manag. Educ. Rev.* **1981**, *128*, 193–201. [CrossRef]
34. Roseen, P.A.; Gubanski, S.M.; Gedde, U.W. External PD resistance of thermoplastic and XLPE containing voltage stabilizers. *IEEE Trans. Dielectr. Electr. Insul.* **1998**, *5*, 189–194. [CrossRef]
35. Hammarström, T.J.Å.; Bengtsson, T.; Gubanski, S.M. Partial discharge characteristics of electrical treeing in XLPE insulation exposed to voltages of different rise times. In Proceedings of the 2017 International Symposium on Electrical Insulating Materials (ISEIM), Toyohashi, Japan, 11–15 September 2017; Volume 1, pp. 407–410. [CrossRef]
36. Chen, X.; Murdany, D.; Liu, D.; Andersson, M.; Gubanski, S.M.; Gedde, U.W.; Suwarno. AC and DC pre-stressed electrical trees in LDPE and its aluminum oxide nanocomposites. *IEEE Trans. Dielectr. Electr. Insul.* **2016**, *23*, 1506–1514. [CrossRef]
37. Chen, X.; Xu, Y.; Cao, X.; Gubanski, S.M. On the conducting and non-conducting electrical trees in XLPE cable insulation specimens. *IEEE Trans. Dielectr. Electr. Insul.* **2016**, *23*, 95–103. [CrossRef]

38. Zheng, Y.; Serdyuk, Y.V.; Gubanski, S.M. Space charge controlled electric field preceding inception of electric tree in XLPE at AC voltage. In Proceedings of the 2015 IEEE 11th International Conference on the Properties and Applications of Dielectric Materials (ICPADM), Sydney, NSW, Australia, 19–22 July 2015; pp. 132–135. [CrossRef]
39. Jarvid, E.M.; Johansson, A.B.; Blennow, J.H.M.; Andersson, M.R.; Gubanski, S.M. Evaluation of the performance of several object types for electrical treeing experiments. *IEEE Trans. Dielectr. Electr. Insul.* **2013**, *20*, 1712–1719. [CrossRef]
40. Chen, X.R.; Hu, L.B.; Xu, Y.; Cao, X.L.; Gubanski, S.M. Investigation of temperature effect on electrical trees in XLPE cable insulation. In Proceedings of the 2012 Annual Report Conference on Electrical Insulation and Dielectric Phenomena, Montreal, QC, Canada, 14–17 October 2012; pp. 612–615. [CrossRef]
41. Jarvid, M.; Johansson, A.; Englund, V.; Gubanski, S.; Andersson, M.R. Electrical tree inhibition by voltage stabilizers. In Proceedings of the 2012 Annual Report Conference on Electrical Insulation and Dielectric Phenomena, Montreal, QC, Canada, 14–17 October 2012; pp. 605–608. [CrossRef]
42. Sonerud, B.; Blennow, J.; Gubanski, S.M.; Nilsson, S.; Bengtsson, T. Continuous monitoring of dielectric properties of LDPE samples during electrical treeing. In Proceedings of the 2010 10th IEEE International Conference on Solid Dielectrics, Potsdam, Germany, 4–9 July 2010; pp. 1–4. [CrossRef]
43. Roseen, P.A.; Reitberger, T.; Gubanski, S.M.; Gedde, U.W. PD resistance of thermally aged polyethylene and carbonyl-containing model polymers. *IEEE Trans. Dielectr. Electr. Insul.* **1999**, *6*, 191–201. [CrossRef]
44. Dang, C.; Fournier, D. Dielectric performance of interfaces in premolded cable joints. *IEEE Tran. Power Deliv.* **1997**, *12*, 29–32. [CrossRef]
45. Fournier, D.; Lamarre, L. Effect of pressure and length on interfacial breakdown between two dielectric surfaces. In Proceedings of the Conference Record of the 1992 IEEE International Symposium on Electrical Insulation, Baltimore, MD, USA, 7–10 June 1992; pp. 270–272. [CrossRef]
46. Fournier, D.; Lamarre, L. Effect of pressure and temperature on interfacial breakdown between two dielectric surfaces. In Proceedings of the 1992 Annual Report: Conference on Electrical Insulation and Dielectric Phenomena, Victoria, BC, Canada, 18–21 October 1992; pp. 229–235.
47. Kantar, E.; Mauseth, F.; Ildstad, E. Effect of pressure and elastic modulus on tangential breakdown strength of solid-solid interfaces. In Proceedings of the 2016 IEEE Electrical Insulation Conference (EIC), Montreal, QC, Canada, 19–22 June 2016; pp. 431–435. [CrossRef]
48. Kantar, E.; Panagiotopoulos, D.; Ildstad, E. Factors Influencing the Tangential AC Breakdown Strength of Solid-Solid Interfaces. *IEEE Trans. Dielectr. Electr. Insul.* **2016**, *23*, 1778–1788. [CrossRef]
49. Kunze, D. *Untersuchungen an Grenzflächen Zwischen Polymerwerkstoffen Unter Elektrischer Hochfeldbeanspruchung in der Garniturentechnik VPE-Isolierter Hochspannungskabel*; Shaker: Düren, Germany, 2000.
50. Du, B.; Gu, L.; Zhang, X.; Zhu, X. Fundamental research on dielectric breakdown between XLPE and silicon rubber interface in HV cable joint. In Proceedings of the 2009 IEEE 9th International Conference on the Properties and Applications of Dielectric Materials, Harbin, China, 19–23 July 2009; pp. 97–100.
51. Chen, X.; Gu, L.; He, X.; Liao, H. Tracking failure process of XLPE–Silicone rubber interface under impulse voltage. In Proceedings of the 2012 International Conference on High Voltage Engineering and Application, Shanghai, China, 17–20 September 2012; pp. 51–54.
52. Hasheminezhad, M.; Ildstad, E. Application of contact analysis on evaluation of breakdown strength and PD inception field strength of solid-solid interfaces. *IEEE Trans. Dielectr. Electr. Insul.* **2012**, *19*, 1–7. [CrossRef]
53. Hasheminezhad, M.; Ildstad, E. Partial discharge inception of interface voids versus mechanical surface pressure. In Proceedings of the 2010 International Conference on High Voltage Engineering and Application, New Orleans, LA, USA, 11–14 October 2010; pp. 397–400. [CrossRef]
54. Hasheminezhad, S.M.; Ildstad, E.; Nysveen, A. Breakdown strength of solid-solid interface. In Proceedings of the 2010 10th IEEE International Conference on Solid Dielectrics, Potsdam, Germany, 4–9 July 2010; pp. 1–4. [CrossRef]
55. Hasheminezhad, S.M. Breakdown strength of solid | solid interfaces. In Proceedings of the Trondheim PowerTech, Trondheim, Norway, 19–23 June 2011; pp. 1–7.
56. Hasheminezhad, M.; Ildstad, E. Breakdown strength of solid-solid interfaces. In Proceedings of the Nordic Insulation Symposium (NORD-IS), Trondheim, Norway, 9–12 June 2011.
57. Kantar, E.; Ildstad, E. Modeling longitudinal breakdown strength of solid-solid interfaces using contact theory. In Proceedings of the 2016 IEEE International Conference on Dielectrics (ICD), Montpellier, France, 3–7 July 2016; Volume 1, pp. 398–401. [CrossRef]
58. Kantar, E.; Hvidsten, S.; Mauseth, F.; Ildstad, E. On the Tangential AC Breakdown Strength of Polymeric Interfaces Considering Elastic Modulus. In Proceedings of the 2017 IEEE Conference on Electrical Insulation and Dielectric Phenomenon (CEIDP), Fort Worth, TX, USA, 22–25 October 2017; pp. 816–819. [CrossRef]
59. Kantar, E.; Hvidsten, S.; Mauseth, F.; Ildstad, E. Tangential AC Breakdown Strength of Solid-Solid Interfaces Considering Surface Roughness. In Proceedings of the 2017 IEEE Conference on Electrical Insulation and Dielectric Phenomenon (CEIDP), Fort Worth, TX, USA, 22–25 October 2017; pp. 580–583. [CrossRef]
60. Kantar, E.; Hvidsten, S.; Mauseth, F.; Ildstad, E. Longitudinal AC Breakdown Voltage of XLPE-XLPE Interfaces Considering Surface Roughness and Pressure. *IEEE Trans. Dielectr. Electr. Insul.* **2017**, *24*, 3047–3054. [CrossRef]

61. Kantar, E.; Hvidsten, S.; Ildstad, E. Effect of Material Elasticity on the Longitudinal AC Breakdown Strength of Solid-Solid Interfaces. *IEEE Trans. Dielectr. Electr. Insul.* **2019**, *26*, 655–663. [CrossRef]
62. Kantar, E.; Hvidsten, S.; Mauseth, F.; Ildstad, E. A Stochastic Model for Contact Surfaces at Polymer Interfaces Subjected to an Electrical Field. *Tribol. Int.* **2018**, *127*, 361–371. [CrossRef]
63. Kantar, E. Mechanisms Governing Longitudinal AC Breakdown at Solid-Solid Interfaces. In Proceedings of the 2020 IEEE Conference on Electrical Insulation and Dielectric Phenomena (CEIDP), East Rutherford, NJ, USA, 18–30 October 2020; pp. 279–283. [CrossRef]
64. Du, B.X.; Ma, Z.L.; Cheng, X.X.; Liu, Y. Hydrophobicity evaluation of silicone rubber insulator using PD-induced electromagnetic wave. *IEEE Trans. Dielectr. Electr. Insul.* **2012**, *19*, 1060–1067. [CrossRef]
65. Illias, H.A. Measurement and Simulation of Partial Discharges within a Spherical Cavity in a Solid Dielectric Material. Ph.D. Thesis, University of Southampton, Southampton, UK, 2011.
66. Illias, H.; Chen, G.; Lewin, P.L. Modeling of partial discharge activity in spherical cavities within a dielectric material. *IEEE Electr. Insul. Mag.* **2011**, *27*, 38–45. [CrossRef]
67. Illias, H.; Yuan, T.S.; Mokhlis, H.; Chen, G.; Lewin, P.L. Partial discharge patterns in high voltage insulation. In Proceedings of the 2012 IEEE International Conference on Power and Energy (PECon), Kota Kinabalu, Malaysia, 2–5 December 2012; pp. 750–755.
68. Adhikari, D.; Hepburn, D.M.; Stewart, B.G. PD characteristics and degradation in PET insulation with vented and unvented internal voids. *Electr. Power Syst. Res.* **2013**, *100*, 65–72. [CrossRef]
69. Adhikari, D.; Hepburn, D.M.; Stewart, B.G. Analysis of partial discharge characteristics in artificially created voids. In Proceedings of the 45th International Universities Power Engineering Conference UPEC2010, Cardiff, UK, 31 August–3 September 2010; pp. 1–4.
70. Mas'ud, A.A.; Stewart, B.G. An investigative study on the influence of correlation of PD statistical features on PD pattern recognition. In Proceedings of the 2018 IEEE 2nd International Conference on Dielectrics (ICD), Budapest, Hungary, 1–5 July 2018; pp. 1–5. [CrossRef]
71. Mohamed, F.P.; Siew, W.H.; Sheng, B.; Stewart, B. Effect of voltage reduction in minimising partial discharge activity in cables—Experimental study. In Proceedings of the 2017 IEEE Conference on Electrical Insulation and Dielectric Phenomenon (CEIDP), Fort Worth, TX, USA, 22–25 October 2017; pp. 331–334. [CrossRef]
72. Adhikari, D.; Hepburn, D.M.; Stewart, B.G. Comparison of partial discharge characteristics and degradation in several polymeric insulators. *IET Sci. Meas. Tech.* **2012**, *6*, 474–484. [CrossRef]
73. Reid, A.J.; Hepburn, D.M.; Stewart, B.G. The influence of external magnetic fields on the partial discharge characteristics of voids. In Proceedings of the 2013 IEEE Electrical Insulation Conference (EIC), Ottawa, ON, Canada, 2–5 June 2013; pp. 147–150. [CrossRef]
74. Kantar, E.; Hvidsten, S. A deterministic breakdown model for dielectric interfaces subjected to tangential electric field. *J. Phys. D Appl. Phys.* **2021**, *54*, 295503. [CrossRef]
75. Gu, L.; He, X. Microcavity on tracking failure of XLPE-SiR interface. In Proceedings of the 2012 Asia-Pacific Power and Energy Engineering Conference, Shanghai, China, 27–29 March 2012; pp. 1–4.
76. Kantar, E.; Ildstad, E. A Novel Methodology to Monitor Partial Discharges in Microvoids at Solid-Solid Interfaces. In Proceedings of the 2019 IEEE Conference on Electrical Insulation and Dielectric Phenomena (CEIDP), Richland, WA, USA, 20–23 October 2019; pp. 113–117. [CrossRef]
77. Kato, M.; Nishimura, Y.; Osawa, N.; Yoshioka, Y.; Yanase, H.; Okamoto, K. Effects of Compressive Force and Dielectric Materials on Contact Area for High-Pressure Region and Interfacial AC Breakdown between Two Solid Dielectrics. In Proceedings of the 21st International Symposium on High Voltage Engineering (ISH 2019), Budapest, Hungary, 26–30 August 2019; pp. 118–129. [CrossRef]
78. Kantar, E. Longitudinal AC Electrical Breakdown Strength of Polymer Interfaces. Ph.D. Thesis, Norwegian University of Science and Technology, Trondheim, Norway, 2019.
79. Gao, Y.; Yuan, Y.; Chen, L.; Li, J.; Huang, S.; Du, B. Direct Fluorination Induced Variation in Interface Discharge Behavior between Polypropylene and Silicone Rubber Under AC Voltage. *IEEE Access* **2018**, *6*, 23907–23917. [CrossRef]
80. Dissado, L.A. Understanding electrical trees in solids: From experiment to theory. *IEEE Trans. Dielectr. Electr. Insul.* **2002**, *9*, 483–497. [CrossRef]
81. Borodich, F.M.; Pepelyshev, A.; Savencu, O. Statistical approaches to description of rough engineering surfaces at nano and microscales. *Tribol. Int.* **2016**, *103*, 197–207. [CrossRef]
82. Almqvist, A. On the Effects of Surface Roughness in Lubrication. Ph.D. Thesis, Luleå Tekniska Universitet, Luleå, Sweden, 2006.
83. Ciavarella, M.; Demelio, G.; Barber, J.; Jang, Y.H. Linear elastic contact of the Weierstrass profile. *Proc. R. Soc. Lond. Ser. A Math. Phys. Eng. Sci.* **2000**, *456*, 387–405. [CrossRef]
84. Majumdar, A.; Bhushan, B. Fractal model of elastic-plastic contact between rough surfaces. *J. Tribol.* **1991**, *113*, 1–11. [CrossRef]
85. Majumdar, A.; Bhushan, B. Role of fractal geometry in roughness characterization and contact mechanics of surfaces. *J. Tribol.* **1990**, *112*, 205–216. [CrossRef]
86. Ganti, S.; Bhushan, B. Generalized fractal analysis and its applications to engineering surfaces. *Wear* **1995**, *180*, 17–34. [CrossRef]
87. Bhushan, B. *Handbook of Micro/Nano Tribology*; CRC Press: Boca Raton, FL, USA, 1998.

88. Zhuravlev, V. On the question of theoretical justification of the Amontons-Coulomb law for friction of unlubricated surfaces. *Proc. IMechE Part J.* **2007**, *221*, 895–898. [CrossRef]
89. Greenwood, J.; Williamson, J. Contact of nominally flat surfaces. *Proc. R. Soc. Lond. Ser. A Math. Phys. Eng. Sci.* **1966**, *295*, 300–319.
90. Bhushan, B. *Principles and Applications of Tribology*; John Wiley & Sons: Hoboken, NJ, USA, 2013.
91. Bhushan, B. Analysis of the Real Area of Contact between a Polymeric Magnetic Medium and a Rigid Surface. *J. Tribol.* **1984**, *106*, 26–34. [CrossRef]
92. Williamson, J.; Pullen, J.; Hunt, R. The shape of solid surfaces. *Surf. Mech. ASME* **1969**, *9*, 334.
93. Greenwood, J. The area of contact between rough surfaces and flats. *J. Lubr. Tech.* **1967**, *89*, 81–87. [CrossRef]
94. Greenwood, J.; Tripp, J. The contact of two nominally flat rough surfaces. *Proc. Instit. Mech. Eng.* **1970**, *185*, 625–633. [CrossRef]
95. Kantar, E. A Deterministic Model for Contact Surfaces at Dielectric Interfaces Subjected to an Electrical Field. In Proceedings of the 2020 IEEE Conference on Electrical Insulation and Dielectric Phenomena (CEIDP), East Rutherford, NJ, USA, 18–30 October 2020; pp. 21–26. [CrossRef]
96. Zhu, B.; Jia, Z.; Hu, H.; Ouyang, X.; Wang, X. Relationship between the Interfacial DC Breakdown Voltage and the Morphology of the XLPE/SiR Interface. *IEEE Trans. Dielectr. Electr. Insul.* **2019**, *26*, 689–697. [CrossRef]

energies

MDPI

Article

Electric Field Distribution and AC Breakdown Characteristics of Polluted Novel Lightning Protection Insulator under Icing Conditions

Jiazheng Lu [1], Jianping Hu [1], Zhen Fang [1], Xinhan Qiao [2] and Zhijin Zhang [2,3,*]

[1] State Key Laboratory of Disaster Prevention & Reduction for Power Grid Transmission and Distribution Equipment, Disaster Prevention and Reduction Center of State Grid Hunan Electric Power Co., Ltd., Changsha 410007, China; babyzhx1@163.com (J.L.); hujianping81016@sina.com (J.H.); policy@139.com (Z.F.)

[2] School of Electrical Engineering, Chongqing University, Chongqing 400044, China; qiaoxinhan@cqu.edu.cn

[3] Chongqing De-Icing Group Technology Co., Ltd., Chongqing 400799, China

* Correspondence: zhangzhijing@cqu.edu.cn; Tel.: +86-13883207915

Citation: Lu, J.; Hu, J.; Fang, Z.; Qiao, X.; Zhang, Z. Electric Field Distribution and AC Breakdown Characteristics of Polluted Novel Lightning Protection Insulator under Icing Conditions. *Energies* **2021**, *14*, 7493. https://doi.org/10.3390/en14227493

Academic Editor: Ryszard Palka

Received: 9 September 2021
Accepted: 1 November 2021
Published: 9 November 2021

Abstract: As a result of lightning strikes, pollution, and ice, overhead distribution wires might be short-circuited and trip. As a result, researchers have developed a new lightning protection composite insulator. There is still a need to test its pollution and icing performance. Based on the finite element and field test method, this paper studies the electric field distribution and AC (Alternating Current) breakdown characteristics of polluted novel lightning protection insulators under icing conditions. Firstly, the finite element calculated results show that this novel insulator's electric field distribution is different from that of a conventional insulator. The locations with sizeable electric fields are located in the insulation section, and the electric field in the arrester section is tiny. In addition, when the insulator surface is covered with ice, there is an increase in the electric field along the surface and pin electrodes. Compared with the dry conditions, when there is an ice layer and icicle, electric field peaks increase by 48.85% and 46.08%, respectively. Secondly, the test results show that there are three types of arc paths in different pollution levels. The arc paths are related to ESDD (equivalent salt deposit density) under icing conditions. U_f shows a downward trend with increased pollution levels, and the maximum flashover voltage is 2.70 times more than the minimum. Finally, four fitting methods are proposed in this paper. After comparing the goodness of fit of different functions, the quadratic function and negative power function with the constant term are recommended as empirical formulas for calculating flashover voltage of novel insulators under icing conditions in different pollution levels. The research results of this paper have a specific guiding role for the selection of the external insulation of transmission lines and structural optimization of novel insulators.

Keywords: AC breakdown characteristics; arresters; lightning protection insulator; electric field calculation; icing conditions

1. Introduction

As a result of lightning strikes, pollution, and ice, overhead distribution wires might be short-circuited and trip [1]. As a result, researchers have developed a new lightning protection composite insulator [2,3]. Lightning accidents can be avoided by using the new insulator. In practical use, it can replace ordinary insulators and lightning arresters at the same time. In terms of length and installation, it is the same as a conventional insulator, but with the added benefits of low cost and ease of installation. However, worries have been raised about its pollution and icing performance. Large-scale power outages caused by pollution-induced flashover accidents might harm social–economic growth [4–6]. In addition, as a unique form of pollution, the harm of insulator surface icing to the power systems is prominent [7–9].

Many research investigations have been carried out on power line insulators to determine their icing [10–13] and pollution [14–17] performance. First and foremost, the pollution level has a significant impact on the flashover performance of composite insulators. A negative power function connection between pollution failure voltage and salt deposit density (SDD) is also discovered [15]. Namely, the flashover voltage decreases with the increase of pollution degree. Because of the narrower distance between composite sheds, icicles may cross it more efficiently, resulting in poor insulation and even power outages [18–20]. In addition, some literature has studied the relationship between electric field and icing [21,22]. The ice accretion performance of 220 kV composite insulators with two shed configurations on occasions of different electric field distributions is analyzed in the study [21]. The test results indicate that the intensity and homogeneous distribution of electric fields positively affect the ice's appearance, density, and weight. In the study [22], based on a kind of 10 kV composite insulator computation model, the electric field distribution magnitude of composite insulator coated by different icicle lengths were calculated, respectively, under icing and de-icing conditions. It could be found that the electric field strength around the icicle increases along with the increasing of the length of the icicle.

The above research has important guiding significance for selecting external insulation of ordinary insulators under pollution and icing conditions. However, there are few studies on the flashover characteristics of this novel lightning protection composite insulator under icing and pollution. Only the literature has made a preliminary study [1,23–25]. In one study [24], the new composite insulator's AC flashover performance was examined in detail. Tests were done at the Natural Ice Test Base to determine the effect of icing thickness on AC flashover performance. Next, to calculate AC flashover voltage, an empirical equation was developed. Literature [25] studied the failure arc paths of the novel insulator. The findings indicate that low, medium, and heavy pollutions caused different arcing under AC voltage. In addition, the arc path was affected by both the pollution level and impulse voltage amplitude. Based on the test results recorded by a high-speed camera, 14 types of arc paths were defined and analyzed. However, the above literature does not consider AC breakdown characteristics of polluted novel lightning protection insulators under icing conditions. In addition, the electric field distribution characteristics on the surface of this novel insulator under icing conditions have not been studied.

Based on the above, this paper studies the electric field distribution and AC breakdown characteristics of polluted novel lightning protection insulators under icing conditions, based on the finite element and field test method. The research results of this paper have a specific guiding role for the selection of the external insulation of transmission lines and the structural optimization of novel insulators.

2. Samples and Methods

2.1. Samples

A novel lightning-protection composite insulator was used as an experimental sample (Figure 1). Its physical drawing is shown in Figure 1a, and its sectional illustration is shown in Figure 1b. An arrester and insulator are combined in this 10 kV lightning protection composite insulator. As shown in Figure 1b, ① (disk-shaped metal electrode) and ② (a pin electrode with a spherical end) constitute the air gap. ③ and ④ are silicone rubber sheds. ⑤ and ⑥ are a zinc oxide resistor and an acid, and high-temperature resistant mandrel. Their geometrical characteristics are shown in Table 1. $d1$ is the arrester section's big-shed diameter, $d2$ is the insulation section's big-shed diameter, l is the leakage distance, and $d3$ is the gap distance between ① and ②. H is the height.

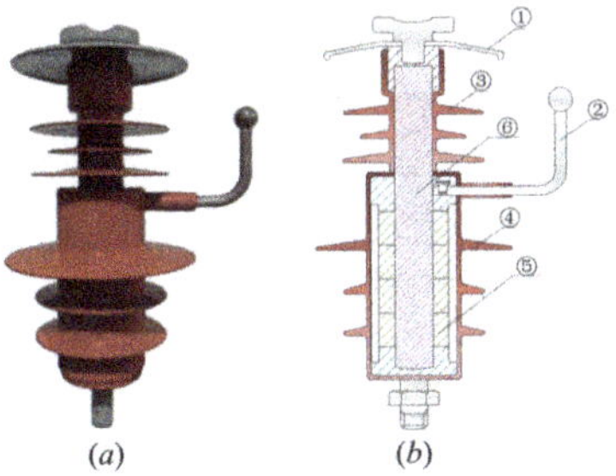

Figure 1. The novel lightning protection composite insulator: (**a**) Physical drawing; (**b**) Sectional illustration.

Table 1. Geometric parameters of the sample.

Sample	Parameters (mm)				
	H	$d1$	$d2$	l	$d3$
Lightning protection composite insulator	293	166	120	686	55

2.2. Electric Field Calculated Methods

This was conducted by building 1:1 two-dimensional models of the clean lightning protection insulator and one with an ice coating on its surface, to determine how the electric field distribution changed under the icing. The state of the ice layer can be divided into two cases, one with icicles and the other without icicles. As for the case with the icicle, seven icicles are set in the model, of which two are 33 mm long, two are 15 mm long, and three are 25 mm long. The icicle is a cone with a circular arc at the tip. FEM is also used to determine the electric potential and field distribution. The auto-generation method divides the computational domain into refined triangular meshes in this model [1]. For the air, a silicone rubber shed, icing layer, zinc oxide resistor, and relative dielectric constants of 1, 3.5, 81, and 600 were utilized in the electric field simulations.

$$\nabla \cdot \mathbf{D} = \rho,\ \mathbf{E} = -\nabla U,\ \mathbf{D} = \varepsilon_0 \varepsilon_1 \mathbf{E} \tag{1}$$

where ρ is charge density, U is the potential, $\mathbf{D}$ is the electric flux density, $\mathbf{E}$ is the electric field strength, and ε_0 is the absolute dielectric constant of the vacuum, ε_1 is the relative dielectric constant.

2.3. Flashover Test Devices and Methods

According to Figure 2, the testing was conducted at the Natural Icing Test Base. Figure 3 depicts the test circuit. The previous paper [22] describes the experimental site and test device parameters in detail. According to the IEC (International Electrotechnical Commission), the power supply satisfies all requirements.

As illustrated in Figure 4, the natural icing test technique is outlined. An initial step was to remove dirt and oil from the composite lightning protection insulators by cleaning them with Na_2PO_3. A solid layer pollution approach was used to pre-treat all lightning protection composite insulators to replicate the surface state of the insulators on running power lines. Insulator surfaces were contaminated with *NaCl* and diatomite to create a corrosive environment. The quality of each was measured, as in (2),

$$\begin{cases} M_{NaCl} = \frac{ESDD}{S} \\ M_{diatomite} = \frac{NSDD}{S} \end{cases} \tag{2}$$

where $M_{diatomite}$ is the mass of diatomite, M_{NaCl} is the mass of *NaCl*, and S is the surface area of composite insulators. *ESDD* and *NSDD* have equivalent salt deposit density and non-soluble deposit density, respectively.

Figure 2. Natural Icing Test Base.

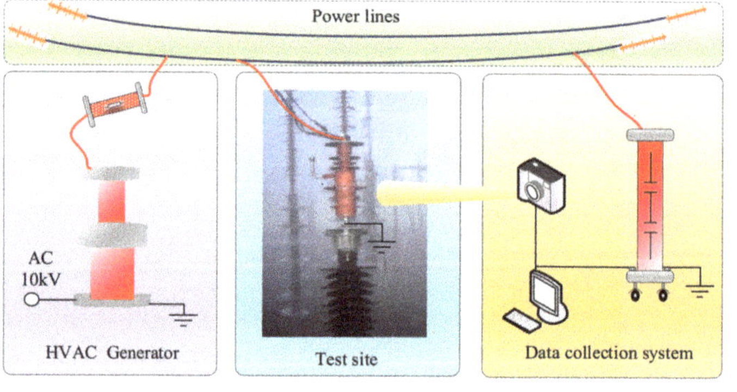

Figure 3. Schematic diagram of natural icing test.

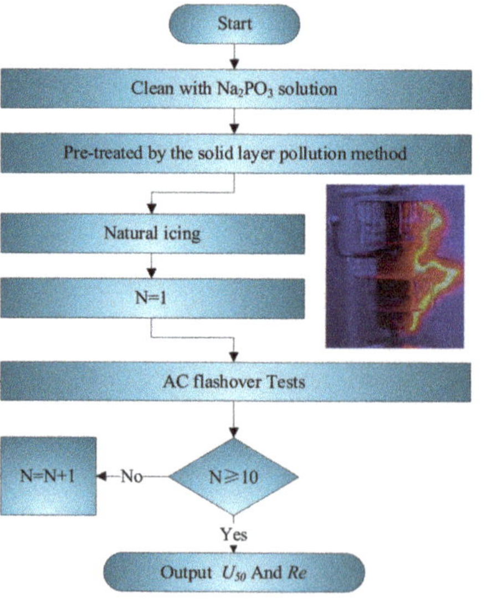

Figure 4. Test process.

Lightning protection composite insulators are installed on the post insulation after it has been contaminated. Then, the test was performed to determine the AC flashover voltage. Lightning protection composite insulators were subjected to at least 10 AC flashover tests under the same conditions. Therefore, it was possible to calculate U_f and Re (relative standard deviation error), as in,

$$U_f = \frac{\Sigma(U_i n_i)}{N} \tag{3}$$

$$Re = \sqrt{\left(\sum_{i=1}^{N}\left(U_i - U_f\right)^2\right)/(N-1)/U_f \times 100\%} \tag{4}$$

where U_i is the applied voltage, n_i is the number of tests, and N is "valid" tests.

3. Results

3.1. Electric Field Distribution of the Insulator

The calculation results of the electric field are shown in Figure 5, from which the following conclusions can be obtained.

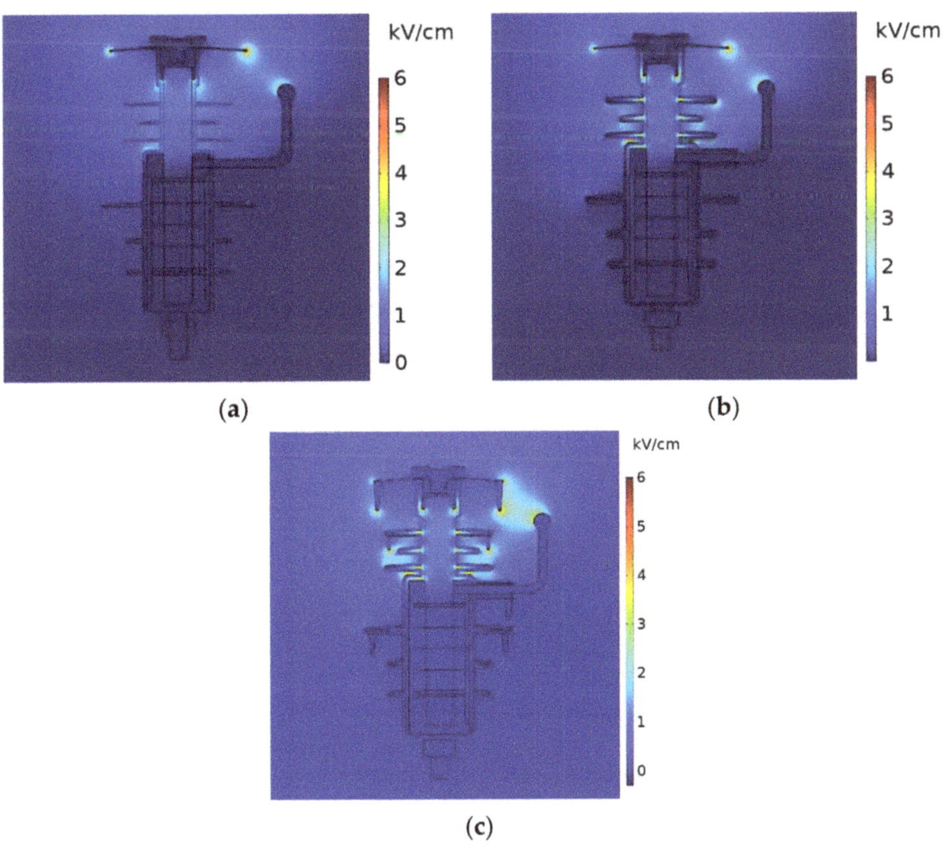

Figure 5. Electric field distribution of the novel lightning protection insulator under icing conditions: (**a**) Electric field distribution of a clean insulator; (**b**) Electric field distribution of an insulator with the ice layer without icicles; (**c**) Electric field distribution of an insulator with the ice layer with icicles.

(1) The electric field distribution of this novel insulator is different from that of a conventional insulator. The locations with sizeable electric fields are located in the insulation section, and the electric field in the arrester section is tiny.

(2) Both the ice layer and icicle affect the electric field distribution on the insulator surface. Specifically, the presence of ice and icicles on the insulator surface increases the electric field intensity on the insulator surface.

(3) When the ice layer is attached to the insulator surface, the icicle will further increase the electric field on the surface. In particular, the electric field of the air gap of the insulator is increased.

To more accurately describe the surface electric field of the insulator, the surface electric field curve along the left side of the insulator is counted in this paper, as shown in Figure 6. In addition, the electric field on the surface of the pin electrode is also counted, as shown in Figure 7.

As can be observed in Figure 6, the electric field peaks of the three conditions on the left side of the insulator surface are located at the creepage distance of 50 mm, and the electric field peaks are 2.17 kV/cm, 3.23 kV/cm, and 3.17 kV/cm, respectively. Compared with the dry conditions, when there is an ice layer and icicle, it increases by 48.85% and 46.08%, respectively. In addition, when there is an icicle on the surface ice layer, an electric field peak (2.78 kV/cm) appears at the tip of the icicle, as shown in Figure 6, at the creepage distance of 125 mm.

As observed in Figure 7, electric field peaks on the surface of the pin electrode vary. The electric field peaks of the conditions with icicles are the largest, reaching 1.92 kV/cm, followed by the dry and icing conditions (1.45 kV/cm and 1.52 kV/cm). Compared with the dry conditions, when there is an ice layer and icicle, it increases by 4.83% and 32.41%, respectively.

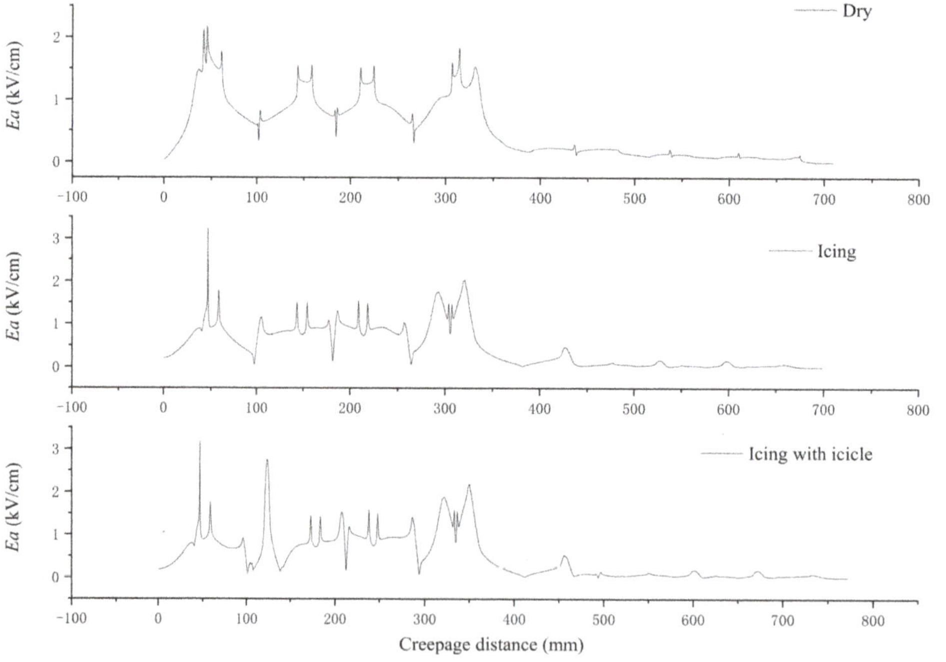

Figure 6. Electric field distribution of the left side.

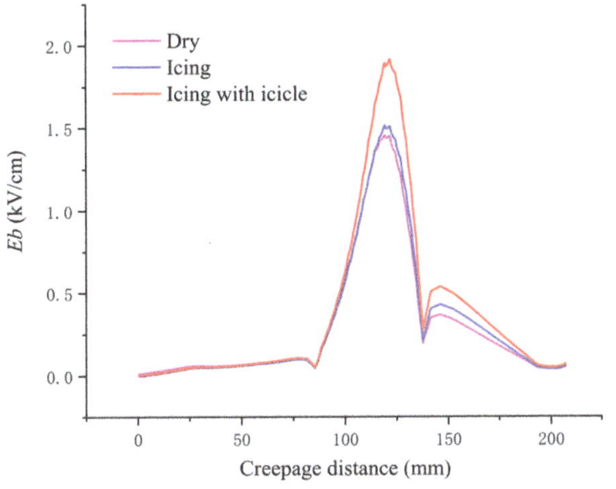

Figure 7. Electric field distribution of the pin electrode.

In sum, when the insulator surface is covered with ice, there is an increase in the electric field along the surface and pin electrode. With icicles in particular, there is a prominent electric field peak at the tip of the icicle. In addition, it can be seen from the calculation results that icing affects the electrical performance of the insulators; therefore, the icing flashover test of insulators is carried out in this paper.

3.2. AC Breakdown Characteristics under Icing Conditions

In the AC flashover test, as is shown in Figure 8, there are three types of arc paths. Figure 8a shows the flashover arc path when ESDD is 0.01–0.02 mg/cm^2 (low pollution). The flashover arc path is between the disk and pin metal gap of the insulation section. Figure 8b shows the flashover arc path when ESDD is 0.03–0.05 mg/cm^2 (medium pollution). The flashover arc path is along the surface of the silicone rubber umbrella skirt of the insulation section. Figure 8c shows the flashover arc path when ESDD is 0.07–0.15 mg/cm^2 (heavy pollution). The flashover arc path is through the whole lightning protection composite insulator, which means that the arrester section composed of zinc oxide resistor did not work in the flashover process.

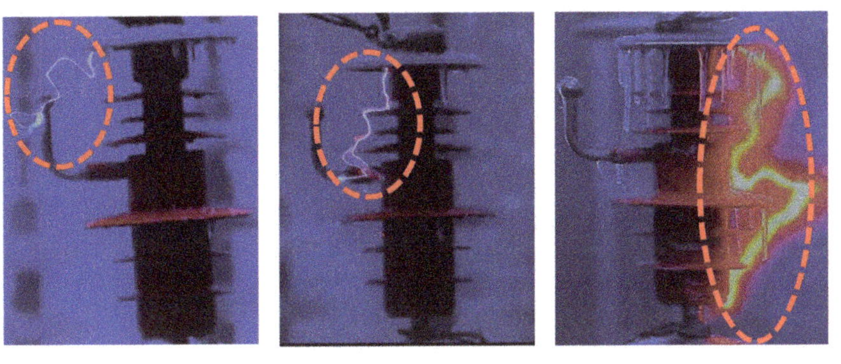

(**a**) (**b**) (**c**)

Figure 8. Different flashover arc paths: (**a**) The flashover arc path when ESDD is 0.01–0.02 mg/cm^2; (**b**) The flashover arc path when ESDD is 0.03–0.05 mg/cm^2; (**c**) The flashover arc path when ESDD is 0.07–0.15 mg/cm^2.

When the insulator surface is covered with ice, the flashover voltage under different pollution degrees is shown in Figure 9. As observed in Figure 9, the following conclusions can be obtained.

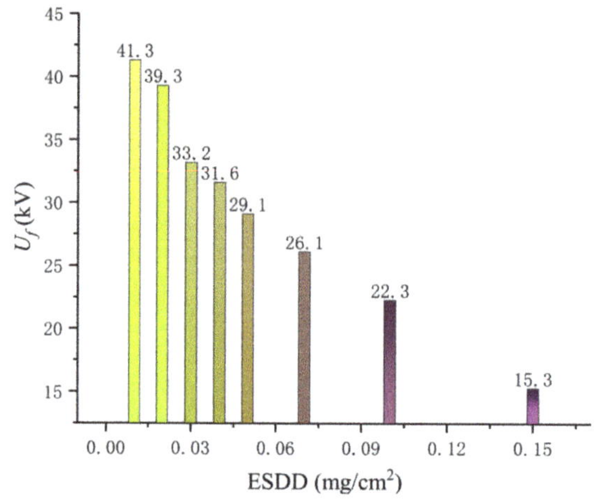

Figure 9. AC flashover test results.

(1) The U_f varies greatly in different pollution levels under icing conditions. The value for U_f (ESDD = 0.01 mg/cm^2) ranks the highest, standing at 41.3 kV. On the contrary, U_f (ESDD = 0.15 mg/cm^2) is the lowest with nearly 15.3 kV. The maximum flashover voltage is 2.70 times more than the minimum. What is more, U_f shows a downward trend with the increase of pollution levels.

(2) With the increase of pollution level, the flashover voltage does not decrease linearly. There is a negative power function relationship between insulator ESDD and flashover voltage under icing conditions for traditional insulators. However, due to the diversification of arc paths during the flashover, the relationship between insulator ESDD and flashover voltage should be discussed again for this novel lightning protection insulator.

In addition, the pollution flashover test of the novel insulator (without the ice layer) was carried out. The tested pollution flashover voltage is 36.9 kV, which is 3.7 kV higher than the insulator with the ice layer (ESDD = 0.03 mg/cm^2). The simulation results show that the electric field at the creepage distance of the insulator with the ice layer is more extensive. Therefore, the electric field of the insulator with the ice layer is easier to reach the air breakdown electric field, resulting in the decrease of the flashover voltage. This tested result is consistent with the findings of the electric field simulation.

4. Discussion

Under the conditions of pollution and icing, the accurate empirical formula of flashover voltage is essential for the design of external insulation. Four fitting methods are proposed in this paper about the calculation formula of flashover voltage of conventional insulators. They are negative power function, quadratic function, negative power function with the constant term, and piecewise function. The functional form is shown in Equations (5)–(8).

$$U_f = a_1 \times ESDD^{b_1} \quad (5)$$

$$U_f = a_2 \times ESDD^2 + b_2 \times ESDD + c_2 \quad (6)$$

$$U_f = a_3 \times ESDD^{b_3} + c_3 \quad (7)$$

$$U_f = \begin{cases} a_4 \times ESDD^{b_4} & 0.03 \le ESDD \le 0.05 \\ a_5 \times ESDD^{b_5} & 0.07 \le ESDD \le 0.15 \end{cases} \quad (8)$$

The fitting results between U_f and ESDD, according to Equations (5)–(8), are shown in Figure 10. Figure 10a–d are the fitting results using negative power function, quadratic function, negative power function with the constant term, and piecewise function, respectively. It can be seen from the fitting results in Figure 10 that the fitting method (negative power function) of the conventional insulator has the lowest goodness of fit (R^2 = 0.9037). The fitting methods of a quadratic function and negative power function with the constant term are better, and the goodness of fit is 0.9798 and 0.9885, respectively. As for the piecewise function, the goodness of fit is also acceptable, which are 0.9537 and 0.9583, respectively.

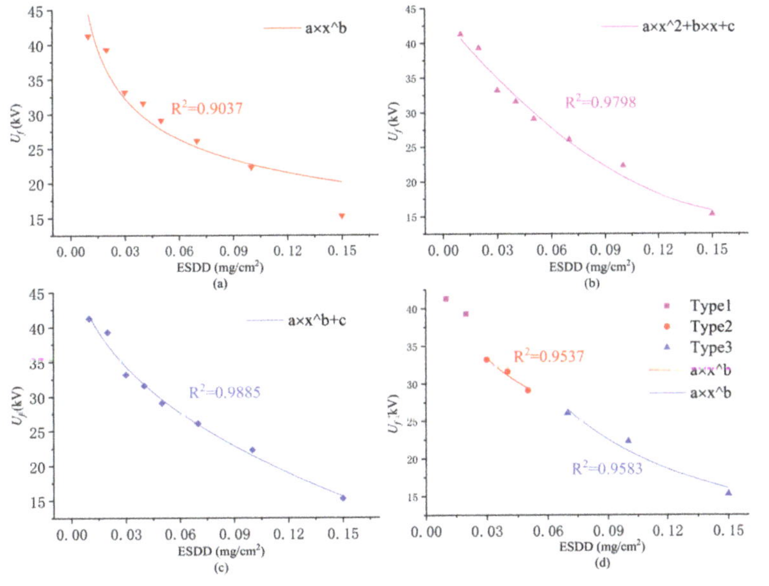

Figure 10. Fitting results of flashover voltage under different pollution degrees: (**a**) The fitting results using negative power function; (**b**) The fitting results using quadratic function; (**c**) The fitting results using negative power function with the constant term; (**d**) The fitting results using piecewise function.

According to the above fitting results, the traditional negative power function is not recommended to calculate the flashover voltage of novel insulators under icing conditions in different pollution levels in this paper. In addition, although the fitting result of the piecewise function is good, it has high requirements for data points, which increases the number of tests required for fitting data. When there are few data points, for example, ESDD is between 0.01 and 0.02, and fitting cannot be performed.

In this paper, the quadratic function and negative power function with the constant term are recommended as empirical formulas for calculating the flashover voltage of novel insulators under icing conditions in different pollution levels. When the icing thickness is between 2.6–2.9 mm, the empirical formula for flashover voltage recommended in this paper is as follows:

$$U_f = 888.7 \times ESDD^2 - 319.3 \times ESDD + 43.8 \quad (9)$$

$$U_f = -84.18 \times ESDD^{0.39} + 55.63 \quad (10)$$

5. Conclusions

Electric field distribution and AC breakdown characteristics of polluted novel lightning protection insulators under icing conditions are investigated. Compared with traditional insulators, we have made some new discoveries. From the findings, the following conclusion can be drawn.

(1) The electric field distribution of this novel insulator is different from that of a conventional insulator. When the insulator surface is covered with ice, there is an increase in the electric field along the surface and pin electrode. Especially with icicles, there is a prominent electric field peak at the tip of the icicle. Compared with dry conditions, when there is an ice layer and icicle, electric field peaks increase by 48.85% and 46.08%, respectively.

(2) Three arc paths in different pollution levels were discovered in the AC flashover test. The arc paths are related to ESDD under icing conditions.

(3) The U_f varies significantly in different pollution levels under icing conditions. Specifically, U_f shows a downward trend with the increase of pollution levels. The maximum flashover voltage is 2.70 times more than the minimum.

(4) This paper puts forward two more accurate empirical formulas. Specifically, after comparing the goodness of fit of different functions, the quadratic function and negative power function with the constant term are recommended as empirical formulas for calculating the flashover voltage of novel insulators under icing conditions.

Author Contributions: Conceptualization, J.L., J.H. and Z.F.; methodology, Z.Z.; investigation, X.Q.; resources, J.H.; writing—original draft preparation, X.Q.; writing—review and editing, Z.Z.; software, X.Q.; validation, Z.Z.; supervision, Z.Z. All authors have read and agreed to the published version of the manuscript.

Funding: This research was funded by State Key Laboratory of Disaster Prevention & Reduction for Power Grid Transmission and Distribution Equipment, grant number [SGHNFZ00ZHJLJS1900208].

Acknowledgments: The authors would like to express their gratitude to Chongqing University for their assistance with this research.

Conflicts of Interest: The authors declare no conflict of interest.

References

1. Qiao, X.; Zhang, Z.; Raji, S.; Jiang, X.; Hu, J.; Fang, Z. AC Breakdown Characteristics of Polluted 10-kV Post Insulator with Concentric Externally Gapped Line Arrester. *IEEE Trans. Power Deliv.* **2021**. [CrossRef]
2. Jiang, Z.; Wu, W.; Wang, B.; Xie, P.; Li, H.; Lin, F. Design and test of 500-kV lightning protection insulator. *IEEE Access* **2019**, *7*, 135957–135963. [CrossRef]
3. Wang, B.; Lu, J.; Fang, Z.; Jiang, Z.; Hu, J. Development of antithunder composite insulator for distribution line. *IEEJ Trans. Electr. Electron. Eng.* **2020**, *15*, 100–107. [CrossRef]
4. Zhang, D.; Chen, S. Intelligent Recognition of Insulator Contamination Grade Based on the Deep Learning of Ultraviolet Discharge Image Information. *Energies* **2020**, *13*, 5221. [CrossRef]
5. Zhang, D.; Xu, H.; Liu, J.; Yang, C.; Huang, X.; Zhang, Z.; Jiang, X. Research on the Non-Contact Pollution Monitoring Method of Composite Insulator Based on Space Electric Field. *Energies* **2021**, *14*, 2116. [CrossRef]
6. Zhang, M.; Wang, R.; Li, L.; Jiang, Y. Size Distribution of Contamination Particulate on Porcelain Insulators. *Coatings* **2018**, *8*, 339. [CrossRef]
7. Li, X.; Zhou, M.; Luo, Y.; Wang, G.; Jia, L. Effect of Ice Shedding on Discharge Characteristics of an Ice-Covered Insulator String during AC Flashover. *Energies* **2018**, *11*, 2440. [CrossRef]
8. Liu, Y.; Li, Q.; Farzaneh, M.; Du, B.X. Image Characteristic Extraction of Ice-Covered Outdoor Insulator for Monitoring Icing Degree. *Energies* **2020**, *13*, 5305. [CrossRef]
9. Xu, J.; Yin, F.; Li, L.; Wen, Q.; Wang, H.; Liu, S.; Jia, Z.; Farzaneh, M. Wet Snow Flashover Characteristics of 500-kV AC Insulator Strings with Different Arrangements. *Appl. Sci.* **2019**, *9*, 930. [CrossRef]
10. Farzaneh, M. Research activities related to insulator flashover and selection criteria under icing conditions. In Proceedings of the 2006 IEEE Power Engineering Society General Meeting, Montreal, QC, Canada, 18–22 June 2006; p. 2.
11. Porkar, B.; Farzaneh, M. In DC flashover performance of an inclined insulator string under heavy icing conditions. In Proceedings of the 2015 IEEE 11th International Conference on the Properties and Applications of Dielectric Materials (ICPADM), Sydney, Australia, 19–22 July 2015; pp. 648–651.

12. Xu, T.; Nan, J.; Wan, X. In Research on icing flashover characteristic of different shed type composite insulators. In Proceedings of the 2014 International Conference on Information Science, Electronics and Electrical Engineering, Sapporo, Japan, 26–28 April 2014; pp. 191–194.
13. Liu, Y.; Farzaneh, M.; Du, B.X. Nonlinear characteristics of leakage current for flashover monitoring of ice-covered suspension insulators. *IEEE Trans. Dielectr. Electr. Insul.* **2016**, *23*, 1242–1250. [CrossRef]
14. Chihani, T.; Mekhaldi, A.; Beroual, A.; Teguar, M.; Madjoudj, D. Model for polluted insulator flashover under AC or DC voltage. *IEEE Trans. Dielectr. Electr. Insul.* **2018**, *25*, 614–622. [CrossRef]
15. Lan, L.; Zhang, G.; Wang, Y.; Wen, X.; Wang, W.; Pei, H. The Influence of Natural Contamination on Pollution Flashover Voltage Waveform of Porcelain Insulators in Heavily Polluted Area. *IEEE Access* **2019**, *7*, 121395–121406. [CrossRef]
16. Maadjoudj, D.; Mekhaldi, A.; Teguar, M. Flashover process and leakage current characteristics of insulator model under desert pollution. *IEEE Trans. Dielectr. Electr. Insul.* **2018**, *25*, 2296–2304. [CrossRef]
17. de Santos, H.; Sanz-Bobi, M.Á. A Cumulative Pollution Index for the Estimation of the Leakage Current on Insulator Strings. *IEEE Trans. Power Deliv.* **2020**, *35*, 2438–2446. [CrossRef]
18. Qiao, X.; Zhang, Z.; Jiang, X.; Sundararajan, R.; Ma, X.; Li, X. AC failure voltage of iced and contaminated composite insulators in different natural environments. *Int. J. Electr. Power Energy Syst.* **2020**, *120*, 105993. [CrossRef]
19. Li, P.; Fan, J.; Li, W.; Su, Z.; Zhou, J. Flashover Performance of HVDC Iced Insulator Strings. *IEEE Trans. Dielectr. Electr. Insul.* **2007**, *14*, 1334–1338. [CrossRef]
20. Jiang, X.; Chao, Y.; Zhang, Z.; Hu, J.; Shu, L. DC flashover performance and effect of sheds configuration on polluted and ice–covered composite insulators at low atmospheric pressure. *IEEE Trans. Dielectr. Electr. Insul.* **2011**, *18*, 97–105. [CrossRef]
21. Hu, Q.; Yuan, W.; Shu, L.; Jiang, X.; Wang, S. Effects of electric field distribution on icing and flashover performance of 220 kV composite insulators. *IEEE Trans. Dielectr. Electr. Insul.* **2014**, *21*, 2181–2189.
22. Zhou, X.; Xiong, J. Computation of a kind of electric field distribution of 10kV composite insulator coated by ice. In Proceedings of the 2008 China International Conference on Electricity Distribution, Guangzhou, China, 10–13 December 2008; pp. 1–5.
23. Lu, J.; Hu, J.; Fang, Z.; Qiao, X.; Zhang, Z.; Jiang, X. AC Flashover Performance and Insulation Coordination of Novel Lightning Protection Composite Insulator. In Proceedings of the 2021 International Conference on Electrical Materials and Power Equipment (ICEMPE), Chongqing, China, 11–15 April 2021; pp. 1–4.
24. Qiao, X.; Hu, J.; Sundararajan, R.; Zhang, Z.; Fang, Z.; Jiang, X. AC Flashover Performance of 10 kV Novel Composite Insulators under Icing Conditions in Natural Environment. In Proceedings of the 2020 IEEE Conference on Electrical Insulation and Dielectric Phenomena (CEIDP), East Rutherford, NJ, USA, 18–30 October 2020; pp. 172–175.
25. Qiao, X.; Zhang, Z.; Sundararajan, R.; Jiang, X.; Hu, J.; Fang, Z. The failure arc paths of the novel device combining an arrester and an insulator under different pollution levels. *Int. J. Electr. Power Energy Syst.* **2021**, *125*, 106549. [CrossRef]

energies

Article

Frequency Resolved Partial Discharges Based on Spectral Pulse Counting

Anderson J. C. Sena [1,†], Rodrigo M. S. de Oliveira [1,*,†] and Júlio A. S. do Nascimento [1,2,†]

1 Electrical Engineering Graduate Program (PPGEE), Federal University of Pará, Belém 66075-110, Brazil; ajcsena@gmail.com (A.C.J.S.); salhebjulio@gmail.com (J.A.S.d.N.)
2 Eletrobras Eletronorte, Rodovia Arthur Bernardes 2175, Belém 66115-000, Brazil
* Correspondence: rmso@ufpa.br
† These authors contributed equally to this work.

Abstract: A partial discharge (PD) classification methodology based counting PD pulses in the spectral domain is proposed and presented in this paper. The spectral counting data are processed using the proposed PD Spectral Pulse Counting Mapping technique (PD-SPCM), which leads to a Frequency-Resolved Partial Discharges (FRPD) map. The proposed map is then used for PD detection and classification. In this work, corona and slot FRPDs are presented in frequency bands up to 500 MHz, obtained from laboratory measurements performed using two hydro-generator stator bars. The electromagnetic signals from the PDs were captured using a patch antenna designed for this purpose and a spectral analyzer. The corona and slot PDs were chosen because one can be mistakenly classified as the other because they may present similar Phase Resolved PD (PRPD) maps and may occupy shared spectral bands. Furthermore, corona and slot PDs can occur concurrently. The obtained results show that the corona and slot PDs can be properly identified using the developed methodology, even when they occur simultaneously. This is possible because, as it is experimentally demonstrated, corona and slot PDs have appreciable levels of spectral pulse counting in particular bands of the frequency spectrum.

Keywords: hydro-generators; partial discharges; PD Spectral Pulse Counting Map (PD-SPCM); microstrip antenna; UHF method; Frequency Resolved Partial Discharges (FRPD)

Citation: Sena, A.J.C.; de Oliveira, R.M.S.; do Nascimento, J.A.S. Frequency Resolved Partial Discharges Based on Spectral Pulse Counting. *Energies* **2021**, *14*, 6864. https://doi.org/10.3390/en14216864

Academic Editor: Zhijin Zhang and Hualong Zheng

Received: 5 September 2021
Accepted: 24 September 2021
Published: 20 October 2021

1. Introduction

Rises in amplitude and occurrence rates of partial discharges (PDs) in insulation dielectrics of hydro-generators stator bars indicate electrical problems that can compromise the operational safety of this kind of high-voltage equipment [1–3]. If the evolution of PD is not investigated, generation outages may take place, along with the consequent huge losses for society, which depends on electric power.

High sinusoidal voltage is produced by a generator, which induces a high electric field in the rotating machine and surroundings, including the air, the electrical insulation materials and surfaces of the generator structure. Therefore, PDs can be associated with several problems, including dielectric rupture of gases confined in stator bar insulation internal voids, the degradation of the semiconductive layer due, for instance, to mechanical or electrical stresses, and the presence of undesired objects in the stator winding [2]. Discharge activities evolve gradually over the years, which can be identified and quantified by measuring acoustic emissions [4,5], electrical PD signals [6,7], and electromagnetic radiation [8]. If the electrical problems PDs indicate are not monitored and treated, they can evolve to reach a state of complete unavailability of the hydro-generator [2,9–11].

In general, several features have been used to extract information from PD signals, such as Weibull probability function [12], cross-wavelet transforms [13], rough sets [14], time-of-pulse travel between sensors, and time-frequency maps [15]. A feature frequently used to identify PD sources from hydro-generators is the PRPD (Phase-Resolved Partial

Discharge) [16,17], which contains PD peaks distributed over the generator 50 Hz or 60 Hz power signal phase, such that the peaks are accumulated over time (at least during 30 s). The patterns seen in PRPD maps are standardized in [6] for each type of hydro-generator PD.

In certain circumstances, however, PRPDs may provide limited classification possibilities [7]. For instance, if slot and corona PDs occur simultaneously, their PRPD clouds will be intertwined or superimposed because the phases they produce are coincident [6,18]. Thus, their PRPDs would overlap, making it difficult to properly identify them only by inspecting the PRPD cloud; i.e., it would be tricky or virtually impossible to determine the presence of both faults or one of them. Furthermore, although specific PD sources are manifested in specific frequency ranges [19], when two faults are simultaneously occurring, their frequency bands may intersect such that one is contained within the spectrum of the other. This is the case of simultaneous slot and corona PDs, in which it is also not possible to distinguish the PD sources just by inspecting the resulting spectrum (or PRPD).

In this paper, a new methodology is proposed in which spectral PD count is introduced, producing a novel tool for statistical analysis of PDs—the Partial Discharge Spectral Pulse Counting Mapping technique (PD-SPCM)—which leads to a Frequency Resolved Partial Discharges (FRPD) map. The proposed method is based on spectral counting of PD pulses, with which one is able to identify the types of simultaneous faults producing the PDs, even if their spectra are in the same frequency bands and their PRPD clouds completely overlap each other. FRPD has been validated experimentally in this work.

2. Slot and Corona Discharges

Slot discharges occur when the semiconductive coating of the bar has high resistance or when friction on the coating, due to machine vibrations and wedge looseness, leaves the high-voltage bare insulation, causing a metallic grounded core [19]. Because of the bare metallic surface of the lamination, the initial availability of free electrons is sufficient to start an electron avalanche in the air gap, regardless of the voltage polarity. After some time, intense slot discharges, along with ozone and nitric acid produced as byproducts, starts to attack both the insulation and the core, causing oxidation. This can affect the availability of free electrons. Thus, the typical asymmetry on the slot PRPD map [6] becomes more pronounced a few hours after the initial slot PDs [2,19].

Corona discharges occur due air ionization produced by high electric fields, originating from the bar surface, usually at the junction between the semi-conductive material (used in the slot) and the stress-grading material (used in the winding output terminals) [19]. This type of activity causes degradation directly at the junction, leaving a deposit of white powder. This activity can evolve slowly over the years. In [19], it is shown experimentally that corona PRPD can change due to degradation of the junction or voltage increases.

In terms of spectral bands, it was determined experimentally in [19] that spectra of corona and slot PDs reaches up to 500 MHz. This means that, although slot PDs tend to produce higher amplitude levels [6], it may be difficult to distinguish slot discharges when they are occurring simultaneously with corona PDs simply by using the measured spectrum (as well as PRPDs [6]). PD energy was registered for frequencies over 1.5 GHz in [20].

Physics of Corona and Slot PDs

Considering a stator bar of a hydro-generator subjected to the 60 Hz voltage produced, the higher the produced voltage, the greater the strength of the electric field in the semiconductive coating and at the junction among the slots and the winding output terminals, causing ionization of the surrounding air. This ionization occurs because, due the mentioned defects on the surfaces, the electric field rises to levels sufficiently high to extract electrons from the valence shells of gas atoms. Each removed electron is accelerated due to the action of the high electric field, causing collisions with the electrons of the other electrically neutral atoms of the gas, generating detachments of more electrons, and so on, leading to the so-called avalanche effect [21].

The phenomena just described produce a plasma, in which the ratio between the electron drift velocity v_d and the electric field strength E is represented as the electron mobility μ_e [22], which is thus given by

$$\mu_e = v_d/E. \quad (1)$$

In addition, the plasma electrical conductivity σ can be obtained by

$$\sigma = e \cdot n_e \cdot \mu_e, \quad (2)$$

in which n_e is the carrier concentration [22] and e is the electron charge. Notice that, as is given by (1), a rise in E leads to reduction of electronic mobility μ_e because the number of free electrons also increases.

In short, during the first stages of the ionization process, the conductivity given by (2) increases because the increase in carrier concentration n_e is dominant over the reduction in μ_e. Thus, while the electric field keeps rising, the number of free electrons capable of ionizing new atoms increases. During this initial ionization period, the current i, in the ionized region, will increase exponentially over time [22,23]. The voltage between the electrodes decreases because of the rapid increase in electrical conductivity of ionized air between the anode and the cathode. This behavior is consistent with what is predicted by Maxwell's equations; i.e., the electric field decreases as conductivity increases [24]. As a consequence, the ohmic resistance R_c of the circuit shown in Figure 1 is subjected to increasing voltage due to the progressive increase in the current with σ [22].

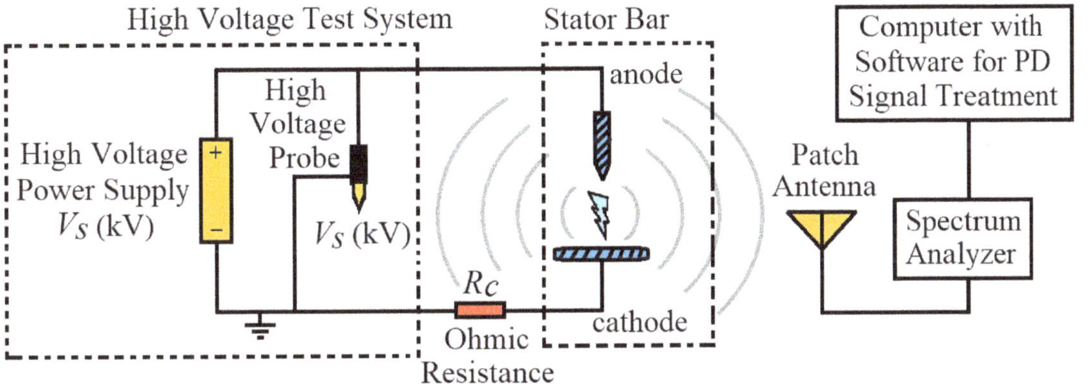

Figure 1. Representation of the discharge circuit proposed in [25]. The radiated electromagnetic field measurement setup (at right) is defined in this work.

Thus, the ionization process begins with intense electric field around the positive electrode, producing a distribution of positive ions around the anode and leading to a phenomenon called corona discharge (see Figure 2a). At this stage, the electric field intensity is high at the anode and decays to zero as one approaches the cathode (Figure 2b) [26]. As ionization continues due to the intense electric field, neutral atoms lose electrons that move towards the anode, causing the field to weaken in the vicinity of that electrode. This mechanism causes the field to increase around the regions farthest from the anode, around the tip of the just-ionized path, favoring the propagation of the ionization region towards the cathode (Figure 2c). As a result, the number of free electrons progressively increases due to the avalanche effect, leading to the consequent increase in current in the channel due to the increase in its conductivity [26]. The region around the ionized channel tip, called the streamer head, propagates towards the cathode as long as ions and electrons emerge more quickly than the weak electrical field of the conducting trail above the streamer head would be capable of. The conducting trail, which is referred to as streamer channel [26], electrically connects the streamer head to the anode.

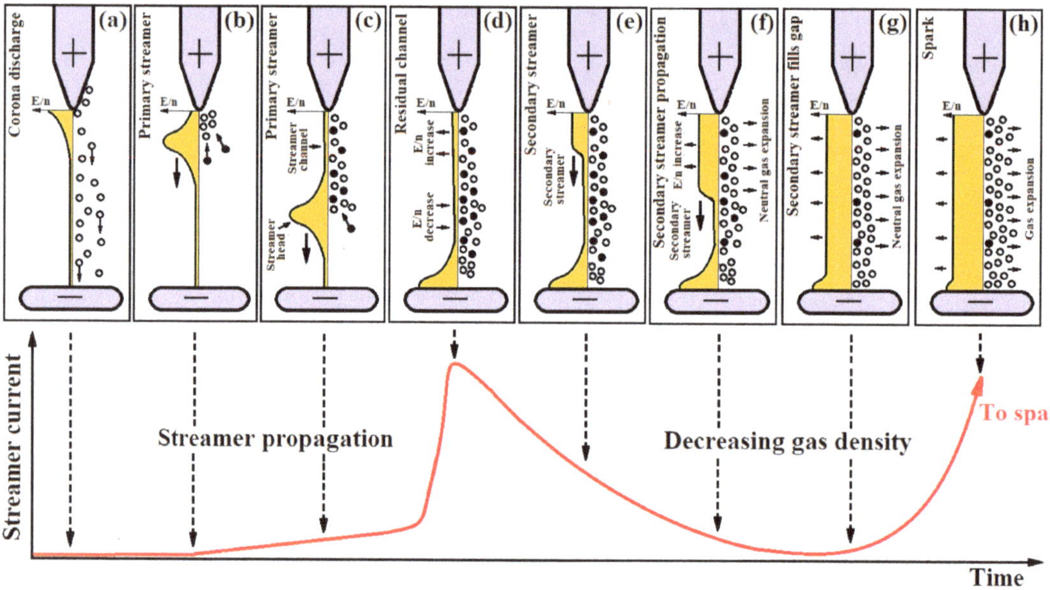

Figure 2. Steps of primary and secondary streamers progressive developments [26]. The black circles represent electrons, while the white circles symbolize positively charged ions.

It should be noted that if the voltage supply level is not sufficient to maintain a high enough electric field in the gap between the electrodes, the streamer head is extinguished before reaching the cathode (as in corona dischargess). Otherwise, the streamer head moves quickly across the channel, closing the circuit between the electrodes (similarly to slot discharges [6]). As the streamer head reaches a distance of approximately one millimeter from the cathode (Figure 2d), a large number of electrons are emitted from the cathode due to intense ultraviolet radiation from the streamer head. This process leads to increases of thousands of volts of channel voltage during about one nanosecond, at least doubling the channel's average field during the process. The number of free electrons increases rapidly, leading to the emergence of a region with great ionization capability that propagates with high speed along the streamer channel towards the anode. This phenomenon favors equipotentialization along the residual streamer channel and produces the so-called return stroke, causing the current to reach its maximum value in this stage [26].

The electrical connection between the electrodes (Figure 2d), due to the ionized air filament, gives rise to a process leading to the so-called spark breakdown discharge [26]. This attachment instability precedes the formation of a secondary streamer, which consists of a progressive enlargement of the channel cross-section area due to a rise in temperature associated to the fast current increase produced by the high voltage from the source fully applied between the two ends of the channel. This process evolves into an equipotentialization state along the residual streamer channel (see Figure 2e–g). The secondary streamer is visually observed as a partial re-illumination of the streamer channel, usually starting at the anode at almost the same time that the streamer head makes contact with the cathode. The secondary streamer starts and moves in the channel created by the previous ionization processes, and although the increase in the channel cross-section area favors reducing n_e and the electrical conductivity σ as a consequence, the concurrent exponential decay of the current in the channel is also significantly dependent on the attachment of electrons to oxygen molecules [26].

When the excitation voltage is sufficient and fully applied across the ionized channel, the gas density n is sufficiently reduced due to the expansion of the gas provided by the heating of the channel, leading to increase in E/n, thus allowing the propagation of the secondary streamer (Figure 2f). The continuous gas expansion and the progressive E/n increase allows the secondary streamer to cross the gap between the electrodes, forming a quasi-uniformly ionized domain (Figure 2g). Since E/n reaches a certain magnitude, an abrupt new increase in the channel current towards a spark breakdown can be observed (see Figure 2h) [26].

3. Discharges Due Sinusoidal Voltage Excitation

In a hydroelectric generator, stator bars are often subjected to 50 Hz or 60 Hz sinusoidal electric fields. As illustrated in Figure 3, since voltage $V(t)$ is changing over time, stator discharges may or may not emerge [1], depending on present phase of the sinusoidal cycle. Discharge amplitude and duration may vary over the cycle with direct dependence on $V(t)$. Formation of corona and slot discharges are governed by the physical mechanisms previously described.

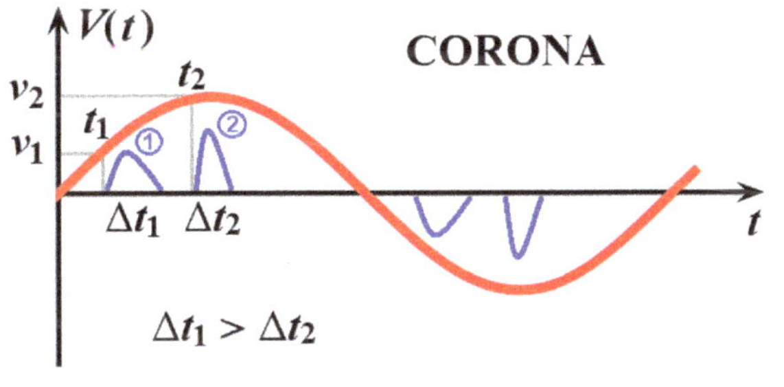

(a)

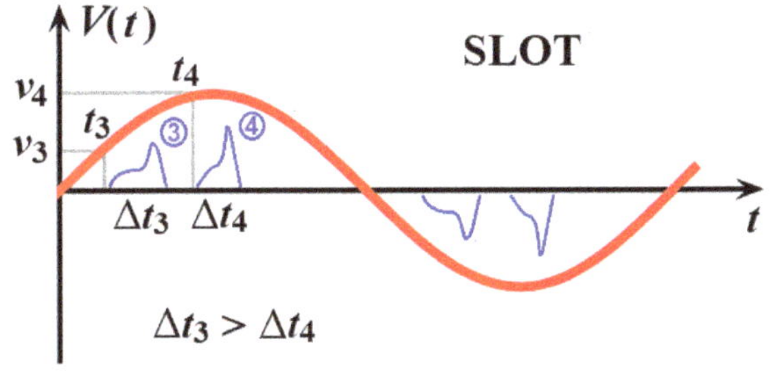

(b)

Figure 3. Illustrations of partial discharges in a sinusoidal voltage signal: (**a**) corona and (**b**) slot.

As voltage magnitude increases, electric field also rises in the plasma channel and, as a consequence, the discharge current rise time is reduced before the spark stage takes place. Additionally, the streamer propagation speed increases [22,25–28]. Deionization takes place, as previously described. Then, with the increase in the generated voltage, the electric field rises again, and, since E exceeds E_c, a new PD will occur at t_2, as shown in Figure 3a. Thus, during the evolution of the positive semicycle of the generated sinusoidal voltage, multiple PDs may occur with different amplitudes and time spams, i.e., the increase in the sinusoidal excitation voltage makes the partial discharge peaks occur in progressively shorter time intervals. As shown in Figure 3a, $\Delta t_1 > \Delta t_2$ [22,25,27]. In short, partial discharge pulses present increasing spectral bands and amplitudes as $V(t)$ increases in magnitude.

Notice that when corona discharges are developed, there is no electrical contact between electrodes due to the discharge channel because this type of discharge occurs between an energized electrode and air [6,29]. However, ionized arcs electrically connect electrodes during slot discharges [6]. For this reason, slot discharges are associated with higher discharge currents and broader frequency spectra because, once the streamer head reaches the cathode, as previously discussed, current rises more rapidly [25], i.e., steep slopes are seen in slot PD signals (occurring as the discharge channel electrically connects anode and cathode). This phenomenon increases frequencies composing the associated fields as a consequence. This is why slot PD functions have been depicted in Figure 3b differently (with the steep slopes) from corona PDs seen in Figure 3a. These phenomena were experimentally observed in [25] using the discharge circuit illustrated in Figure 1 (excluding the antenna receiving setup).

Figure 4a shows the discharge currents obtained in [25] using 8.2, 7.2, and 6.2 kV voltage excitation levels. Figure 4b–d shows wavelet transforms, calculated in this work, of PD current pulses in Figure 4a. Note that the higher the excitation voltage, the higher are the discharge current pulses and the shorter the time intervals necessary for current to reach its peak. Consequently, their spectral bands become larger, especially during the time periods of steep current slopes, as can be easily seen by comparing the wavelet transforms to each other and by observing their respective current signals evolving over time in Figure 4. Wavelet transforms $W(x(t), a, t')$ in Figure 4 were calculated by numerically evaluating $W(x(t), a, t') = \frac{1}{\sqrt{a}} \int_{-\infty}^{\infty} \psi(\frac{t-t'}{a}) x(t) \mathrm{d}t$, in which t is time, $a = 1/f$, f is frequency, t' is the time shift factor, and $\psi(t) = \exp(-\frac{1}{2}t^2)\cos(2\pi t)$ (the real-valued Morlet wavelet) [30].

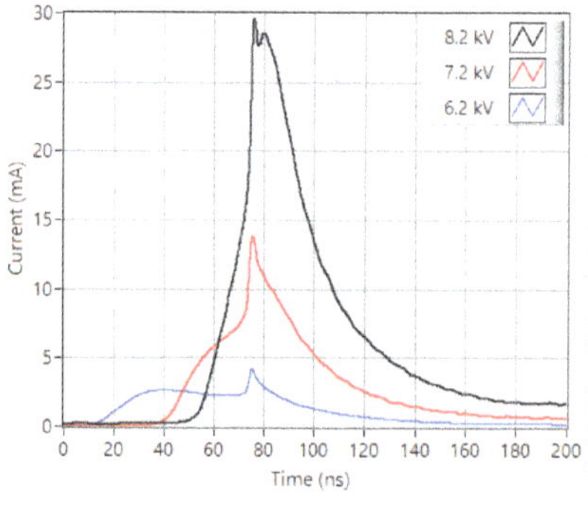

(a)

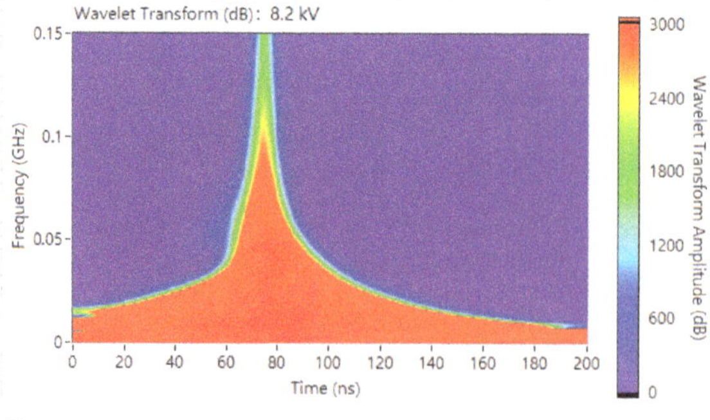

(b)

Figure 4. *Cont.*

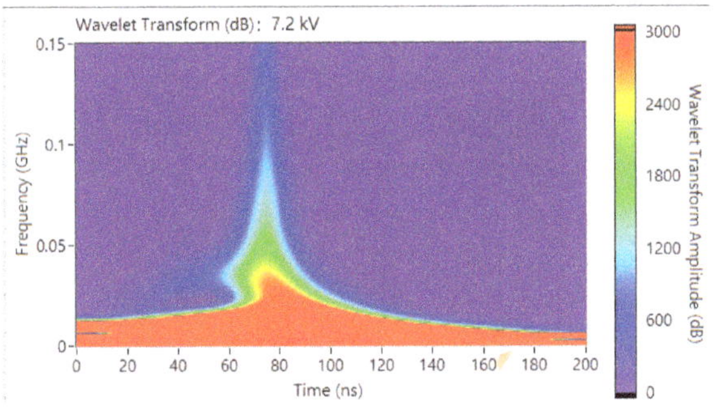

(c)

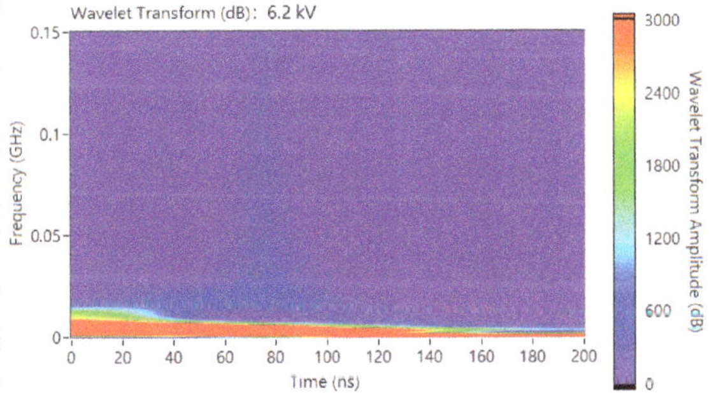

(d)

Figure 4. Discharge current pulses: (**a**) evolution of waveforms over time (8.2 kV, 7.2 kV and 6.2 kV pulses) [25], (**b**) wavelet transform of the 8.2 kV pulse, (**c**) wavelet transform of the 7.2 kV pulse, and (**d**) wavelet transform of the 6.2 kV pulse.

4. Frequency Resolved Partial Discharges (FRPD)

In this section, basic concepts necessary for defining the proposed Partial Discharge Spectral Pulse Counting mapping technique, which leads to FRPD maps, are discussed and experimentally demonstrated. Furthermore, computer implementation of the proposed method is also examined.

4.1. FRPD: Basic Concepts

Since generated voltage sinusoidally oscillates in hydroelectric generators, amplitude and especially PD duration are not constant over the generator's 50 Hz or 60 Hz cycle. Because pulse duration directly affects spectral responses, PD count can be analyzed as a function of time and frequency, as shown by the experimental results obtained in this work, as shown in Figure 5. The spectrograms in Figure 5 show corona and slot PD pulses over the time-frequency domain, each registered within 8 s over a band defined between 20 MHz and 160 MHz. The graphs show that the spectral PD pulses occur with different repetition levels when different frequencies are analyzed over time. This way, we have performed PD pulse count over time and frequency domains using data extracted from an spectrum analyzer employed to perform experimental measurements. The spectrum analyzer is connected to a patch antenna designed in this work for receiving high frequency PD radiated signals. The antenna design is detailed in Appendix A.

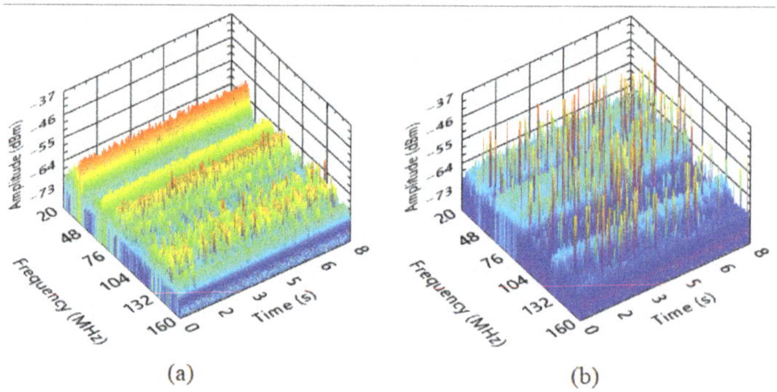

Figure 5. Spectrograms of (**a**) corona and (**b**) slot partial discharges. The spectrograms were registered in this work within eight seconds between 20 MHz and 160 MHz.

Based on this experimental confirmation of the theoretical prediction, we developed in this work the Frequency Resolved Partial Dischage (FRPD) map (which is based on the developed Partial Discharge Spectral Pulse Counting technique), of which an illustrative example is shown in Figure 6. FRPD and its classification capabilities are the main contributions of this paper. In a short explanation, we may say that the map in Figure 6 is constructed by detecting pulses with amplitudes that are higher than a pre-established power threshold in the time-frequency domain (Figure 5) by counting the repeating pulses over time for each defined small registering frequency window and by averaging the amplitudes of the registered pulses in each registering frequency window. Normalized (percentage) spectral PD counting levels are represented by the colors of the dots in Figure 6.

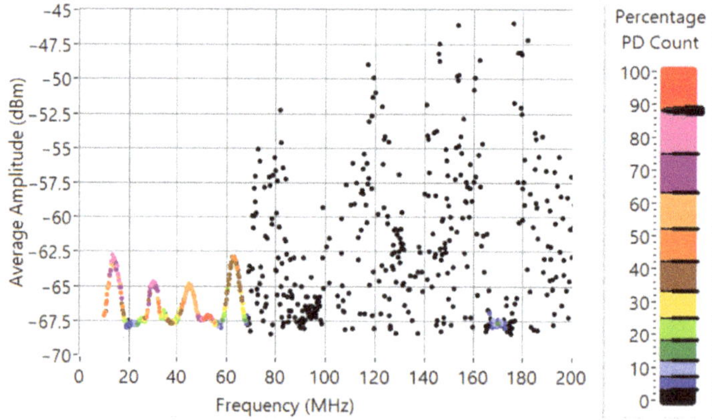

Figure 6. An illustrative example of the proposed FRPD map.

For the sake of clarification of the ideas just introduced, we show Figures 7 and 8. Figure 7a shows that the pre-established noise threshold is set at −60 dBm, which is considerably higher than the laboratory ambient power noise (noise can also be seen in Figure 7a). The graph in Figure 7b shows the first occurrence of a pulse at 387 MHz. During this occurrence, PD frequency and amplitude are registered, and the number of PD occurrences at 387 MHz is increased (from zero to one in this case) and recorded. In a third time window in which the spectrum is measured, depicted by the graph in Figure 7c, the power noise threshold is once more exceeded at 387 MHz, and it is also crossed at 417 MHz (for the first time). Then, PD frequency and amplitude recordings are made, and the occurrence count at 387 MHz is once more incremented (from one to two). For

417 MHz, recordings of PD frequency and amplitude are also made, and the number of PD occurrences is set to one for that frequency (since, up to this time, a single PD occurence has been registered for 417 MHz). The PD frequency and amplitude recordings and spectral PD counting processes are repeated every time the signal amplitude is higher than the threshold for a given frequency (or frequencies). To obtain statistically consistent results, measurements need to be performed during a large number of sinusoidal cycles, i.e., over several seconds.

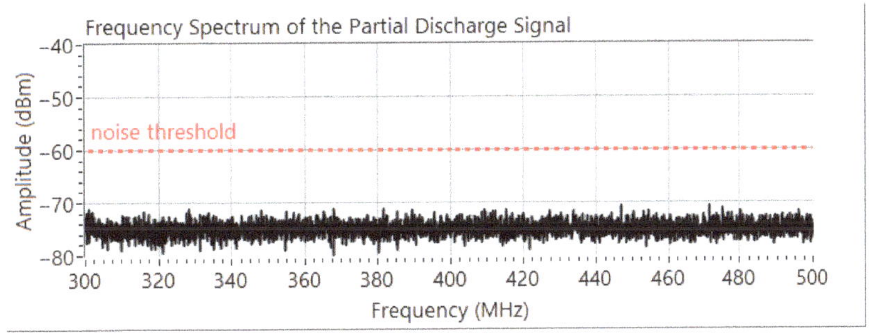

(**a**)

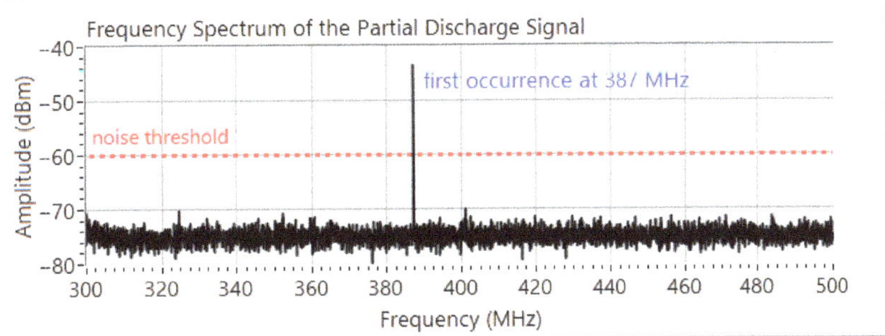

(**b**)

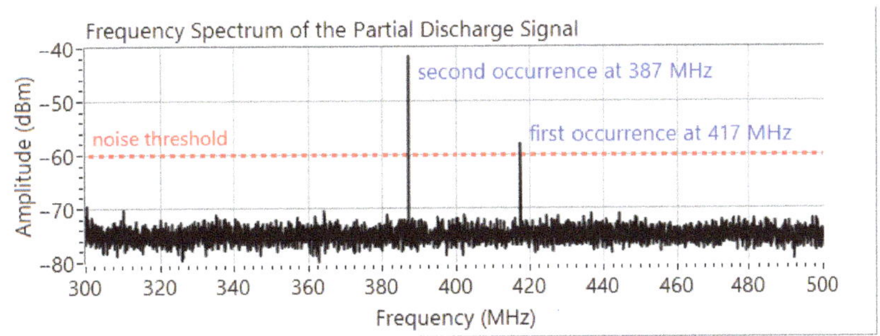

(**c**)

Figure 7. Spectral registering of PD pulses obtained at different times: (**a**) measurement with no occurrence records, at $t = t_0$; (**b**) detection of the first occurrence at 387 MHz, at $t = t_1$; and (**c**) detection of the second occurrence at 387 MHz and of the first occurrence at 417 MHz, at $t = t_2$.

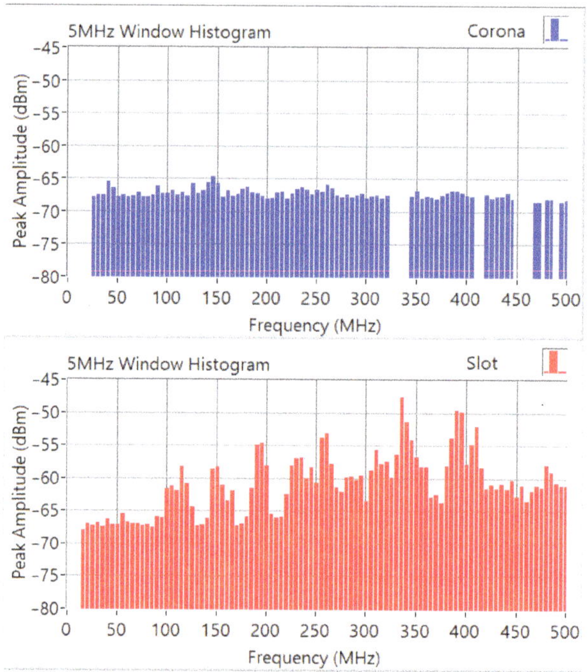

(**a**)

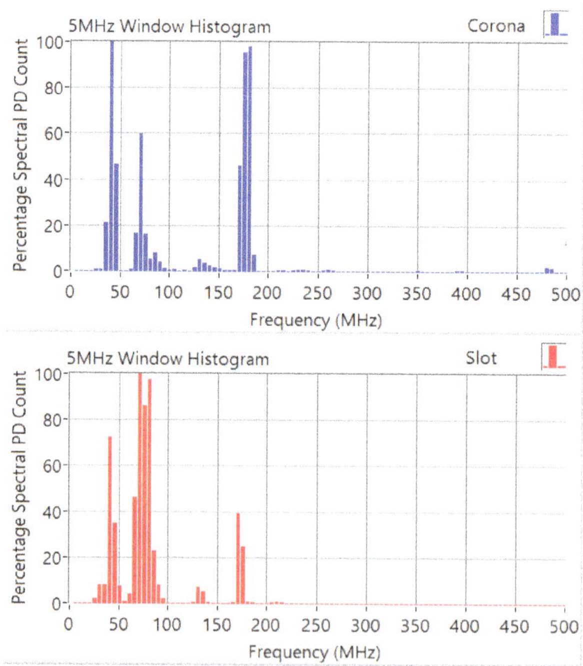

(**b**)

Figure 8. Spectral histograms of corona and slot PDs: (**a**) peak amplitude histograms and (**b**) normalized (percentage) pulse count histograms.

In order to build FRPD maps, such as the illustrative example of Figure 6, since frequencies at which of PD pulses occur can slightly fluctuate for a given sinusoidal phase, we employed measuring frequency band windows, similarly to the 5 MHz wide windows seen in the histograms of Figure 8. Thus, powers of pulses crossing the power noise threshold in a given measuring frequency window are averaged over time. Furthermore, the number of PD occurrences are individually computed for each frequency window. Therefore, every time a pulse crosses the threshold within a given frequency window, the number of PD occurrences is incremented for that window.

Peak amplitude histograms of Figure 8a show that spectral amplitudes of corona PDs tend to be almost constant over the frequency range shown, while slot PDs amplitudes oscillate over the frequency spectrum, with power peaks considerably higher than the power levels reached by corona PDs. This feature is associated with the electrically closing arcs between the circuit poles during slot PD occurrence. Notice that simultaneous occurrences of corona and slot PDs would be difficult to detect using peak spectral amplitudes. On the other hand, the pulse count histograms in Figure 8b show 100% for the normalized percentage count for slot PD at 70 MHz, which corresponds to 451,875 spectral peaks recorded in this frequency window, the counting of which is used as the reference level to set the percentage for the discussed measurements in other frequency bands. The pulse count histogram for corona PD shows 100% for the normalized percentage count in the 40 MHz window, which corresponds to 394,956 recorded spectral peaks. Thus, corona PDs have higher PD count rates at 40 MHz (with 100%), 45 MHz (with 46.6%), 70 MHz (with 59.9%), 170 MHz (with 46.2%), 175 MHz (with 95.5%), and 180 MHz (with 97.9%). Slot PDs have their higher PD count rates at 40 MHz (with 72.4%), 45 MHz (with 34.9%), 65 MHz (with 45.9%), 70 MHz (with 100%), 75 MHz (with 86.0%), 80 MHz (with 97.1%), and 170 MHz (with 39.2%). Corona discharges tend to have higher counting rates distributed over the spectrum because they occur between the energized electrode and the air, with plasma channel length highly dependent on the 50 or 60 Hz instantaneous signal amplitude. In contrast, slot PD counting strongly depends on arcs connecting the circuit poles, producing characteristic resonance frequency, which is associated to the length of the produced ionized electric arcs (which tends to be the distance between the poles) and especially on the total length of the metallic parts shortened by the arcs (the lengths of hydro-generator bars [31], for instance). This behavior explains the fact that slot PDs have, in our test measurements, their highest registered PD repetition rate around 70 MHz. Complementary Figure 9 shows spectral pulse counting per second, obtained by sampling the spectra 5000 times over 150 s.

An important matter is related to the spectral peaks seen in Figure 5. Although spectra of current PD pulses can be smooth functions, with a certain bandwidth, it is demonstrated in detail in [31], using full-wave simulations (by solving Maxwell's equations numerically), that electromagnetic waves produced by the PD currents resonate inside the winding bar while propagating in the structure, producing resonating electromagnetic fields propagating outwards from the bar. The fundamental resonance frequencies depend on the dimensions of the bar and on the wave propagation velocity inside the structure. In [31], an analytical formula from [32] is also used to confirm the resonance frequencies obtained from the simulations. Higher harmonics are also seen in simulation results and confirmed using the analytical formula. This physical phenomenon produces electric field spectra with well-defined spectral peaks, due to the influence of the bar and of spectral band of the PD current pulse. It is interesting to see the effect of increasing the pulse rise time in [31], which reduces spectral amplitudes in higher frequencies, but resonance peaks are still seen if excitation pulse has energy levels at those higher frequencies. The mentioned resonance effects are stronger in slot PDs because metallic slot structure reflects an electromagnetic field, forcing it to penetrate the bar (deterioration of semiconductive coating facilitates wave penetration). Corona PDs can occur in a face of the bar with no contact with slot structure, and, thus, a smaller part of electromagnetic wave power refracts into the bar, creating different spectrogram patterns (see Figure 5a,b).

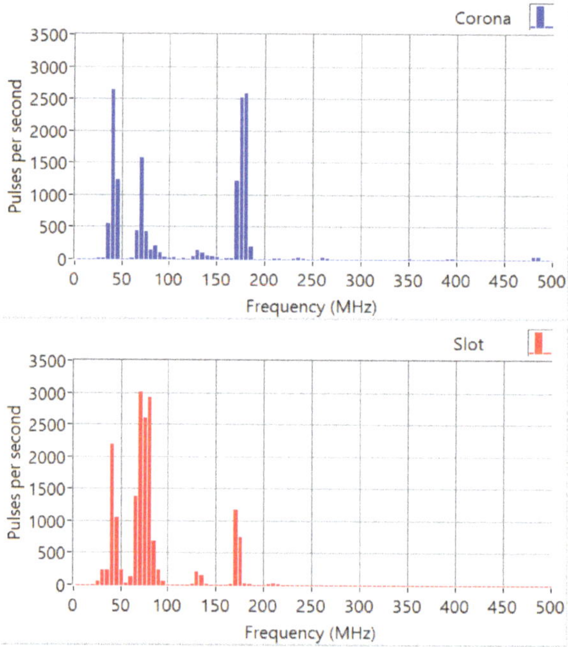

Figure 9. Spectral pulses per second registered in this work, during 150 s, for corona and slot PDs.

Temporal analysis of simulated PD radiated signals in [31] show that resonating frequencies may occur asynchronously, as transient signals clearly contain different frequencies at different time windows. This feature was confirmed in this work in the laboratory, when radiated signals' spectra were measured 5000 times over 150 s. Spectral peaks occur asynchronously, with particular repetition rates over time.

Thus, the peaks seen in Figure 5 are due the electromagnetic phenomenon just described. Notice that the spectrum is actually continuous, but amplitudes attenuated by the bar (at non-resonating frequencies) are not visible in Figure 5 because they are mixed with environmental noise.

Thus, an advantage of using the proposed FRPD methodology is that it allows the identification of classes of partial discharges occurring concurrently, especially those of which PRPDs or spectra are difficult to distinguish when they are superimposed, as is the case involving concurrent corona and slot PDs.

4.2. Computer Implementation of FRPD

Computer implementation of FRPD is based on two one-dimensional arrays: one array for PD count and another for average PD amplitude, computed using peaks crossing the power threshold for each frequency window individually. The dimensions of both arrays are coincident with the number of employed frequency windows.

The measurements of spectra of PD signals are made in time windows of 5 ms with a spectrum analyzer logically connected to a personal computer. The FRPD maps are formed with six frequency sub-bands, namely 7–20 MHz, 20–100 MHz, 100–200 MHz, 200–300 MHz, 300–400 MHz, and 400–500 MHz. For each frequency sub-band, 5000 time windows are acquired and 2000 measuring frequency windows are employed. Thus, for the 7 MHz–20 MHz sub-band, measuring frequency windows are 6.5 kHz wide. In the 20 MHz–100 MHz band, measuring frequency windows are 40 kHz wide. Finally, in the sub-bands 100–200 MHz, 200–300 MHz, 300–400 MHz, and 400–500 MHz, measuring frequency windows are 50 kHz wide.

As previously discussed, the methodology consists of detecting the PD pulses in a frequency spectrum exceeding the noise power threshold (in dBm). The threshold is determined to minimize the influence of background noise. The threshold is determined automatically and experimentally by the software. First, the noise signal is measured and divided into 10 frequency sub-bands. Each of them has its average power value calculated within 10 s. The highest average power is selected, and then, the noise threshold corresponding to 108% of the highest calculated average noise power is established.

During PD measurements, when any pulse in the spectrum exceeds the threshold, the pulse amplitude value, its frequency value, and the number of times that pulses occurred will be stored for each of the measuring frequency windows. Once the measurements and data recordings are complete, the total count of the occurrences of the spectral PDs pulses is determined, and the average PD amplitude is calculated for each measuring frequency window.

More specifically, for each measuring frequency window f_w, the number of pulse occurrences $P_o(f_w)$ is divided by the highest number of pulse occurrences found, $\max\{P_o\}$, which is obtained among all the sub-bands. Then, the so-called percentage PD count function $P_c(f_w)$ is recorded, of which values are mapped by using a color scale. Thus, the function $P_c(f_w)$ is given by

$$P_c(f_w) = \frac{P_o(f_w)}{\max\{P_o\}} \times 100\%. \tag{3}$$

Finally, results obtained from (3) are used to plot FRPDs such as that in Figure 6 and normalized (percentage) pulse count histograms such as those in Figure 8b. Thus, 100% pulse count indicates the frequency window in which maximum P_o was obtained in a given FRPD map or pulse count histogram. The algorithm for the proposed methodology is graphically shown in Figure 10.

By analogy with the PRPD (Phase Resolved Partial Discharge) method [17], which associates the PD peak occurrence with the sinusoidal signal phase, we may say that this new spectral statistical map is a Frequency Resolved Partial Discharge map because, with the presented technique, maps are assembled from PD peak counting derived from time evolving frequency spectrum of PD signals. The main advantage of FRPD over PRPD is the fact that FRPD can be easily used to detect more than one concurrent PD type, even with overlapping PRPDs and spectra, as is shown experimentally in the next sections.

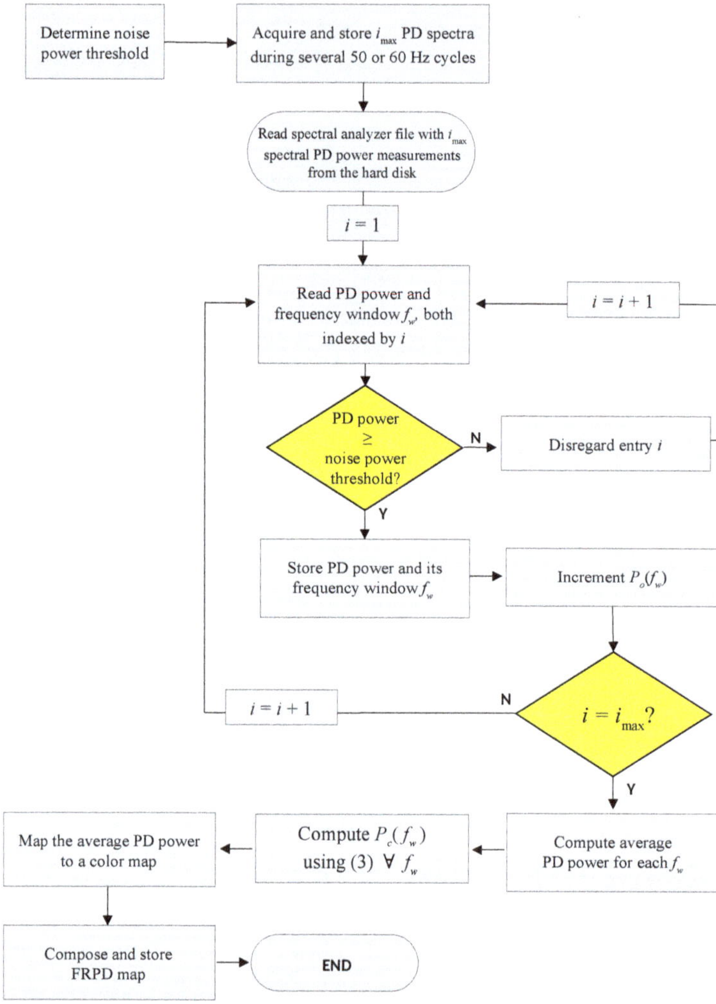

Figure 10. Algorithm for the proposed FRPD methodology. In this work, $i_{max} = 5000$.

5. Experimental Measurements and Results

5.1. The Designed Golden Ratio Patch Antenna

Since partial discharges radiate high-frequency electromagnetic signals, recently, microstrip patch antennas (consisting of a dielectric substrate with a face containing a specifically designed metallic patch and a ground plane of metal deposited on the opposite substrate face) have been used for PD detection [20,33–35]. Ultra-wide-band (UWB) patch antennas for partial discharge detection were proposed in [33] (0.5 GHz to 1.5 GHz), ref. [34] (0.305 GHz to 1.495 GHz), and [35] (0.35 GHz to 1.38 GHz). An appropriate band for detecting PD radiation is from 0.3 GHz to 1.5 GHz [35]. However, PD frequencies over 1.5 GHz were found in [20].

The well-known golden ratio can be observed not only in biological systems such as shell formation and flower petal formation, but also at the atomic scale in the magnetic resonance of spins in cobalt niobate crystals [36]. In telecommunications engineering, the golden ratio has been used in antenna design [37,38]. Therefore, in this work, we design a golden ratio patch antenna operating between 0.33 GHz and 2.02 GHz. The patch is formed by three circles that are centered on the borders of smaller intercepting circles, such as is illustrated by the proposed design shown in Figure 11. For the designed antenna, $R_1/R_2 = R_2/R_3 = r_a$ = 1.618, which is the golden ratio. The developed antenna has been used to receive high frequency electromagnetic PD signals in this work. Figure 12a properly illustrates the antenna being used in our labaratorial setup.

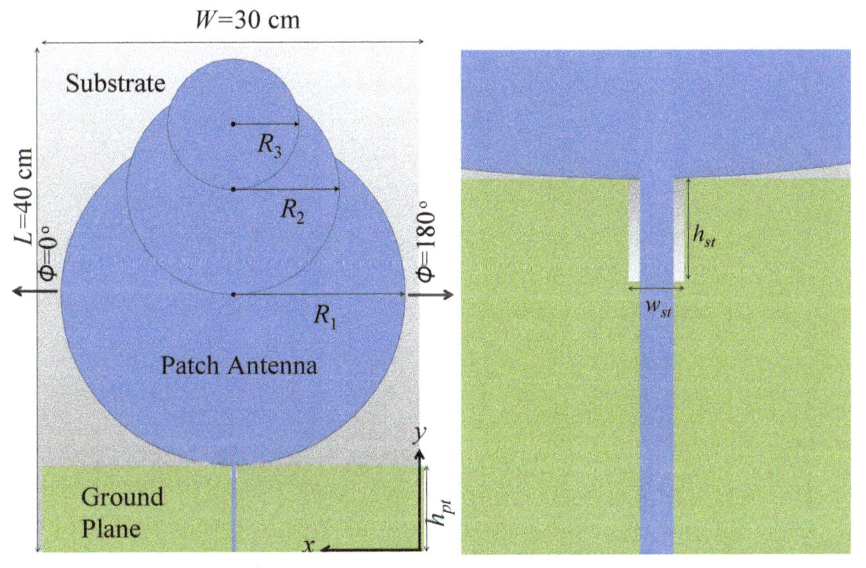

(a) (b)

Figure 11. Preliminary antenna model simulated using COMSOL Multiphysics: (**a**) overview and (**b**) details of the ground plane slot (the ground plane and patch are on opposite sides of the substrate).

For sake of comparison to other antennas, spiral antennas are low-gain wide-band antennas [39], while rod antennas can be high-gain narrow-band devices (e.g., Yagi antennas) [40]. Spiral antennas (with reflector) can achieve 2 dBi gain [39] and Yagi antennas can easily achive 12 dBi or more [40]. Our patch antenna lies in between the two: it is a wide-band antenna with the moderate maximum gain of approximately 3.8 dBi at 450 MHz. Further details, parameters, and engineering characteristics of the device are described in detail in the Appendix A.

5.2. Laboratory Setup

The designed golden ratio patch antenna was used to perform high-frequency PD measurement with the setup illustrated by Figure 1. In this work, the process of measuring PD signals to produce FRPD maps involved the following instruments and apparatus: a spectrum analyzer, the patch antenna, hydro-generator stator bars, a 60 Hz high voltage source, and a laptop computer with the measurement and analysis software developed in this work. The experimental setup can be seen in Figure 12a.

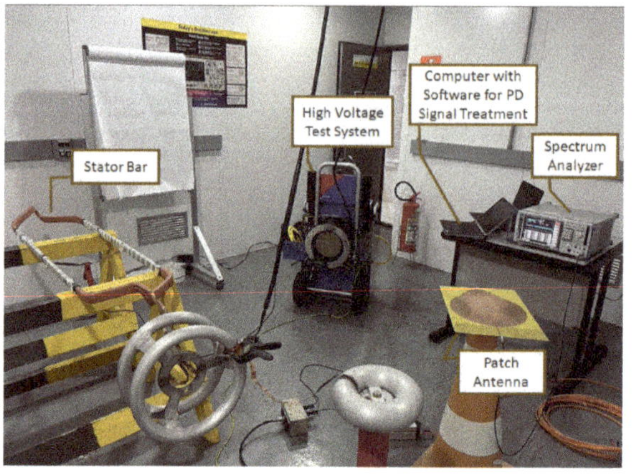

(**a**)

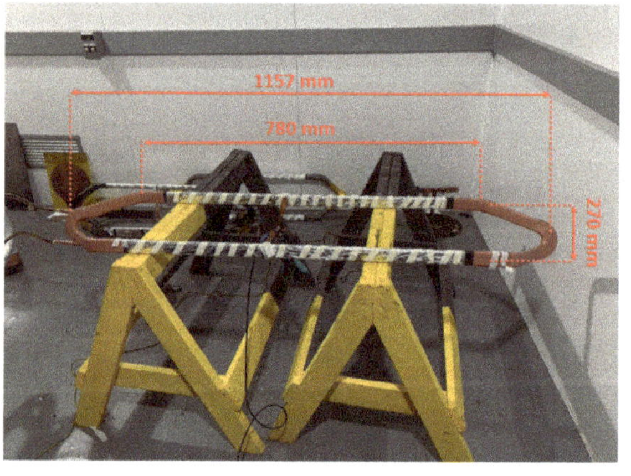

(**b**)

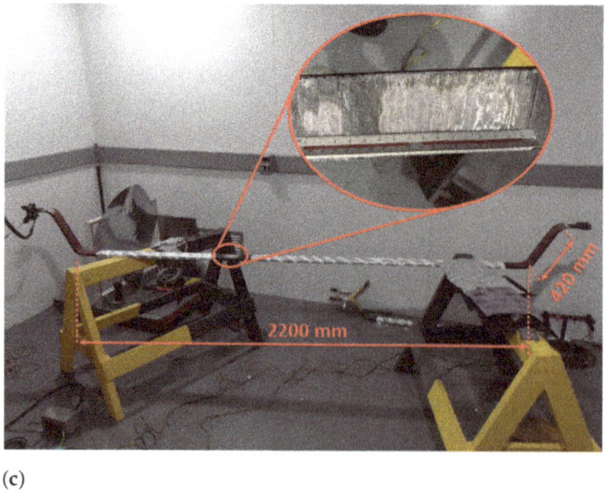

(**c**)

Figure 12. *Cont.*

(d)

Figure 12. The measurement setup: (**a**) overview of the measurement setup for acquiring high frequency PD radiated signals, (**b**) General Electric bar from Curuá-Una hydroelectric power plant (with corona PDs), (**c**) Jeumont-Schneider bar from Tucuruí hydroelectric power plant (with slot PDs), and (**d**) parallel-connected bars for producing concurrent corona and slot PDs.

The hydro-generator bars were energized with 12 kV by the 60 Hz high voltage system. This voltage level was sufficient to capture the PD signals from the bars using the patch antenna. The patch antenna was connected to the spectrum analyzer using a 50 Ω impedance coaxial cable. PD spectral signals, acquired over time, were registered by the developed software in which the proposed FRPD methodology is implemented.

For the measurement of the corona discharges, a bar with mica insulation from the Curuá-Una Hydroelectric Plant (Brazil) was used, which is shown in Figure 12b. For the slot discharges, a bar with mica insulation from the Tucuruí Hydroelectric Plant (Brazil) swas used, which can be seen in Figure 12c. The slot defect produced in the bar is highlighted in Figure 12c. The measurements of concurrent corona and slot PD discharges were performed by implementing a parallel circuit connection between the Curuá-Una and the Tucuruí Hydroelectric Plant bars, as is seen in Figure 12d.

5.3. Results

Figure 13 shows PRPDs obtained in this work for the experiments of Figure 12. The measured PRPDs of corona, slot, and the simultaneous corona and slots PDs are shown by Figure 13a–c, respectively. Note that PRPD profiles shown in Figure 13a,b are compatible with standardized PRPD patterns in [6] for corona and slot PDs, respectively, validating our experimental PD sources. By comparing Figure 13a,b, we can see that corona and slot PDs occur practically on the same phase ranges for both the positive and the negative 60 Hz semi-cycles. For this reason, concurrent corona and slot PRPDs overlap each other in Figure 13c, as one would expect. Thus, distinguishing these defects patterns by inspecting solely Figure 13c is a remarkably difficult task.

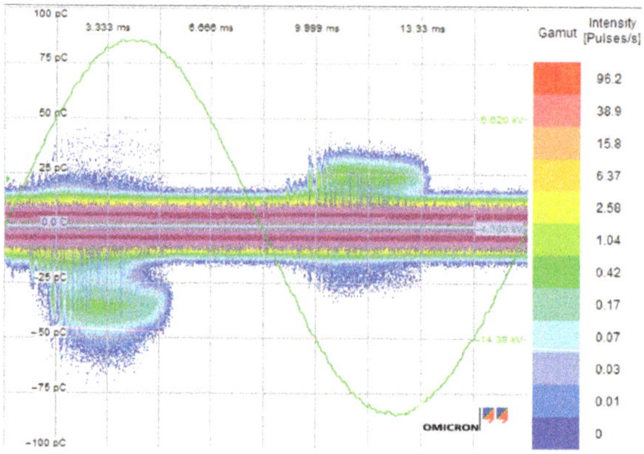

(a)

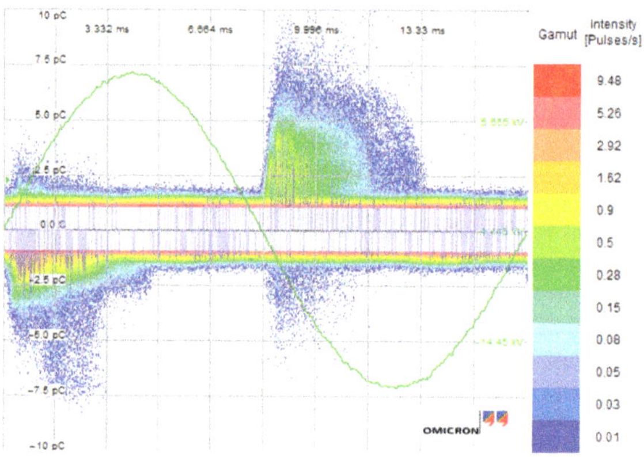

(b)

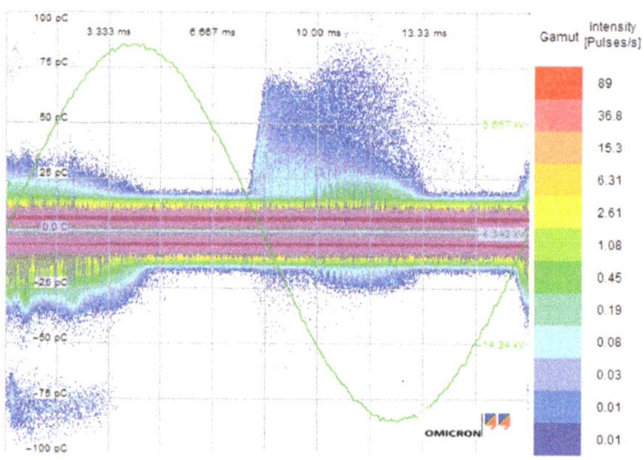

(c)

Figure 13. PRPDs experimentally obtained in this work for: (**a**) corona (12 kV—General Electric bar), (**b**) slot (10 kV—Jeumont-Schneider bar), and (**c**) simultaneous corona and slot (12 kV) obtained usingboth bars.

The situation is different when FRPDs are used for identifying the measured corona and slot PDs. FRPDs seen in Figure 14 show spectral counting computed from 0 to 500 MHz for different cases: corona discharges in Figure 14a, slot discharges in Figure 14b, and corona PDs with simultaneous slot PDs in Figure 14c. In order to quantify background noise and to map radiated signals from the high voltage source, we have registered the background noise with non-energized and energized high voltage power source, as shown by Figure 14d,e, respectively.

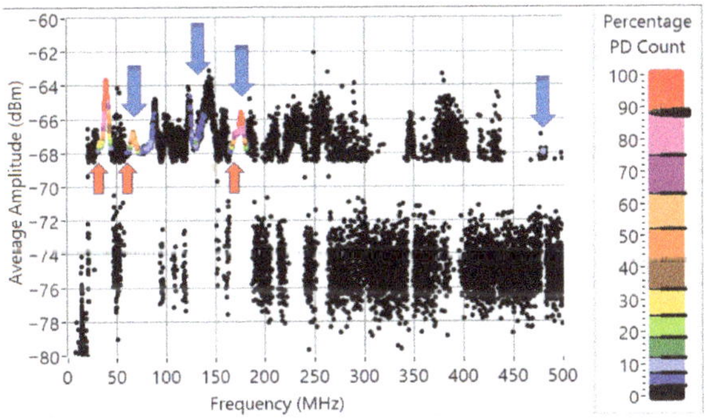

(**a**)

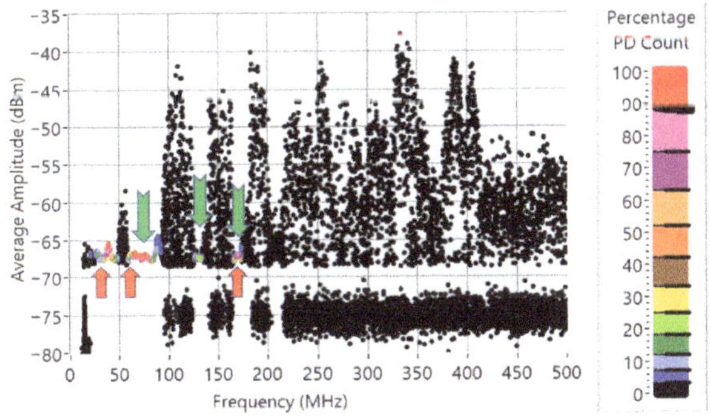

(**b**)

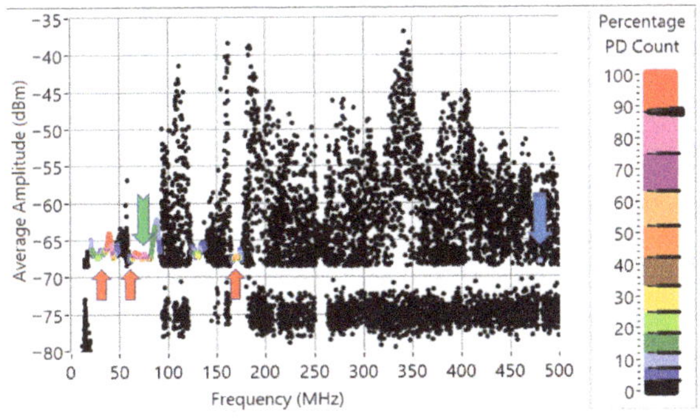

(**c**)

Figure 14. *Cont.*

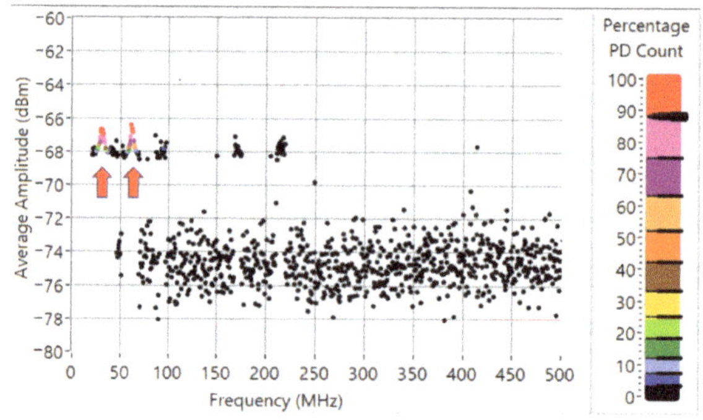

(**d**)

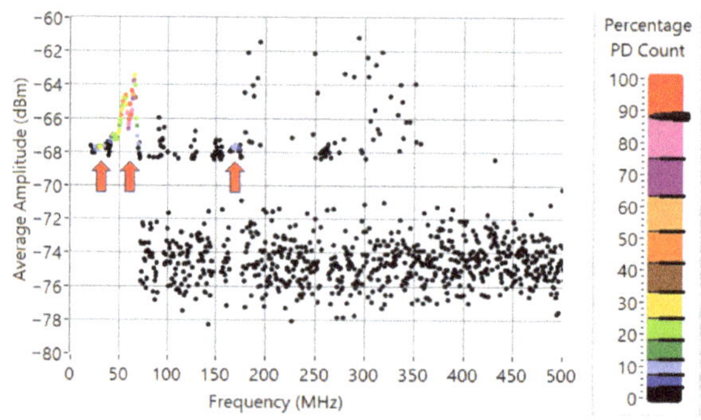

(**e**)

Figure 14. FRPD obtained from 0 to 500 MHz for different cases: (**a**) corona PD, (**b**) slot PD, (**c**) corona PD with concurrent slot PD, (**d**) background noise (high voltage power source not energized), and (**e**) background noise with an energized high-voltage power source.

Once measurements of Figure 14d,e had been made, we considered the noise power threshold of −68.5 dBm for performing PD measurements. Percentage PD count, which is mapped using a color scale, was calculated with respect to the highest measured counting, which is represented by the color red. Black points indicate low-percentage counting pulses, i.e., up to 3% of the highest measured counting.

In Figure 14, red arrows (pointing up) indicate high counting noise or high counting signals from the high voltage power source, which are not related to PDs. These external signals were acquired and identified in measurements whose results are shown in Figure 14d,e. In Figure 14a, the blue arrows (pointing down) indicate high-counting spectral regions associated to corona PDs. Similarly, in Figure 14b, the green arrows (pointing down) indicate high-counting pulses associated with slot PDs. Additionally, note that low counting pulses (represented by black points) with amplitudes over −60 dBm from slot discharges are widely spread across the measuring spectrum, as shown in Figure 14b. This distinct feature is not seen in Figure 14a for corona PDs.

Figure 14c, which was obtained using the measuring setup shown in Figure 12d, shows the simultaneous presence of the corona and slot PDs. Indications are made by following the arrow representations defined above. The low count pulses (black points) are widely spread across the spectrum, indicating the presence of slot discharges in Figure 14c. One may also observe that the spectral regions of high counting pulses corresponding exclusively to corona and slot PDs (around 475 MHz and 75 MHz, respectively), which were previously detected in the measurements shown in Figure 14a,b, are also visible and properly indicated in Figure 14c. Thus, with the proposed methodology, it is possible to identify, undoubtedly, simultaneous slot and corona PDs.

FRPD maps seen in Figure 15 highlight the results from 7 to 50 MHz. Figure 15d,e shows repetitive noise between 25 and 37 MHz and between 41 and 50 MHz (influence of energized power supply), with amplitudes around −66 dBm. Figure 15a, Figure 15b, and Figure 15c, respectively, show FRPDs for corona, slot, and for corona with slot, with amplitudes reaching −64 dBm. However, frequency bands where high counting corona and slot PDs are observed overlap each other (besides overlapping the high counting band of noise). Thus, in this band, although PDs can be seen standing out from the noise, it is difficult to distinguish concurrent corona and slot PDs.

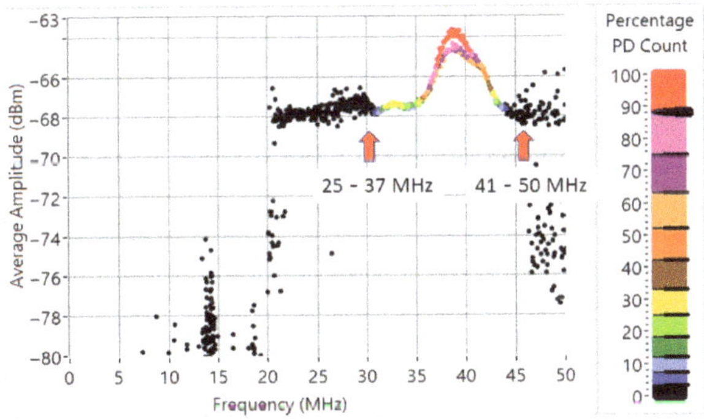

(a)

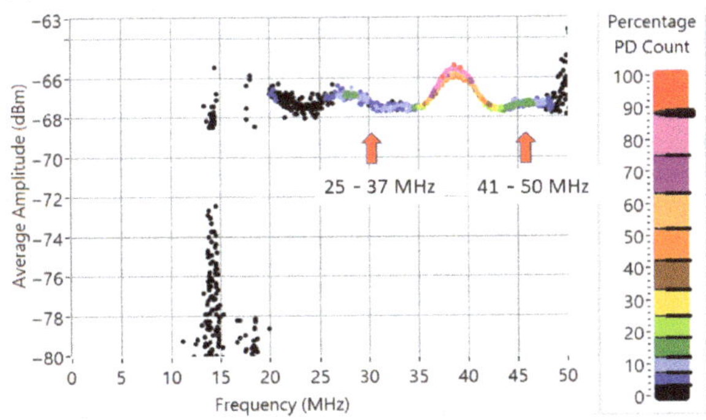

(b)

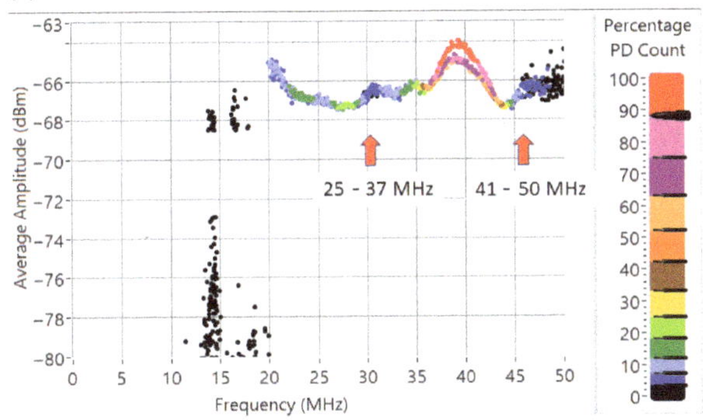

(c)

Figure 15. *Cont.*

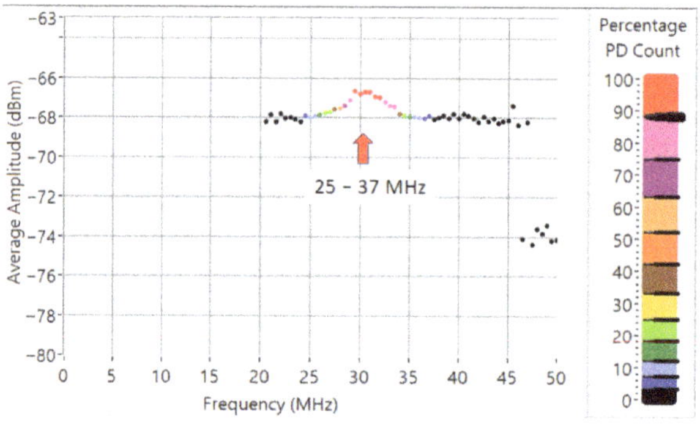

(d)

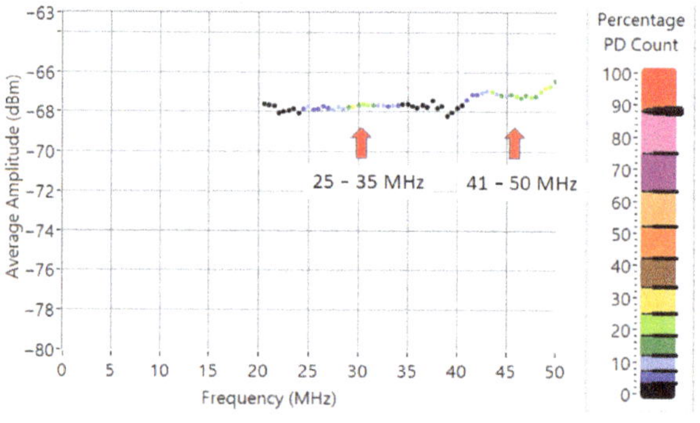

(e)

Figure 15. FRPD for different cases, from 0 to 50 MHz: (**a**) corona, (**b**) slot, (**c**) corona with slot, (**d**) noise with the power supply not energized, and (**e**) noise with energized power supply.

Figure 16 shows FRPD results from 50 to 100 MHz. Figure 16d,e shows noise between 50 and 70 MHz, in such a way that the high voltage source has an important influence on registered counting, with amplitude reaching −63 dBm. Figure 16a shows corona PDs between 65 and 70 MHz, with PD counting up to 63% and amplitudes reaching −67 dBm and low counting registered in noise band (50–70 MHz). Figure 16b shows slot PDs between 65 and 80 MHz, with PD counts up to 100% and amplitudes that reach −67.5 dBm. In Figure 16b, the noise region indicated by the red arrow has a low count. Figure 16c shows simultaneous corona and slot PDs between 63 and 80 MHz with PD counts up to 100%. The region between 65 and 70 MHz is a region of high PD count shared by corona and slot PDs, as one can see by inspecting Figure 16a,b. The green arrow in Figure 16c highlights the high PD count band, between 70 and 80 MHz, where solely slot PDs occur. The band between 80 and 90 MHz is a communal PD band for both phenomena. Thus, in the FRPD of Figure 16c, corona PDs cannot be identified with confidence, but slot PDs can be clearly seen in Figure 16c.

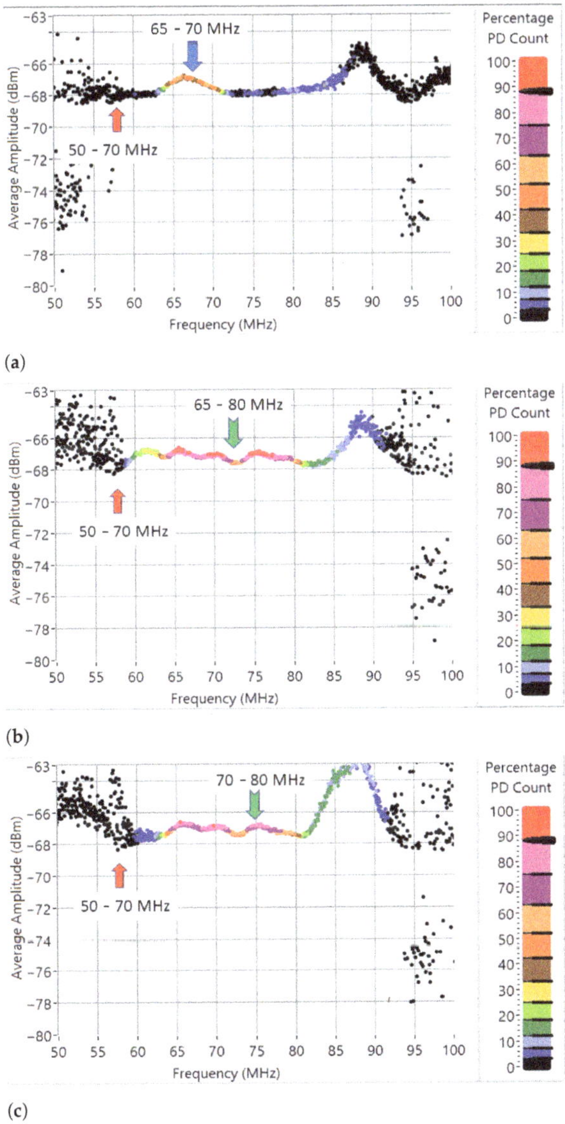

Figure 16. *Cont.*

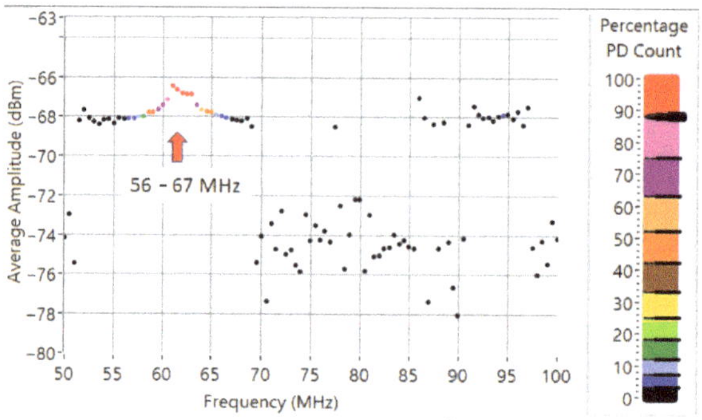

(**d**)

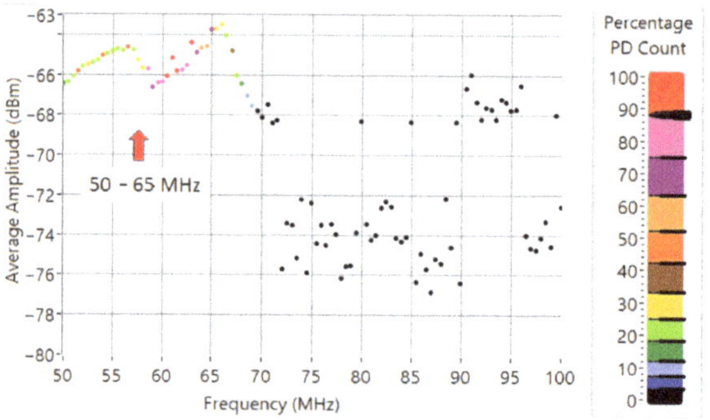

(**e**)

Figure 16. FRPD for different cases, from 50 to 100 MHz: (**a**) corona, (**b**) slot, (**c**) corona with slot, (**d**) noise with the power supply not energized, and (**e**) noise with energized power supply.

FRPDs obtained from 100 to 150 MHz are shown in Figure 17. High counting noise is not seen in Figure 17d,e. Figure 17a shows corona PDs between 124 and 137 MHz, with PD counts up to 12% and amplitudes that reach −65.5 dBm, as the blue arrow indicates. Slot PDs are seen between 125 and 133 MHz in Figure 17b, with PD counts up to 25% and amplitudes reaching −67 dBm, as the green arrow points out. Figure 17c shows coronas with slot PDs between 125 and 135 MHz with PD counts up to 33%. In this case, since the corona's and slot's high counting bands completely intercept each other, it is difficult to distinguish the phenomena by inspecting exclusively the 100–150 MHz band. A very similar situation is seen by inspecting the 150–200 MHz range. As Figure 18a,b show, corona and slot high counting bands are completely superimposed on each other, making discrimination of patterns an intricate process when inspecting Figure 18c.

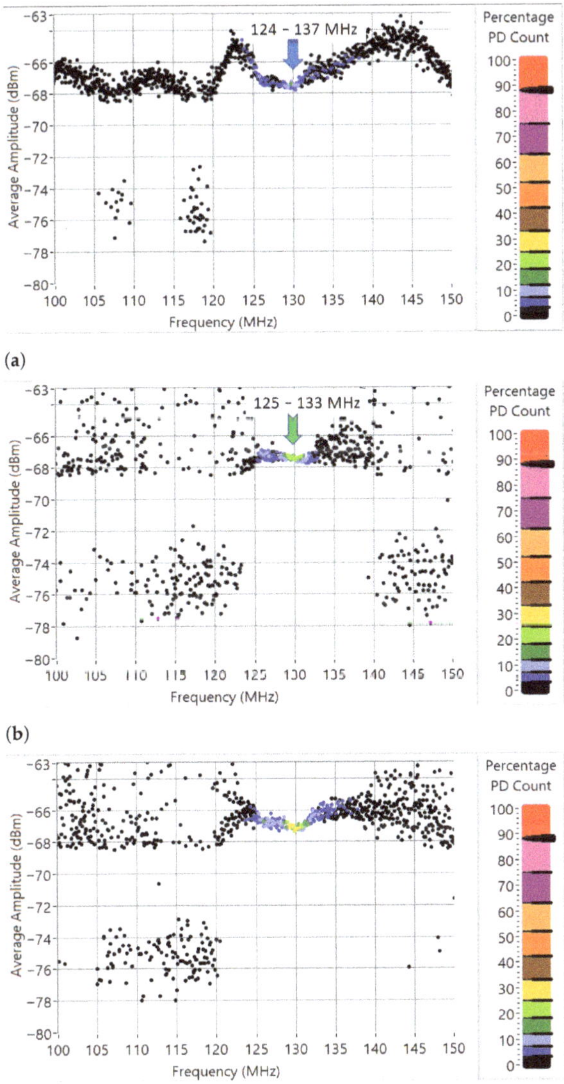

Figure 17. *Cont.*

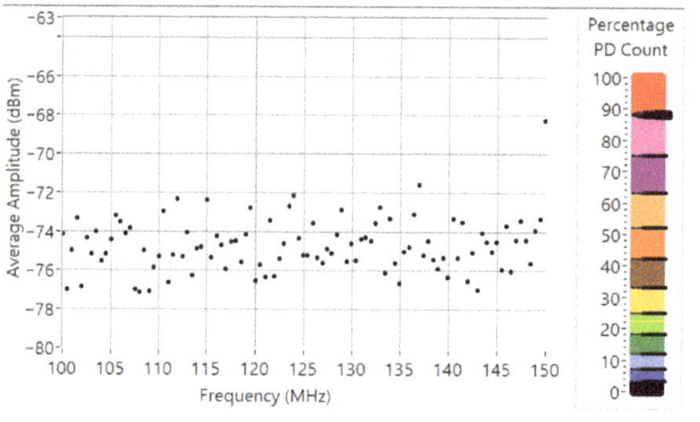

(**d**)

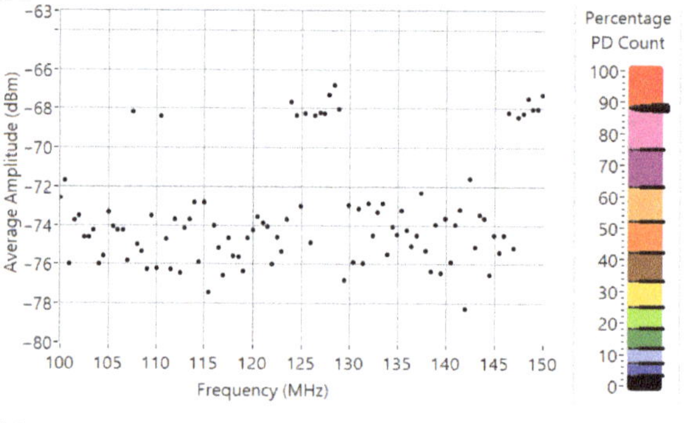

(**e**)

Figure 17. FRPD for different cases, from 100 to 150 MHz: (**a**) corona, (**b**) slot, (**c**) corona with slot, (**d**) noise with the power supply not energized, and (**e**) noise with energized power supply.

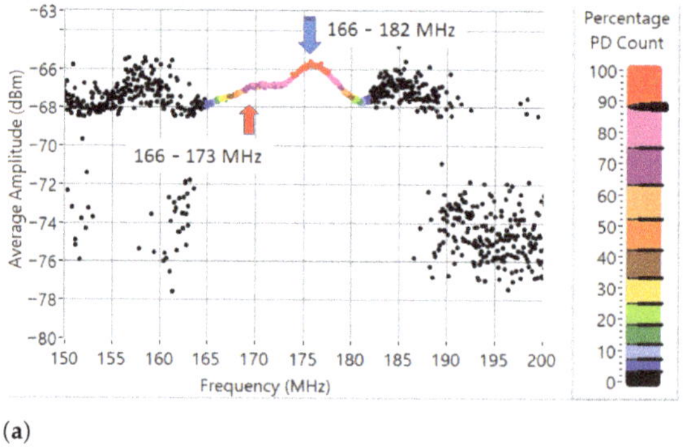

(**a**)

Figure 18. *Cont.*

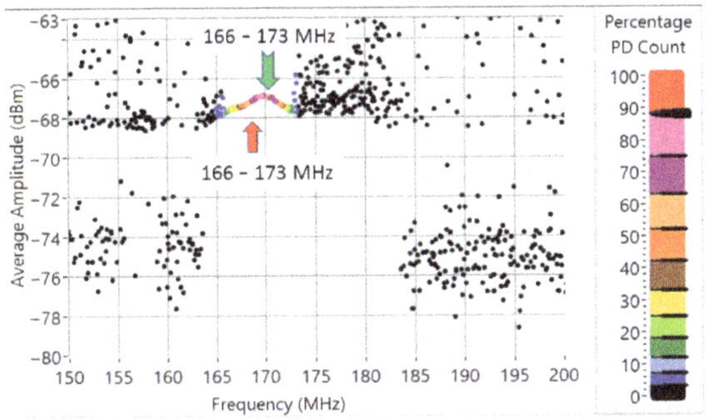

(**b**)

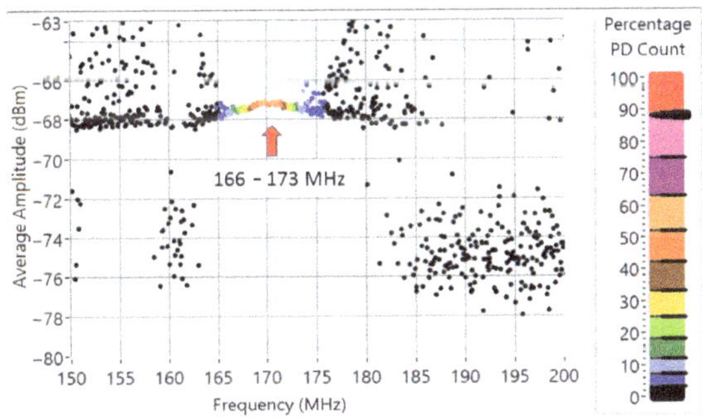

(**c**)

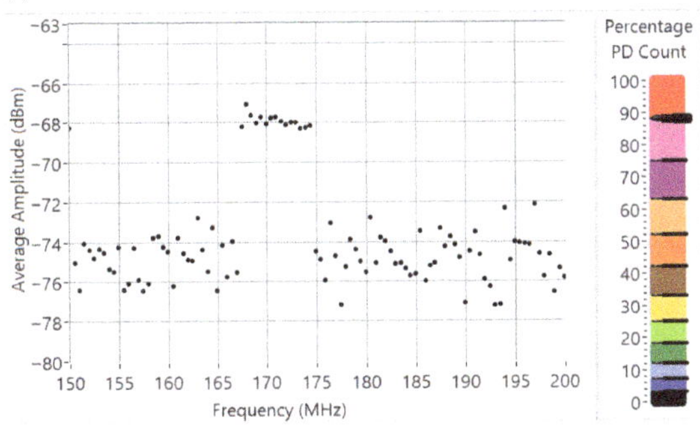

(**d**)

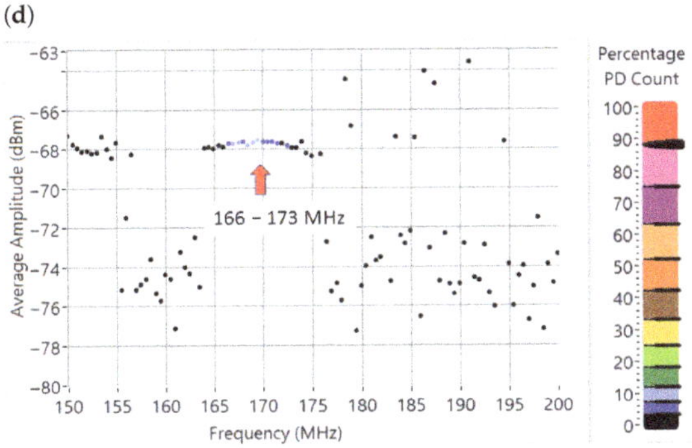

(**e**)

Figure 18. FRPD for different cases, from 150 to 200 MHz: (**a**) corona, (**b**) slot, (**c**) corona with slot, (**d**) noise with the power supply not energized, and (**e**) noise with energized power supply.

Finally, the FRPDs seen in Figure 19 highlight results from 450 to 500 MHz. Figure 19d,e does not show high counting noise in this sub-band. Figure 19a shows corona PDs between 479 and 481 MHz, with PD counts up to 12% and amplitudes that reach −67 dBm, as the blue arrow pinpoints. Figure 19b shows occurrence of slot PDs with low PD counts in this sub-band, of which power ranges from −68.75 to −65 dBm. Figure 19c shows concurrent corona and slot PDs, in which we can see significant PD counts between 479 and 481 MHz (up to 12%) with amplitudes reaching up −68 dBm, clearly indicating corona PDs. Furthermore, low count discharges with amplitudes ranging between −68.75 and −65 dBm are also seen spread across the frequency band in Figure 19c, suggesting the concurrent influence of slot discharges. Therefore, one can properly identify corona and slot PDs in the 450–500 MHz band by inspecting the FRPD map in Figure 19c.

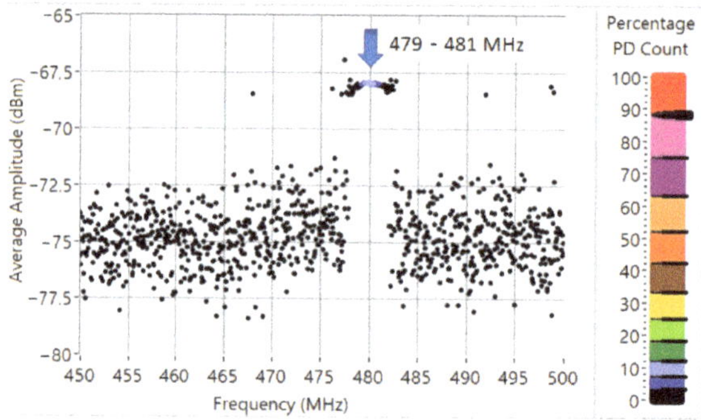

(a)

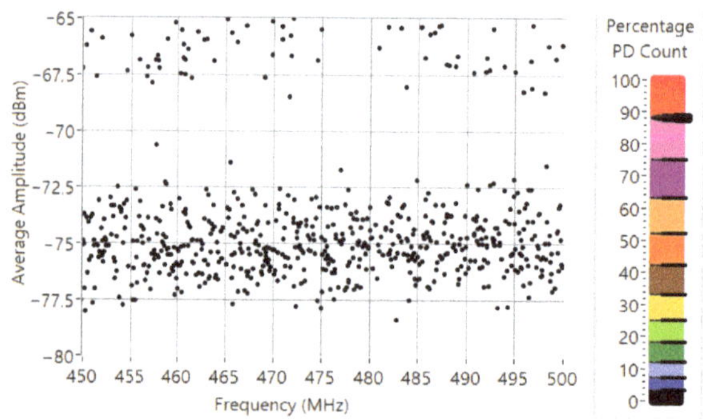

(b)

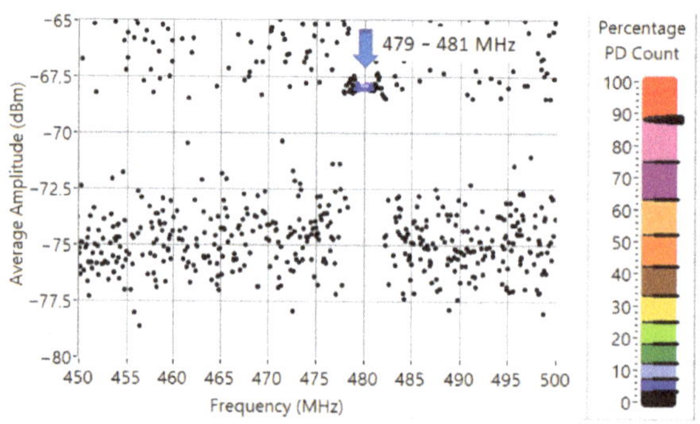

(c)

Figure 19. *Cont.*

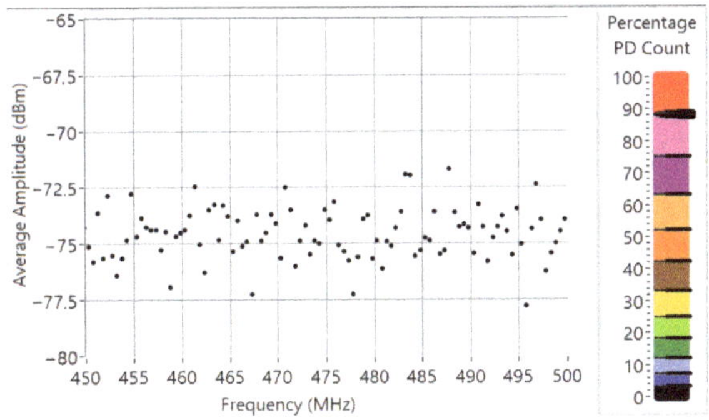

(**d**)

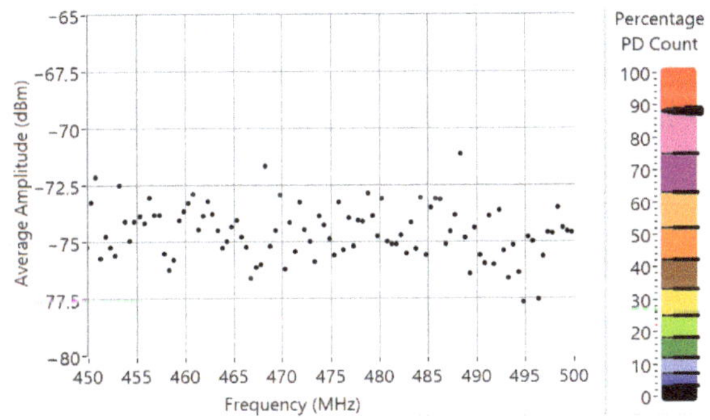

(**e**)

Figure 19. FRPD for different cases, from 450 to 500 MHz: (**a**) corona, (**b**) slot, (**c**) corona with slot, (**d**) noise with the power supply not energized, and (**e**) noise with energized power supply.

6. Conclusions

This paper proposes a new methodology for spectral statistical analysis of partial discharge pulses: frequency resolved partial discharge (FRPD). The proposed methodology is based on counting pulses in spectral domain (which evolves over time), allowing the identification of faults that give rise to discharges, even when the signals occupy shared frequency bands and communal phases of the sinusoidal high voltage excitation signal.

In order to measure the electromagnetic signals radiated from partial discharges, a patch antenna was designed, built and employed in laboratory measurements. The antenna was connected to a RF frequency analyzer. Besides that, a software was developed to analyze the discharge signals using the proposed FRPD methodology. The measurements were made in an electromagnetically shielded room, in a high-voltage laboratory, with two bars from hydro-generator stators. One of the bars has a confirmed corona fault and the other one has certified slot PDs, as measured PRPDs show. In order to obtain corona and slot PDs simultaneously, the two bars were electrically connected in a parallel circuit configuration.

With the proposed methodology, one can generate a FRPD graph showing percentage pulse count rates distributed over the frequency-amplitude plane. Thus, specific bands display high counting PDs for each type of partial discharge source, and each type has its own spectral PD counting map. As was shown experimentally in this work, this feature allows the proper identification of simultaneous PD classes with overlapping PRPDs and spectra, such as corona and slot PDs. In particular, the band from 450 MHz to 500 MHz seems to be important for undoubtedly detecting corona and slot FRPD signatures. Since the proposed FRPD methodology seems to be a strong complement for previously developed techniques, such as PRPD and classical spectral analysis, i future works, further investigation and possibly standardization of FRPDs are suggested for all known classes of PD sources in rotating machines and other HV systems.

Author Contributions: Conceptualization, R.M.S.d.O.; Methodology, R.M.S.d.O., A.J.C.S. and J.A.S.d.N.; software, A.J.C.S.; validation, A.J.C.S. and J.A.S.d.N.; formal analysis, R.M.S.d.O., A.J.C.S. and J.A.S.d.N.; investigation, R.M.S.d.O., A.J.C.S. and J.A.S.d.N.; resources, R.M.S.d.O. and J.A.S.d.N.; data curation, A.J.C.S.; writing—original draft preparation, A.J.C.S.; writing—review and editing, R.M.S.d.O., A.J.C.S. and J.A.S.d.N.; visualization, A.J.C.S. and J.A.S.d.N.; supervision, R.M.S.d.O.; project administration, R.M.S.d.O.; funding acquisition, R.M.S.d.O. All authors have read and agreed to the published version of the manuscript.

Funding: This paper publication fee was funded by PROPESP/UFPA (*Edital* PAPQ).

Institutional Review Board Statement: Not applicable.

Informed Consent Statement: Not applicable.

Data Availability Statement: The authors reserve the right to not disclose the Eletronorte's private data sets of laboratorial PD measurements used in this study.

Acknowledgments: Authors would like to thank the *Universidade Federal do Pará* (UFPA) for financial support and *Centrais Elétricas do Norte do Brasil* (Eletronorte/Eletrobras) for granting permission to use their high voltage laboratory for developing the experimental part of this work.

Conflicts of Interest: The authors declare no conflicts of interest.

Appendix A. Antenna Design

In this work, we designed and used a patch antenna for covering the band between 0.33 GHz and 2.02 GHz, since PD signals with frequencies over 1.5 GHz were measured in [20]. Initially, a preliminary model of the a golden ratio patch antenna was developed, as is presented in Figure 11. The patch is represented using the blue color, the ground plane is depicted in green, and the dielectric substrate is outlined in gray. The substrate is manufactured using a Flame Retardant epoxy material (type 4), of which relative electric permittivity is 4.4 and its loss tangent is 0.02. The antenna dimensions are L = 400 mm (length), W = 300 mm (width), and H = 1.5 mm (thickness).

The patch is formed by circles that are centered on the borders of smaller circles. The ratio of the radius of any circle to its immediately smaller circle is given by the golden ratio r_a = 1.618. The geometric parameters of the proposed antenna are indicated in Figure 11. For the preliminary antenna model, the geometric parameters are h_{pt} = 57 mm, h_{st} = 10 mm, w_{st} = 6 mm, R_1 = 140 mm, $R_2 = R_1/r_a$= 86.52 mm, and $R_3 = R_2/r_a$ = 53.47 mm. The length of the microstrip feed line is $h_{pt} + 1$ mm, and its width is 3 mm. The parameter $L = 40$ cm was chosen in order to produce resonance at approximately 300 MHz. The three circles were used to achieve additional multiple resonance bands at higher frequencies. The ground plane also contributes with its high frequency resonance band. Consequently, a broadband antenna can be obtained by optimizing its geometric parameters, aiming to achieve proper merging of the resonance bands produced by each circular element, by the half perimeter of the patch, by the length of the feeding line, and by the ground plane dimensions.

Reflection coefficient S_{11}, obtained at the antenna terminals [41], is calculated in this work considering the reference impedance $Z_0 = 50\ \Omega$. The reflection coefficients (in dB) of the preliminary antenna model, calculated from COMSOL Multiphysics simulation, are presented in Figure A1 (curve C_1). As one can see in curve C_1, for the initial (preliminary) antenna model, S_{11} undesirably crosses the reference level of -10 dB around 0.8 GHz, near 1.3 GHz, and at 1.8 GHz. For this reason, optimization of the dimensional parameters of the initial antenna model were necessary to allow full coverage of the aimed PD frequency range (300–2000 MHz) [20].

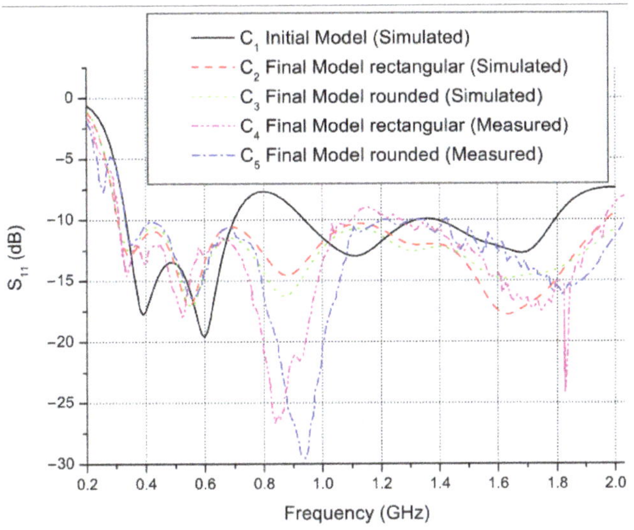

Figure A1. Simulated and measured reflection coefficients of the antennas.

Initial adjustments to the dimensional parameters were done by preserving other dimensions of the initial model; i.e., tuning was performed, at first, by varying a single parameter at a time. The results of these individual parametric sweeps of h_{pt}, R_1, h_{st}, and w_{st} are presented in Figure A2. By setting R_1 to 135 mm, reflection coefficient S_{11} was greatly reduced around 0.8 GHz with respect to the initial model of the antenna. However, with this adjustment on R_1, the parameter S_{11} was increased to levels over -10 dB in the band between 1.1 GHz and 1.45 GHz. On the other hand, by setting h_{st} to 8 mm, levels of S_{11} in the band between 1.1 GHz and 1.45 GHz are well below -10 dB, as Figure A2 shows. Thus, an optimized antenna was obtained by performing a combination of parameters producing reflection coefficients under -10 dB in frequency bands, in which the initial antenna model does not provide that characteristic and in regions of the spectrum in which a desired parameter modification produces undesired S_{11} values in other frequency bands of interest. Thus, we have obtained the following optimized set of parameters: h_{pt} = 68 mm, h_{st} = 9 mm, w_{st} = 5 mm, R_1 = 135 mm, $R_2 = R_1/r_a$ = 83.43 mm, and $R_3 = R_2/r_a$ = 51.56 mm.

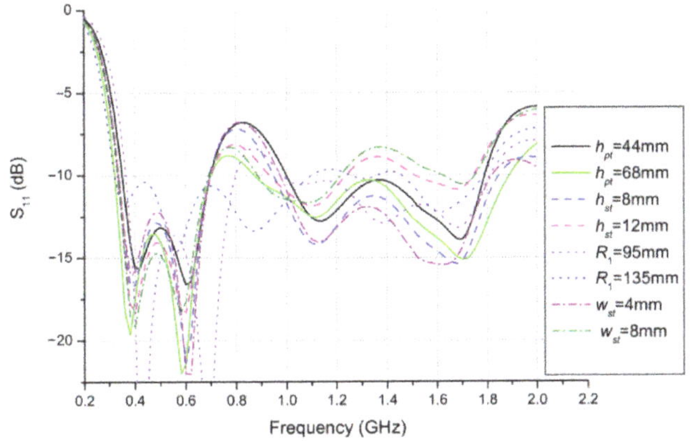

Figure A2. Simulated reflection coefficients regarding adjustments of dimensional parameters.

Finally, the optimized antenna has been built, as Figure A3 illustrates. A vector network analyzer (VNA) has been used for measuring the parameter S_{11} of the antenna, and the experimental results are compared with the numerical curve in Figure A1 (see curves C_2 and C_4). As one can observe, even with the obtained similarity between the measured and the simulated reflection coefficients, measured signal is above -10 dB in the frequency band between approximately 1.1 GHz and 1.25 GHz. As long as the ground plane width is $W = 0.3$ m, this dimension is close to $\lambda/2$ in the dielectric substrate at 1.25 GHz. As a consequence, the ground plane has an important influence on the antenna response around 1.25 GHz. Thus, in order to reduce the reflection coefficient in the frequency range of interest, the edges of the ground plane were modified to be rounded, as is shown in Figure A4. The comparison between the measured and simulated S_{11} curves are presented in Figure A1 (see curves C_3 and C_5). This final improvement introduced to the antenna geometry produces reflection coefficients measured under -10 dB over the entire targeted PD frequency band, between 0.33 GHz and 2.02 GHz. The obtained fractional bandwidth (FBW) is 143.83%.

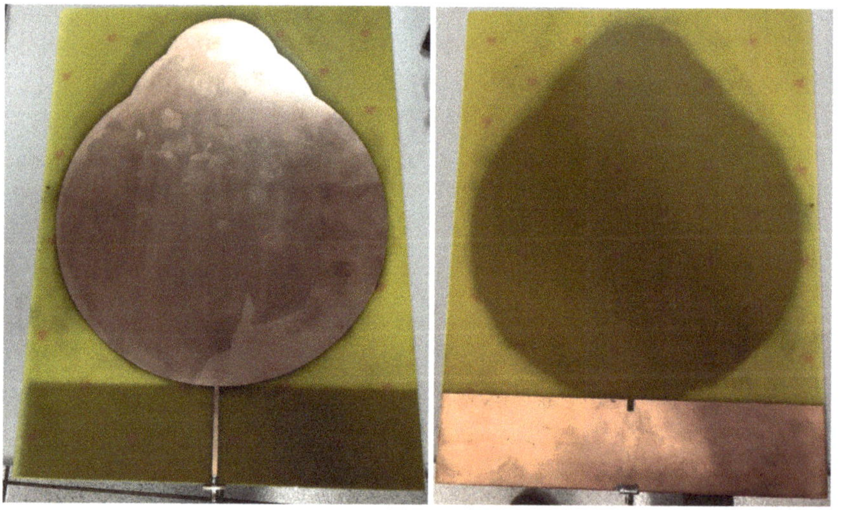

(a) (b)

Figure A3. Optimized antenna designed with right angles ground plane: (**a**) front view and (**b**) back view.

Figure A5 shows real and imaginary parts of the final antenna's impedance $Z = R + jX$, where R and X are the resistance and the reactance of the antenna. One can confirm that the antenna satisfies desired impedance matching goals, since the imaginary part of the impedance is around 0 Ω, reaching maximum and minimum values of 25.16 and -26.43 Ω, while the real part is around $Z_0 = 50$ Ω, with maximum and minimum values of 73.33 and 27.75 Ω, respectively, in the frequency range of interest. This feature is confirmed by inspecting the curve C_5 Figure A1, which is under -10 dB in the aimed frequency range.

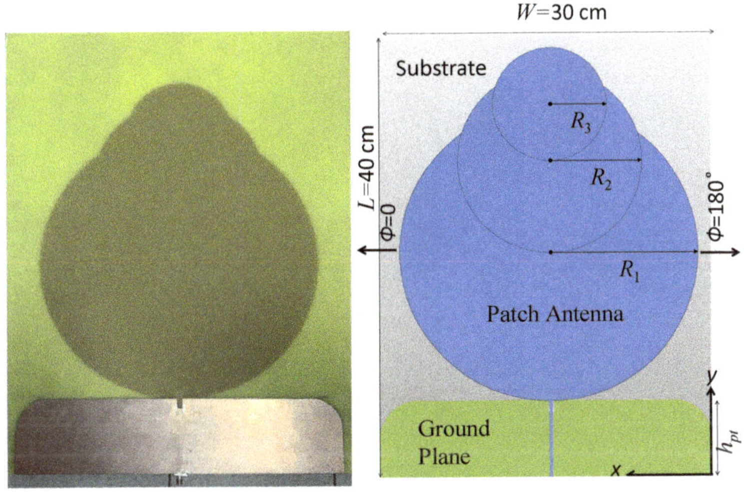

(a) (b)

Figure A4. The final antenna, with rounded ground plane: (**a**) manufactured and (**b**) final geometric design.

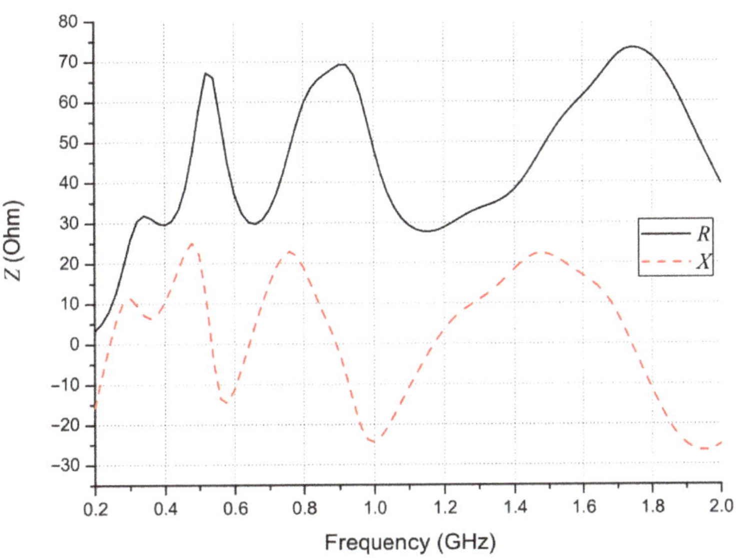

Figure A5. Simulated input impedance of the final antenna model.

Finally, for the sake of illustration, the three-dimensional radiation pattern of the antenna is shown in Figure A6, which was calculated at 450 MHz. Figure A7 shows maximum gain calculated as a function of frequency.

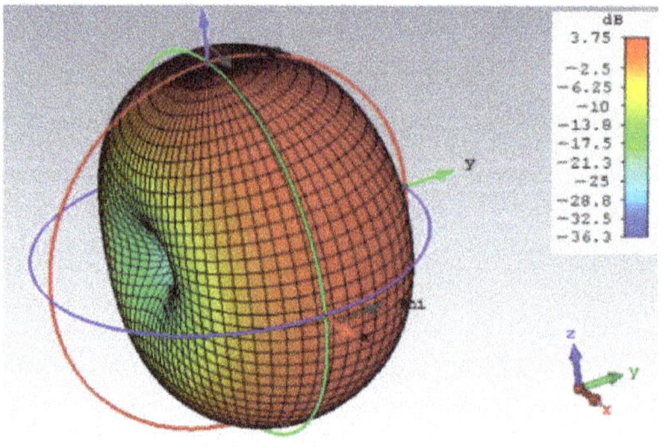

(a)

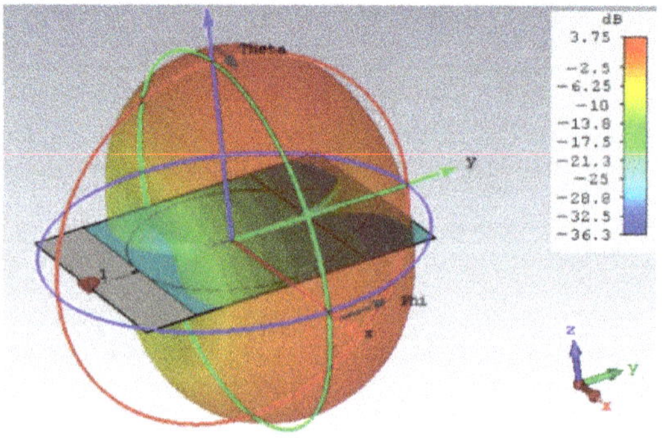

(b)

Figure A6. Simulated 3D radiation pattern of the antenna obtained at 450 MHz: (**a**) standard diagram and (**b**) radiation pattern and the patch antenna.

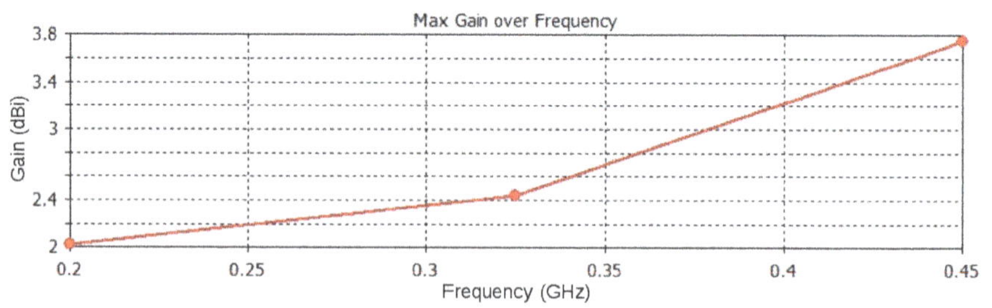

Figure A7. Maximum gain as a function of frequency.

References

1. Stone, G. Condition Monitoring and Diagnostics of Motor and Stator Windings A Review. *IEEE Trans. Dielectr. Electr. Insul.* **2013**, *20*, 2073–2080. [CrossRef]
2. Stone, G.C.; Culbert, I.; Boulter, E.A.; Dhirani, H. *Electrical Insulation for Rotating Machines*, 2nd ed.; Wiley: Hoboken, NJ, USA, 2004.
3. Dehlinger, N.; Stone, G. Surface partial discharge in hydrogenerator stator windings: Causes, symptoms, and remedies. *IEEE Electr. Insul. Mag.* **2020**, *36*, 7–18. [CrossRef]
4. Cselko, R.; Tamus, Z.A.; Szabo, A.; Berta, I. Comparison of acoustic and electrical partial discharge measurements on cable terminations. In Proceedings of the 2010 IEEE International Symposium on Electrical Insulation, San Diego, CA, USA, 6–9 June 2010; pp. 1–5. [CrossRef]
5. Ilkhechi, H.D.; Samimi, M.H. Applications of the Acoustic Method in Partial Discharge Measurement: A Review. *IEEE Trans. Dielectr. Electr. Insul.* **2021**, *28*, 42–51. [CrossRef]
6. IEC/TS60034-27-2. *Rotating Electrical Machines—Part 27-2: On-line Partial Discharge Measurements on the Stator Winding Insulation of Rotating Electrical Machines*, 1st ed.; International Electrotechnical Commission: Geneva, Switzerland, 2012.
7. Araujo, R.C.F.; de Oliveira, R.M.S.; Brasil, F.S.; Barros, F.J.B. Novel Features and PRPD Image Denoising Method for Improved Single-Source Partial Discharges Classification in On-Line Hydro-Generators. *Energies* **2021**, *14*, 3267. [CrossRef]
8. Yadam, Y.R.; Ramanujam, S.; Arunachalam, K. An Ultrawideband Conical Monopole With Radome for Detection of Partial Discharges. *IEEE Sens. J.* **2021**, *21*, 18764–18772. [CrossRef]
9. Raymond, W.J.K.; Illias, H.A.; Bakar, A.H.A.; Mokhlis, H. Partial discharge classifications: Review of recent progress. *Measurement* **2015**, *68*, 164–181. [CrossRef]
10. Luo, Y.; Li, Z.; Wang, H. A Review of Online Partial Discharge Measurement of Large Generators. *Energies* **2017**, *10*, 1694. [CrossRef]
11. Lee, S.B.; Stone, G.C.; Antonino-Daviu, J.; Gyftakis, K.N.; Strangas, E.G.; Maussion, P.; Platero, C.A. Condition Monitoring of Industrial Electric Machines: State of the Art and Future Challenges. *IEEE Ind. Electron. Mag.* **2020**, *14*, 158–167. [CrossRef]
12. Contin, A.; Montanari, G.; Ferraro, C. PD source recognition by Weibull processing of pulse height distributions. *IEEE Trans. Dielectr. Electr. Insul.* **2000**, *7*, 48–58. [CrossRef]
13. Dey, D.; Chatterjee, B.; Chakravorti, S.; Munshi, S. Rough-granular approach for impulse fault classification of transformers using cross-wavelet transform. *IEEE Trans. Dielectr. Electr. Insul.* **2008**, *15*, 1297–1304. [CrossRef]
14. Peng, X.; Wen, J.; Li, Z.; Yang, G.; Zhou, C.; Reid, A.; Hepburn, D.M.; Judd, M.D.; Siew, W.H. Rough set theory applied to pattern recognition of Partial Discharge in noise affected cable data. *IEEE Trans. Dielectr. Electr. Insul.* **2017**, *24*, 147–156. [CrossRef]
15. Stone, G.; Culbert, I.; Bolter, E.; Dhirani, H. *Electrical Insulation for Rotating Machines: Design, Evaluation, Aging, Testing, and Repair*, 2nd ed.; IEEE Press: Piscataway, NJ, USA; Wiley: Hoboken, NJ, USA, 2014.
16. Dakin, T.W. The Relation of Capacitance Increase with High Voltages to Internal Electric Discharges and Discharging Void Volume. *Trans. Am. Inst. Electr. Eng. Part III Power Appar. Syst.* **1959**, *78*, 790–794. [CrossRef]
17. Fruth, B.; Niemeyer, L.; Hassig, M.; Fuhr, J.; Dunz, T. Phase Resolved Partial Discharge Measurements and Computer Aided Partial Discharge Analysis Performed on Different High Voltage Apparatus. In Proceedings of Sixth Int. Symp. on High Voltage Engineering, 6th ISH, New Orleans, LA, USA, 28 August–1 September 1989; pp. 15.03–15.06.
18. Oliveira, R.M.S.; Araujo, R.C.F.; Segundo, A.P.M.; Zampolo, R.F.; Fonseca, W.; Dmitriev, V.; Brasil, F.S. A System Based on Artificial Neural Networks for Automatic Classification of Hydro-generator Stator Windings Partial Discharges. *J. Microwaves Optoelectron. Electromagn. Appl.* **2017**, *16*, 628–645. [CrossRef]
19. Hudon, C.; Belec, M. Partial Discharges Signal Interpretation for Generator Diagnostics. *IEEE Trans. Dielectr. Electr. Insul.* **2005**, *12*, 297–319. [CrossRef]
20. Sako, H.; Mio, K.; Okada, S. Analysis of Phase Resolved Partial Discharge patterns with microstrip antenna. In Proceedings of the 2015 IEEE Electrical Insulation Conference (EIC), Seattle, WA, USA, 27 August 2015; pp. 346–357. [CrossRef]
21. Bartnikas, R. Corona Discharge Processes in Voids. In *Engineering Dielectrics: Corona Measurement and Interpretation*, 2nd ed.; Bartinikas, R., Macmahon, E.J., Eds.; ASTM: West Conshohocken, PA, USA, 1979; Chapter 2, Volume I.
22. Raizer, Y.P. *Gas Discharge Physics*; Springer-Verlag: Berlin/Heidelberg, Germany, 1991.
23. Meek, J.M.; Craggs, J.D. *Electrical Breakdown of Gases*; Clarendon Press: Oxford, UK, 1953; pp. 251–290.
24. Balanis, C.A. *Advanced Engineering Eletromagnetics*; John Wiley and Sons: New York, NY, USA, 1989.
25. Eichwald, O.; Ducasse, O.; Dubois, D.; Abahazem, A.; Merbahi, N.; Benhenni, M.; Yousfi, M. Experimental analysis and modelling of positive streamer in air: towards an estimation of O and N radical production. *J. Phys. Appl. Phys.* **2008**. [CrossRef]
26. Sigmond, R.S. The residual streamer channel: return strokes and secondary streamers. *J. Appl. Phys.* **1984**, *56*, 1355–1370. [CrossRef]
27. Huiskamp, T.; Sengers, W.; Beckers, F.J.C.M.; Nijdam, S.; Ebert, U.; van Heesch, E.J.M.; Pemen, A.J.M. Spatiotemporally resolved imaging of streamer discharges in air generated in a wire-cylinder reactor with (sub)nanosecond voltage pulses. *Plasma Sources Sci. Technol.* **2017**, *26*, 075009. [CrossRef]
28. Hudson, G.G.; Loeb, L.B. Streamer Mechanism and Main Stroke in the Filamentary Spark Breakdown in Air as Revealed by Photomultipliers and Fast Oscilloscopic Techniques. *Phys. Rev.* **1961**, *123*, 1. [CrossRef]

29. IEC60270. *High Voltage Test Techniques—Partial Discharge Measurements*, 3rd ed.; International Electrotechnical Commission—British Standard: London, UK, 2001.
30. Schneider, K.; Farge, M. Wavelets: Mathematical Theory. In *Encyclopedia of Mathematical Physics*; Francoise, J.P.; Naber, G.L.; Tsun, T.S., Eds.; Academic Press: Oxford, UK, 2006; pp. 426–438. [CrossRef]
31. Oliveira, R.M.S.; Modesto, J.F.M.; Dmitriev, V.; Brasil, F.S.; Vilhena, P.R.M. Spectral Method for Localization of Multiple Partial Discharges in Dielectric Insulation of Hydro-Generator Coils: Simulation and Experimental Results. *J. Microwaves Optoelectron. Electromagn. Appl.* **2016**, *15*, 170–190. [CrossRef]
32. Sadiku, M. *Elements of Electromagnetics*, 3rd ed.; Oxford University Press: Oxford, UK, 2000.
33. Yang, F.; Peng, C.; Yang, Q.; Luo, H.; Ullah, I.; Yang, Y. An UWB printed antenna for partial discharge UHF detection in high voltage switchgears. *Prog. Electromagn. Res. C* **2016**, *69*, 105–114. [CrossRef]
34. Xavier, G.; Costa, E.; Serres, A.; Nobrega, L.; Oliveira, A.; Sousa, H. Design and Application of a Circular Printed Monopole Antenna in Partial Discharge Detection. *IEEE Sens. J.* **2019**, *19*, 3718–3725. [CrossRef]
35. Zhang, Y.; Lazaridis, P.; Abd-Alhameed, R.; Glover, I. A compact wideband printed antenna for free-space radiometricdetection of partial discharge. *Turk. J. Electr. Eng. Comput. Sci.* **2017**, *25*, 1291–1299. [CrossRef]
36. Dunlap, R.A. *The Golden Ratio and Fibonacci Numbers*; World Scientific Publishing Company: Toh Tuck Link, Singapore, 1998.
37. Gupta, S.; Arora, T.; Singh, D.; Kumar Singh, K. Nature Inspired Golden Spiral Super-Ultra Wideband Microstrip Antenna. In Proceedings of the 2018 Asia-Pacific Microwave Conference (APMC), Kyoto, Japan, 6 November 2018; pp. 1603–1605. [CrossRef]
38. Meena, A.; Garhwal, A.; Ray, K. Nature Inspired Fibonacci Sequence Microstrip Patch Antenna For Energy Harvesting Applications. *Int. J. Recent Technol. Eng. (IJRTE)* **2019**, *8*, 6154–6160.
39. Tanabe, M.; Nakano, H. Low-Profile Wideband Spiral Antenna With a Circular HIS Reflector Composed of Homogenous Fan-Shaped Patch Elements. *IEEE Trans. Antennas Propag.* **2020**, *68*, 7219–7222. [CrossRef]
40. Soboll, P.; Wienstroer, V.; Kronberger, R. Smooth Moves in Power Transition: New Yagi-Uda Antenna Design for Wireless Energy. *IEEE Microw. Mag.* **2016**, *17*, 75–80. [CrossRef]
41. Balanis, C.A. Fundamental Parameters of Antennas. In *Antenna Theory: Analysis and Design*;Wiley-Interscience: Hoboken, NJ, USA, 2005; Chapter 2, pp. 27–132.

energies

MDPI

Article

Failure Diagnosis and Root-Cause Analysis of In-Service and Defective Distribution Transformers

Saravanakumar Arumugam

Institute of Marine Engineering, University of Rostock, Justus-von-Liebig-Weg 2, 18059 Rostock, Germany; saravanakumar.arumugam@uni-rostock.de

Abstract: Diagnostic data of transformers are essential in determining their integrity and to estimate their reliability. Such diagnostic data is not available for distribution transformers as only meagre attempts are made to accumulate this information. The reason for lack of such attention is attributed to the compromise made to meet financial constraints and cost-cutting measures. It is needless to emphasize the significance of these diagnostic data in evading premature failures and enhancing the accuracy of the maintenance program. Overall, the concepts of diagnostic methods are well defined regarding power transformers, but cannot be extended to distribution transformers. This paper attempts to fill this gap by measuring the diagnostic data from in-service and defective distribution transformers. For this purpose, eight transformers (six in-service, two defective) showing deviations in their diagnostic status were selected from a group. As per their kVA rating and operating status, the chosen transformers were categorized into three sets and their integrity of insulation and winding-core arrangement were analyzed. Later, the root-cause analysis was performed on defective transformers to understand their failure reasons and possibilities to evade them. The pertinent data obtained are quite valuable for making decisions regarding reliability-oriented maintenance and for extending its potential in the distributed generation.

Keywords: distribution transformers; failure analysis; monitoring; diagnostic method; frequency response; open and short-circuit response

Citation: Arumugam, S. Failure Diagnosis and Root-Cause Analysis of In-Service and Defective Distribution Transformers. *Energies* **2021**, *14*, 4997. https://doi.org/10.3390/en14164997

Academic Editors: Zhijin Zhang and Hualong Zheng

Received: 12 July 2021
Accepted: 12 August 2021
Published: 14 August 2021

1. Introduction

Distribution transformers are essential in securing a power supply to the industries and household applications [1]. The selection criteria of these transformers depend on energy requirements at the receiving end and the budget of the user and the service provider [2]. Naturally, they emerge as an important asset to the utilities and industries and requires undivided attention to ensure their operating condition and reliability [3]. Present energy requirements, environmental restrictions and financial constraints have put forth the notion to reduce the budget and life-cycle costs. In addition, there is a constant demand for stringent maintenance procedures and sensitive measurement techniques [3,4]. This implies inclusion of sensitive monitoring tools and diagnostic test methods to track their operating condition and efficiency of distribution transformers. Nevertheless, the modern distribution transformers are not provided with such monitoring or diagnostic support favouring the cost cutting measures [5]. It is also mentioned to be unlikely to outweigh the benefit of replacing a transformer than employing diagnostic tests and/or monitoring support [3–5]. Nonetheless, it is critical for the service provider to know the failure reasons not only to evade its re-occurrence but also to always ensure the quality and reliability of the supplied electric power [6,7]. Pertinent monitoring and diagnostic solutions create a decisive data for effective management of the network of transformers and enhance their efficiency, performance, reliability [3]. So, any attempt to improve the existing maintenance procedures remains highly desirable.

The concepts of detecting premature failures and exercising diagnostic methods are well defined on power transformers, while extending the same to a distribution transformer may not reveal similar results. The corresponding reason may be due to the differences in geometry, construction, winding design, etc., which usually reflects in the measured responses as natural frequencies, resonant peaks, and so on [5]. For instance, the open and short-circuit natural frequencies of the power transformers appear as several resonant peaks within 700 to 900 MHz while the same for a distribution transformer will be entirely different and may appear (if any) above 1 MHz. So, the usual shift in the resonant peaks initiated by the terminal conditions and faulty situations which are observable in a power transformer may not appear in the distribution transformer [6]. In this context, the complete spectrum of the distribution transformer and the respective individual influence of winding, core, and insulation geometry must be studied. There are several literatures and technical reports that discusses the diagnostic concepts, methods, etc., with focus on its application on transformers, bushings, power cables, and rotating machines. Surprisingly, such concepts are not extended to distribution transformers only for the reasons of meeting financial constraints and cost-cutting measures [3,4]. The utilities report that the benefit of replacing a distribution transformer outweighs the cost of exercising a diagnostic method, which is incorrect [4]. Availing such diagnostic data of distribution transformers and failure root cause analysis are very important to retain their service time and widening their operating potential when venturing their application into renewable energy and distributed generation plants. Such diagnostic data is also very useful while enhancing the sensitivity and accuracy of reliability-oriented maintenance program.

A distribution network feeding a city predominantly contains a fleet of distribution transformers manufactured from different factories and are designed to operate under minimal supervision and maintenance [1,5,6]. In recent years, the distribution transformers have occupied a significant role in the distributed generation [7]. Pertinent transformers have multiple purposes such as maintaining inverter output voltage to the desired level, interconnecting dispersed generators to the grid or utility, isolates faulty side and retains voltage regulation [8]. Although made from different factories, their design and construction are expected to be similar, as their voltage/current/kVA ratings are identical [5,9]. Despite design excel, robust construction and better-quality materials, a specific number of distribution transformers in a fleet inevitably fail during its service time [10–16]. Pertinent failures are mostly internal and are attributed to overloading, over voltage, transients, and internal winding faults [11,13]. So, the usual practice is to limit the load of the transformer within 80% of its full capacity so as to minimize the ageing and deterioration [11,13,14]. Alternatively, an adequate maintenance program that would constantly track, monitor, quantify the diagnostic status of the distribution transformers at all times may be exercised, which forms the subject matter of this paper [3–5].

Modern asset management program involves three strategies viz., correction, prediction, and reliability-oriented maintenance, respectively [1]. Pertinent tests (electrical, visual inspection, etc.) are similar and depending on their application time, condition, etc., their objectives change. Amongst all, the corrective maintenance tests are performed only after the fault has occurred in the transformer [1,2]. Naturally, the corresponding repair tasks are too expensive and require skilled labour. The preventive maintenance tests involve time and/or condition-based monitoring tools and diagnostic testing methods that are slightly expensive, nevertheless ensures the normal life of the transformer [1,2]. In the meantime, the reliability centered maintenance tests have evolved recently and drawn wide attention of utilities and industries. This reliability centered approach not only evades the risk of premature failures but also provides the health index and quantifies the lifetime and reliability of the transformer in numerical terms [1]. Nevertheless, employing the reliability centered maintenance requires a large number of onsite/field data of in-service condition, rate of failures, mode of failures, etc. [4]. Such data is abundantly available for power transformers while the same is not true for distribution transformers [1,2].

2. Existing Literature

The importance of distribution transformers in a network and their significance in maintaining voltage regulation has been well known for several decades. In recent years, the distribution transformers in distributed generation have found a variety of applications. In this regard, there are several literatures that are available and describe all the aspects of design, construction, voltage regulation, etc. Despite all, the literature that are relevant to this study focusing on incipient fault conditions and premature failures are alone gathered and discussed in this paper. Pertinent literatures are grouped into five major categories and their inferences related to the present context are explained.

2.1. Significance of Distribution Transformers in Distributed Generation

The primary significance of distribution transformers in distributed generation lies in their ability to interconnect several distributed sources to grid and utility system [7]. Depending on their application, they are addressed as grid, interconnecting or distribution transformers [8]. In general, the grid and distribution transformers occupy a quite significant place in both conventional and distributed generation network [17]. During this, the distribution transformers step up the output voltage of the inverter sufficiently enough to meet the distribution level [18,19]. At the same time, the grid transformers, similar in construction to distribution transformers, are used to inject power into the grid by interconnecting the distributed generation network [19]. So, naturally, the operating cost and efficiency are one of the important factors that directly influence the plant income. In most cases, the operating conditions such as power rating, system voltage, duty cycle, site conditions, etc., are important criteria that decide the functionality of distribution transformers [19]. The corresponding average failure time of such distribution transformers used in the distributed generation network is estimated to be approximately 30 years, provided there are no external factors that cause premature failures [20–22]. Nevertheless, these distribution transformers suffer premature failure caused by natural events, line to ground faults, etc., despite the fault current being equally handled by the distributed generation source [21,22]. In all, the probability of failures due to line to ground are higher than the other reasons.

2.2. Significance of Diagnostic Measurements

Currently available literatures mainly focus on reporting the sensitivity of diagnostic measurements on revealing the dielectric and mechanical status of power transformers [23–28]. The mechanical strength of the windings [23], the influence of temperature and moisture on the transformer insulation [24], and the impact of terminal connections in improvising the sensitivity of pertinent diagnostic method [25,26] and using non-conventional methods on determining the interwinding capacitance [27–29] have received wide attention. In addition, availing the diagnostic data, the lifetime model of a transformer model under combined electrical and thermal stresses are reported [30]. However, extending the same to a distribution transformer is yet to be investigated [1,5,13,28]. The currently practiced test methods on distribution transformers are not diagnostic but only primitive and indicate only after the failure has occurred [13]. Instead, the focus of currently practiced tests on distribution transformers are related to load factors [1,2].

2.3. Current Focus on Distribution Transformers

It is reported that the distribution transformers are operated with a low load factor, i.e., 40% to 60% of loading during normal conditions [1,2]. Additionally, only 25% of the distribution transformers in a network carries peak load that too for a limited duration [2]. So, naturally it is believed that due to the low load factor, the distribution transformers are immune to any failure situation [1]. Consequently, corrective maintenance strategies are popularly used on the current distribution transformers [3–5]. Nevertheless, it is well known that a certain percentage of distribution transformers suffer internal faults due to several factors such as overloading, transient condition or internal winding faults and so on [1,2,10,11]. Most of these internal failures may go unnoticed under the initial phase and appear only after they reach an adverse situation [13]. So, it is important to employ diagnostic tests to monitor distribution transformers.

2.4. Failures in Distribution Transformers

The failures in the distribution transformers can be broadly grouped into two categories viz., incipient, and developed faults [6–13]. The developed fault appears once the event has occurred and caused significant damage to the system [11]. As opposed to this, the incipient faults are nothing but evolving and have not created damage and can henceforth be rectified if identified within proper time [5,6,11,13]. The failures in a distribution transformer over their service time follows a curve which resembles the shape of a 'bath-tub' [10]. This 'bath-tub' curve indicates that the failures are at the highest at the initial (installation and commissioning) and final stages (normal end of service life) of a distribution transformer while the same is at its minimum within the intermediate stage [10,11,13]. The failures in the intermediate phase are known as premature in nature and usually occur due to overloading, transients (lightning, switching, etc.), and short-circuit faults [13]. Literature indicates that around 17% of failures in distribution transformers are due to the insulation, and the same due to overloading and winding-core are within 24% and 25%, respectively [11,13,15,16,23]. In any case, these failures are internal and broadly affect either the insulation and/or the winding-core arrangement.

2.5. Currently Practiced Maintenance Test Methods

The tests that are popularly employed on distribution transformers can be broadly grouped into two viz field tests (insulation resistance, polarization index, turns-ratio, etc.) and diagnostic tests (dielectric response analysis (DRA), frequency response analysis (FRA), dissolved gas analysis (DGA), low-voltage impulse, etc.), respectively [29,31,32]. In all, the field tests are used to determine and quantify the developed faults in the distribution transformers. A subsequent step would be to bring the distribution transformer offline and initiate corrective measures such as oil change, refurbishment, etc., which only adds up as additional costs to the investment [3,4]. However, it is expected that the implementation of monitoring methods would be too expensive [3,4]. The corrective maintenance would cost up to 40% of the total investment and the pertinent aftermath would result in refurbishment and/or scrapping, which is non-profitable [3,4,13]. At the same time, the predictive maintenance tests would reduce costs by 15%, avoid the expensive refurbishment, scrapping process and help in retaining the normal lifetime of the distribution transformers. In addition, the utilization of the distribution transformer is enhanced, which generates more revenue as intended [2].

In summary, it appears from these literatures that the present distribution transformers are provided with field tests and corrective maintenance strategies are suited for determining faults that have developed after an event. Pertinent maintenance strategies are most correctly involving procedures that add additional costs to the investments. At the same time, diagnostic tests such as DGA, DRA, FRA, etc., are much more sensitive in detecting even the incipient faults yet are disregarded to be used on distribution transformers. It is due to the belief that the inclusion of such tests might increase the operating cost, which is not correct. The predictive maintenance is only 15% of the total costs and provides an excellent opportunity to retain their lifetime and reliability. The utilization of the transformer is further prolonged, which eventually generates more revenue [2]. Furthermore, the data obtained would be valuable for the reliability-oriented maintenance strategy, as it requires many onsite/field data of in-service conditions, rate of failures, mode of failures, etc. [3,4]. Such data is abundantly available for power transformers, while the same is not true for distribution transformers [3,4]. In this context, a group of in-service and defective distribution transformers are selected and subjected to diagnostic measurements.

3. Onsite Transformers

Eight distribution transformers that were in-service at the residential and industrial areas were selected and subjected to diagnostic measurements. Table 1 shows the list of in-service and defective distribution transformers involved in the present study. The chosen distribution transformers were of the same vector group (Dyn5), containing the same winding connection (Δ/Y-n) and were constructed at different factories on a different timeline or year. Nevertheless, their design and constructional features were expected to be similar, making them comparable.

Table 1. List of distribution transformers involved.

Set	Transformer	Year	Ratings			Tap/Total
			kVA	kV	A	
1 [a]	3φ, 2-winding, Δ/Y-n	2015 1994 1994	400	20/0.4	20/900	3/5
2 [b]	3φ, 2-winding, Δ/Y-n	2015 2002 1983	630	20/0.4	18/900	3/5
3 [c]	3φ, 2-winding, Δ/Y-n	1994 © 1998 §	630	20/0.4	18/900	3/5

a, b—In-service transformer, c—defective transformer, ©—Industrial area (Thermal defect), §—Residential area (Winding defect).

So, the chosen transformers were classified as per their kVA rating into three groups viz., set-1, set-2, and set-3, and identified using their group and timeline. In all, the distribution transformers in set-1 and set-3 are in-service energizing residential area. In all, the distribution transformers in set-1 and set-3 are in-service energizing residential area. The transformers in set-3 were in-service and emerged to be defective during their service time. Two defective transformers manufactured in 1994 and 1998 were enlisted in the set-3 category. The first defective transformer, manufactured in set-3/1994 had suffered thermal failure while the second, manufactured on set-3/1998, suffered winding failure. Naturally, the DRA and FRA methods were employed to evaluate their complete diagnostic status. The transformer with thermal failure was energizing an industrial site and had failed at least two times in 2016 and 2017. The reason for the failure was identified as a thermal problem, as the same caused the oil to vent from its hermetically sealed tank.

4. Measurements and Analysis

Figure 1a–d shows the picture of test setups, in-service transformers, terminal connections, etc., adopted during DRA and FRA tests, while the terminal connection was adopted, and the parameters were measured. The excitation approach is shown in Figure 2a–h, respectively. It is well known that the DRA tests assess the insulation and its geometry [24,29,31,32] while the FRA identifies defects in the winding and core of the transformers [25–29]. The results obtained from DRA and FRA methods collectively describe the complete diagnostic status of a distribution transformer. The basic test procedure of DRA and FRA methods are the same while the test frequencies applied are different. The DRA setup involves a dielectric response analyzer (Figure 1a) that generates sinusoidal test voltages up to 200 V with frequencies ranging from 100 μHz to 5 kHz in discrete steps. This study uses sinusoidal test voltage of 200 V in amplitude and frequencies ranging 500 μHz to 5 kHz. The frequency dependent insulation parameters such as the loss factor, permittivity, etc., are obtained. The same is compared with the model curve (using the XY model) generated using an inbuilt curve fitting algorithm [27]. The inbuilt algorithm then fits the measured data with the model curve and calculates the moisture level and oil conductivity by referencing to database [27]. The FRA test setup shown in Figure 1c,d comprises of a frequency response analyzer popularly used for diagnosing the windings of a transformer. The frequency response analyzer generates spectrally pure sinusoidal signal of 1 V_{rms} amplitude with frequency sweep from 10 Hz to 20 MHz. Initially, a suitable terminal connection and system function pair as recommended by the current literature and IEEE/IEC standards are adopted [12,25,26]. Pertinent mode of excitation, terminal connection, measurable input, and output parameters, etc., are shown in Figure 2a–h. Following this, sinusoidal signals of 1 V_{rms} in amplitude with frequencies sweeping from 10 Hz to 2 MHz are injected into the terminals and the pertinent input, reference and output signals are recorded. The amplitude response is obtained from the measured data which is further analyzed to understand the diagnostic status of the transformer windings.

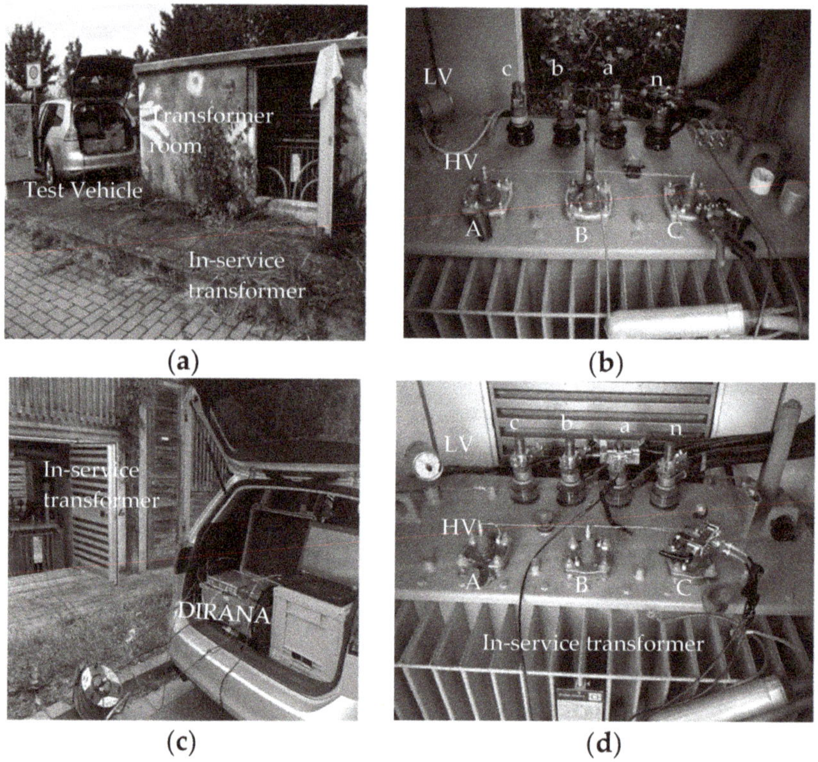

Figure 1. Test setup used for assessing the diagnostic status of the chosen distribution transformers. (**a**) DRA method; (**b**) terminal connection (DRA); (**c**) FRA method; (**d**) terminal connection (FRA).

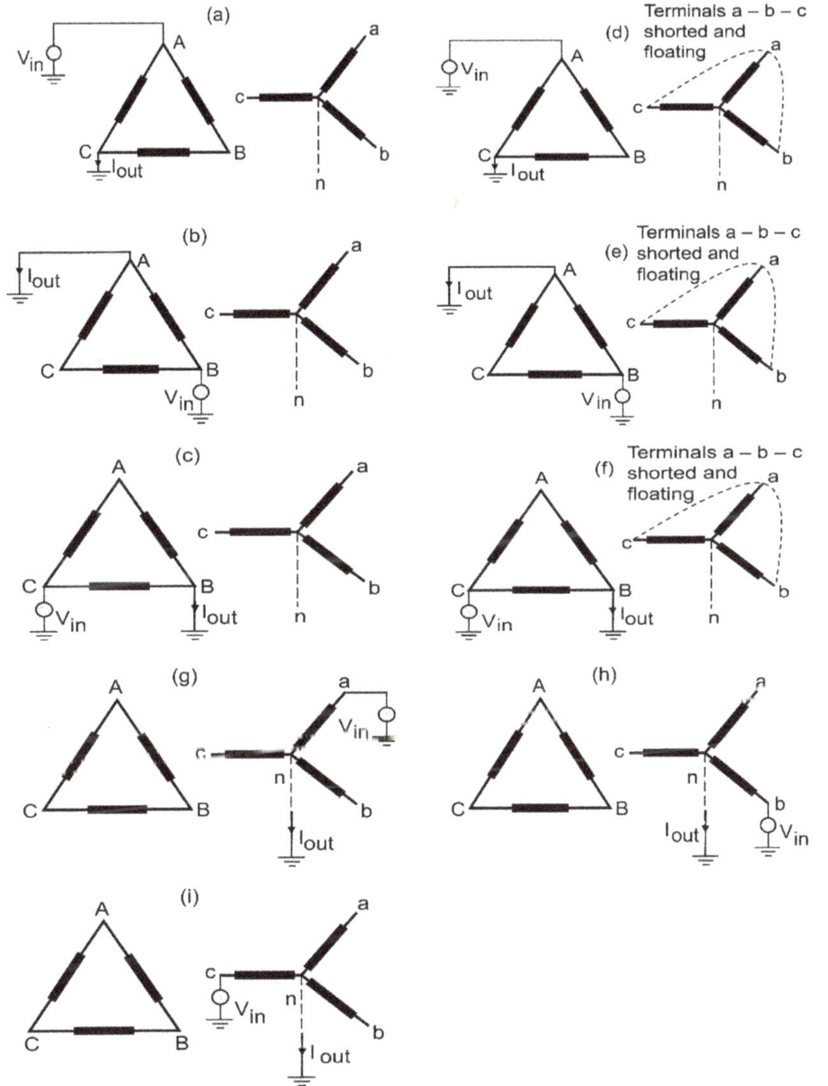

Figure 2. Terminal connection and measurable parameters adopted as recommended by the standard IEEE PC57.149 and literature [25,26]. (**a**) High voltage open-circuit response: A-C; (**b**) B-A; (**c**) C-B; (**d**) high-voltage short-circuit response: A-C; (**e**) B-A; (**f**) C-B; (**g**) low-voltage open-circuit response: a-n; (**h**) b-n; (**i**) c-n.

4.1. In-Service Transformer Units

The DRA tests on in-service transformer units employ two important analysis strategies. First is the comparative analysis of measured data with the model curve developed by an algorithm inbuilt in the device. Later, using the geometry, temperature, etc., a response that fits to the measured data is calculated and the corresponding values of oil conductivity, moisture content, etc., are estimated [29]. The second strategy employs comparative analysis of measured data with the new or refurbished sister-unit transformers that are identical to the test object. Table 2 shows the loss factor, capacitance, oil conductivity and estimated level of moisture ingress in the chosen set-1 and set-2 in-service distribution transformers.

It can be seen from the Table 2 that the insulation of the in-service transformers (set-1 and set-2) are in better status. The measured loss factor of all in-service units (set-1 and 2) lies within 0.002 to 0.003, which is well within the acceptable limit. The conductivity of the oil-paper insulation of set-1 distribution transformers remained within 1.3 pS to 3.7 pS, which is also within the acceptable limit. The same for the set-2 transformer manifested slightly higher values, i.e., within 6.1 pS to 7.2 pS. So, these two parameters confirm that there are no ageing or deterioration in the insulation system of set-1 and set-2 distribution transformers. However, the moisture levels of these transformers yielded different results.

Table 2. Insulation parameters of set-1, set-2 in-service transformers.

Set	Year	Insulation	Loss Factor (50 Hz)	Oil Conductivity		Moisture	
				σ	Status	Estimation	Status
			p.u.	pS		%	
1 a	2015	C_{HL}	0.003	1.3	Good	1.1	Dry
	1994		0.002	1.7	Good	1.3	Dry
	1994		0.003	3.7	Good	3.6	Moderately wet
2 b	2015	C_{HL}	0.003	6.1	Good	2.3	Wet
	2002		0.003	7.2	Good	2.0	Moderately wet
	1983		0.003	6.1	Good	3.6	Moderately wet

a, b—In-service transformer.

Comparatively, the set-1 in-service transformers (2015 and 1994-1) emerged as 'dry' with a low level of moisture in its insulation. Pertinent values of moisture were estimated as 1.1% to 1.3%, respectively. At the same time, the in-service unit, identified as set-1/1994-2 emerged with a higher level of moisture in its insulation. The level of moisture is estimated to be around 3.6%, therefore marked as 'moderately-wet'. The set-2 distribution transformers emerged as either 'wet' or 'moderately wet' as their estimated moisture levels remain within 2.0% to 3.6%. Particularly, the in-service unit identified as set-2/1983 manifested slightly higher level of moisture ingress, i.e., around 3.6%, henceforth marked as 'wet'. The reason for such increased level of moisture might be their operating condition and location. The set-1 in-service transformer units (other than set-1/1994-2) are installed in residential areas while the set-2 transformers are installed and feeding households outside the city. Being exterior to the city limits, their constant exposure to the open weather conditions and lesser loads may be the reason for such minor deviations.

Figure 3a–d shows the insulation parameters (i.e., loss factor, capacitance, resistance, and impedance) of in-service distribution transformers grouped under set-1 category. It becomes clear from these figures that the insulation parameters of in-service transformers identified as set-1/2015 and set-1/1994-1 are within the acceptable level as their pertinent traces match each other. Pertinent insulation parameters such as loss factor (Figure 3a), capacitance (Figure 3b), resistance (Figure 3c), and impedance (Figure 3d) remain in-line to each other. Such observations are not surprising as these in-service units, grouped in set-1 category are either new (set-1/2015) or newly refurbished (set-1/1994-1). However, the transformer identified as set-1/1994-2 emerged with mild deviations in its respective insulation parameters. The loss factor of this set-1/1994-2 unit emerged with slight deviations throughout the frequency span. These deviations in the loss factor (Figure 3a), capacitance (Figure 3b) and impedance (Figure 3c) appeared more dominant at very lower frequencies (i.e., say from 100 mHz to μHz) indicating moisture ingress in the insulation. Such deviations in the resistance (Figure 3b) appeared predominantly from few 100 Hz to μHz, respectively. These deviations in insulation parameters collectively indicate that this transformer (set-1/1994-2) relatively has higher moisture ingress in insulation. This finding is in consonance with observations (moisture 3.6%) from Table 2.

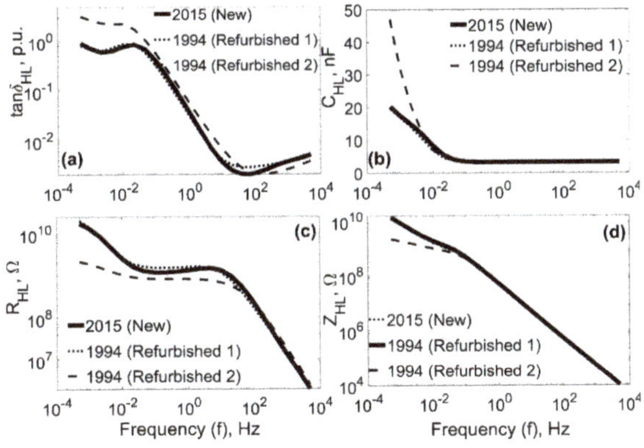

Figure 3. Insulation parameters of set-1 in-service transformers (**a**) loss factor; (**b**) capacitance; (**c**) resistance; (**d**) impedance.

Figure 4a–d shows the insulation parameters measured from the in-service distribution transformers grouped under set-2 category. As opposed to the set-1 units, the insulation parameters of the set-2 distribution transformer manifested notable deviations in the response function. The loss factor of all the three units shown in Figure 4a manifested deviations throughout the frequency span. In particular, the deviations in the mid-frequency range (i.e., within 100 mHz to 100 Hz) of transformer unit set-2/2002 indicate a slightly higher level of oil conductivity. The same observation can also be made from Table 2. Nevertheless, the percentage of change in the oil conductivity is relatively smaller, and can therefore be deemed an acceptable level. At the same time, the deviations in the lower frequencies of the loss factor (Figure 4a), capacitance (Figure 4b), and resistance (Figure 4c) of set-2/1983 in-service unit indicates moisture ingress in the insulation, thereby positioning it in 'wet' state. This is expected as the in-service transformer unit identified as set-2/1983, is relatively older than the others and had been operating in the outer-skirts of residential area.

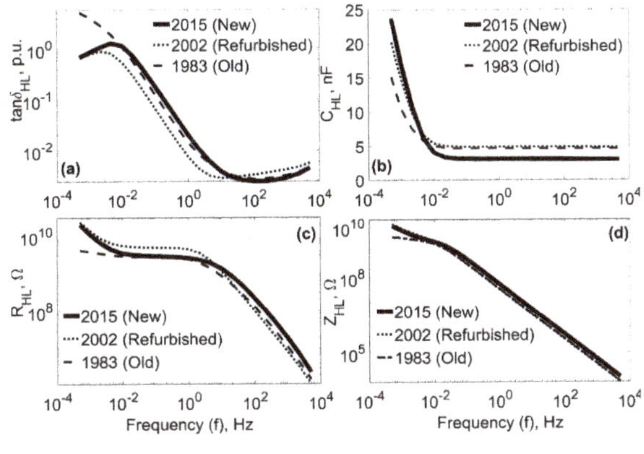

Figure 4. Insulation parameters of set-2 in-service transformers; (**a**) loss factor; (**b**) capacitance; (**c**) resistance; (**d**) impedance.

Figure 5a–i shows the open and short-circuit responses of set-1 in-service distribution transformers. At first glance, it becomes clear from these figures that the in-service units identified as set-1/2015 and set-1/1994-1 manifested no significant deviations in their respective open and short-circuit response functions. The high voltage open-circuit (HVOC) response of phases A-C (Figure 5a), B-A (Figure 5d), and C-B (Figure 5g) exhibits an exact match between each other. Similarly, the high voltage short-circuits (HVSC) response of phases A-C (Figure 5b), B-A (Figure 5e), and C-B (Figure 5h) also exhibit an exact match between each other. Naturally, there are no significant changes observable in the low voltage open-circuit (LVOC) response of all the three phases viz., a-n (Figure 5c), b-n (Figure 5f) and c-n (Figure 5i). So, the outcome of the FRA tests altogether confirms that the windings of all the in-service units identified as set-1/2015 and set-1/1994-1 are intact without any major deviations from its diagnostic condition. However, the open and short-circuit response of the other in-service unit identified as set-1/1994-2 manifested a small deviation in its amplitude response function. These changes are insignificant and follows the pattern of other phases. It hence can be concluded that this transformer unit is also in good condition.

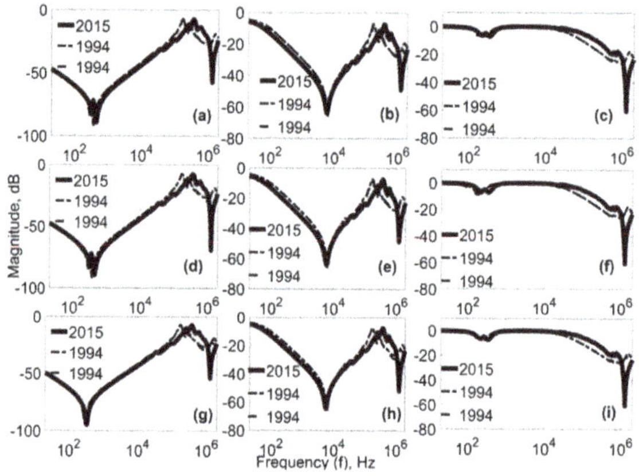

Figure 5. Open and short-circuit responses of the set-1 in-service distribution transformers: HVOC (**a**) A-C; (**b**) B-A; (**c**) C-B, HVSC: (**d**) A-C; (**e**) B-A; (**f**) C-B, LVOC: (**g**) a-n; (**h**) b-n; (**i**) c-n.

Further FRA tests (Figure 6a–i) on set-2 in-service distribution transformers revealed similar results. Figure 6a–i shows the open and short circuit responses of set-2 in-service transformers adopted in this study. The open and short-circuit responses of units set-2/2015 and set-2/2002 manifested no notable deviations in their open and short-circuit response function. All their traces of HVOC and HVSC response of phases A-C (Figure 6a,b), B-A (Figure 6d,e) and C-B (Figure 6g,h) and their LVOC responses of phases a-n (Figure 6c), b-n (Figure 6f) and c-n (Figure 6i) match each other without any deviations. So, it can be confirmed that the windings of these in-service units (set-2/2015, set-2/2002) are in good condition. The in-service unit identified as set-2/1983 manifested mild deviations in their respective HVOC, HVSC, and LVOC response function. In all, the deviations appear more prominent at the low voltage phases, i.e., a-n, b-n, and c-n, respectively. As the deviations are quite smaller in magnitude and the respective pattern followed is similar to the other phases, it can be once again concluded that this in-service unit is in good condition.

Thus, it becomes clear from these experiments that there are no significant changes in the respective open and short circuit response functions. This confirms the fact that the transformers grouped in this category are in good condition and can be further employed in the distribution network. With this information, the defective distribution transformers grouped under set-3 category are subjected to diagnostic investigation.

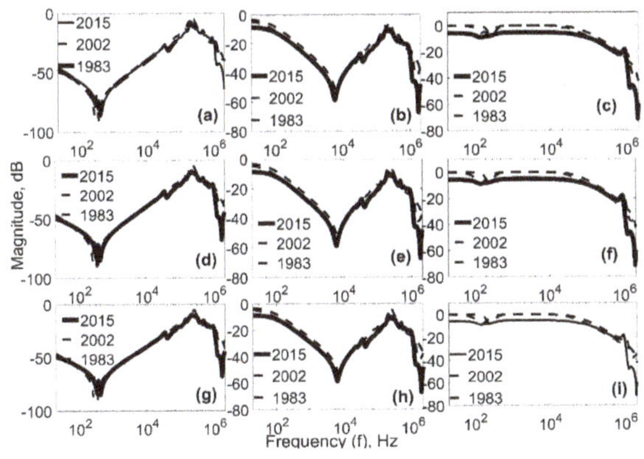

Figure 6. Open and short-circuit responses of the set-2 in-service distribution transformers: HVOC (**a**) A-C; (**b**) B-A; (**c**) C-B, HVSC: (**d**) A-C; (**e**) B-A; (**f**) C-B, LVOC: (**g**) a-n; (**h**) b-n; (**i**) c-n.

4.2. Defective Distribution Transformer Units

Two distribution transformers (Table 1) that had several interruptions during its service time are selected and subjected to diagnostic tests. The first defective transformer (set-3/1994) grouped under set-3 category, was in-service at an industrial site and was tripped out at least by three intervals due to elevated temperature. During each interruption, an oil sample was extracted, and the gas dissolved in oil is estimated. After the third interruption, the transformer suffered a failure causing the oil to vent, forcing it to go offline. After gathering this information, the dielectric, mechanical and thermal integrity of this defective transformer (set-3/1994) is studied and the results that reveal interesting information are alone discussed. The second defective transformer (set-3/1998) was feeding a residential area in the outskirts of the city. Figure 7a,b shows the picture of set-3/1998, the second defective transformer that has winding failure initiated by the localized hot spot temperature and arcing phenomena. The oil system and thermal indicators of this transformer (set-3/1998) were initially normal, however the relay has tripped repeatedly at peak loads. Eventually, this transformer suffered an untimely failure, so it was removed from service and subjected to tests. After these tests, this transformer was opened to visually inspect the failure locations.

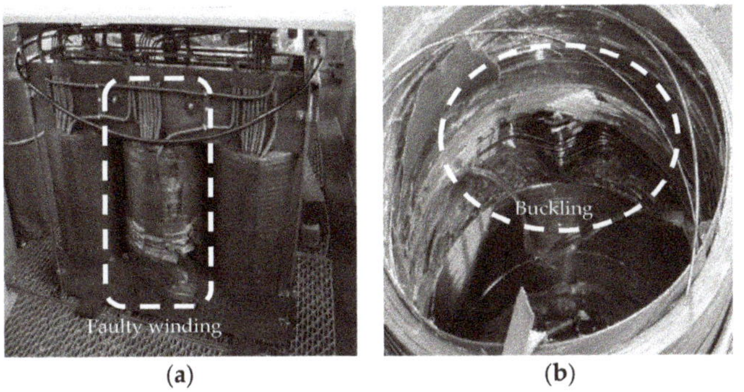

Figure 7. Picture of 630 kVA, 20 kV/400 V, distribution transformers with defective windings grouped under set-3 category. (**a**) Frontal view showing defective winding; (**b**)internal bucking winding defect (interior view).

Figure 7a,b pictorially describes the failure locations in the windings. Table 3 lists the values of gases detected in the oil sample and the possible reasons for their development. The oil samples extracted during three interruptions are mentioned in the Table 3 as sampling intervals 1, 2, and actual. The actual being the present case and the intervals 1 and 2 are data from previous samples. It can be observed from Table 3 that the gases hydrogen (H_2), carbon monoxide (CO), and carbon-dioxide (CO_2) have gradually evolved to a higher value. In all, the H_2 evolved from 14 ppm to 150 ppm, indicating minor interior discharge activities. The CO increased from 57 to 1000, indicating carbonization problems. The CO_2 gas that increased from 4979 ppm to a higher value of 12,000 ppm indicated possible moisture content in the transformer. In addition, the presence and increase of methane (CH_4) (up to 100 ppm) dissolved in the gas indicated interior sparking problems. Additionally, the ethane (C_2H_6) and ethylene (C_2H_4) gases were identified, which increased to higher values indicating localized heating in the transformer. So, acetylene (C_2H_2) gas evolved indicating winding hot spot.

Table 3. Gas dissolved in oil of thermally defective transformer.

Gas in Oil	Sampling Interval			Reference Values		Remarks
	1	2	Actual	Normal	Caution	
	ppm	ppm	ppm	<ppm	<ppm	
H_2	14	103	150	150	1000	Discharge
O_2	5024	2266	n. a.	–	–	Leakage in gasket
N_2	79,985	85,683	n. a.	–	–	–
CO	57	74	1000	10,000	15,000	Carbon
CO_2	4979	5385	12,000	10,000	15,000	Moisture
CH_4	32	65	100	25	80	Sparking
C_2H_6	208	228	300	10	35	Localized heating
C_2H_4	6	47	30	20	150	Surface heating
C_2H_2	<1	<1	500	15	70	Hot spot

n. a.—Not applicable.

In all, the presence of H_2, CO does not pose a serious threat as their values when compared with the reference data are well within the acceptable limit. The level of CO_2 when compared with the reference values indicated minor moisture content, which is also not a serious threat. At the same time, the level of gases such as methane, C_2H_6, and C_2H_4 indicates sparking, localized heating and hot spot problems, which is a serious issue. This might be the reason for the thermal failure of this distribution transformer. Following this, the insulation condition of these defective transformers is studied. Table 4 shows the values of insulation parameters of defective transformers (set-3/1994, set-2/1998) measured using the DRA method. It can be observed from Table 4 that the status of the oil of both the defective transformers are in acceptable condition. However, a noticeable level of moisture has ingress into the insulation of both the defective transformers. The transformer with thermal fault emerged with 4.6% of moisture while the same for the winding fault was close to 4.5%, respectively. So, to make a detailed analysis, the measured insulation parameters are compared with a sister-unit transformer.

Table 4. Measured insulation parameters of set-3 defective distribution transformers.

Set	Year	Insulation	Loss Factor	Oil Conductivity		Moisture	
				σ	Status	Estim.	Status
			p.u.	pS	–	%	–
3 c	1994 *	C_{HL}	0.003	3.2	Good	4.6	Wet
	1998 §	C_{HL}	0.002	28	Satisfy	4.5	Wet

*—Thermal fault; §—Winding fault; c—defective transformer.

Figure 8a–d shows the insulation parameters of the thermally defective (set-3/1994) distribution transformer. It becomes clear from these figures that the insulation condition of the thermally defective transformer remains in acceptable condition. The loss factor of the thermally defective transformer within the frequency span of 10^{-2} Hz to 10^2 Hz appears to remain in-line. The loss factor beyond 100 Hz appears higher than its sister-unit counterpart. Similar observations can be made at frequencies below 10^{-2} Hz. The reason for the increase in loss factor can be attributed to the moisture ingress in the insulation. The other insulation parameters such as capacitance (Figure 8b) and impedance (Figure 8d), manifests a close match with the sister-unit data throughout the frequency. However, the resistance of the thermally defective seems to be slightly higher than the sister-unit data at very higher frequencies. This confirms the fact that the insulation system of the thermally defective transformer is intact, but suffers moisture problems. This finding is in consonance with the observation made from the Table 4.

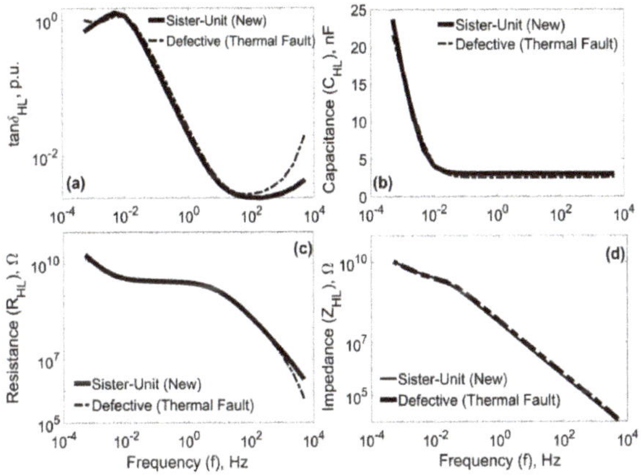

Figure 8. Measured insulation parameters of thermally defective distribution transformer grouped under set-3 category (**a**) loss factor; (**b**) terminal capacitance; (**c**) resistance (**d**) impedance.

Figure 9a–d shows the insulation parameters measured from the set-3/1998 winding-defective transformer. It becomes clear from these figures that the condition of the insulation of the winding-defective transformer unit (set-3/1998) remains compromised. The values of the loss factor shown in Figure 9b are relatively higher, indicating a drastic change in the condition of the oil-paper insulation. Additionally, the changes of the loss factor in the mid-frequency span, i.e., between 10^{-2} Hz and 10^{2} Hz indicate deviations in the condition of oil conductivity, which is in consonance with the model-curve algorithmic analysis (refer to Table 4).

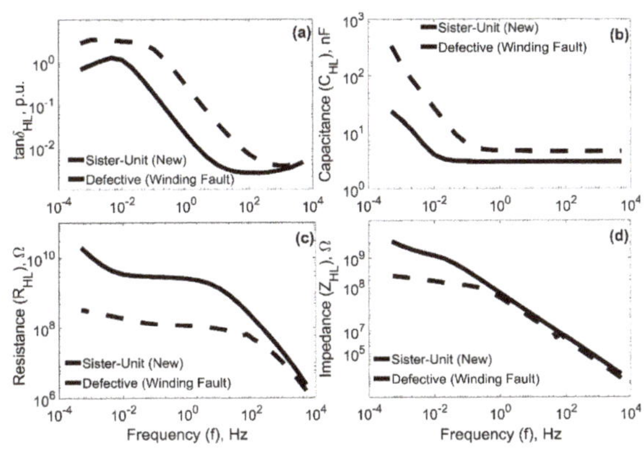

Figure 9. Measured insulation parameters of winding defective distribution transformer grouped under set-3 category. (**a**) Loss factor; (**b**) terminal capacitance; (**c**) resistance (**d**) impedance.

So, these observations collectively confirm that the integrity of the insulation arrangement of the winding-defective transformer remains compromised. This is reasonable, as it can be observed from Figure 9a,b that the winding failure was so severe that it has disturbed the insulation arrangement at this location. The reason for such a failure is due to the lack of timely diagnostic support and inadequate maintenance strategy. Understanding this, detailed FRA tests are conducted on the set-3/1998 winding-defective transformer.

Figure 10a–c shows phase-to-phase comparison of open and short-circuit responses of set-3/1998 winding-defective transformer. It becomes clear from these figures that the windings in the outer phases A-C and C-B (Figure 10a,b) of high voltage and a-n, b-n, and c-n in the low voltage (Figure 10c) have no significant defect. Pertinent magnitude function match each other, indicating that there is no change in impedance. At the same time, the impedance of the winding in the middle phase has drastically changed. These changes appear more prominent at the lower and mid frequencies of open and short-circuit response functions (Figure 10a,b) indicating stronger displacement between the winding-core and individual turns, respectively. These changes altogether confirm that the winding in the middle phase (L_2 or phase B-A of the high voltage winding) has suffered severe damage. This observation matches with the findings made during visual inspection (Figure 10a,b) of the winding defective (set-3/1998) defective transformer. In order to confirm this, the open and short-circuit responses of the faulty unit is compared with the sister-unit transformer. Figure 11a–i shows the magnitude function of the open and short-circuit responses of the defective transformer. It becomes clear from these figures that the winding in middle phase B-A (refer Figure 11d,e) has severe damage. Pertinent open and short-circuit impedance manifest stronger deviations against the sister-unit measurements. Additionally, the corresponding damages are so severe that they have induced or displaced the outer phase windings (phase A-C, C-B) that are in proximity. These induced damages in the outer phase windings appear as deviations in the magnitude responses shown in Figure 11a, b, g, and h, respectively. At the same time, it appears from Figure 11c,f,i that the low voltage windings are free from any defects as there are no changes appeared in its magnitude response. The same is ensured during visual inspection of the faulty transformer.

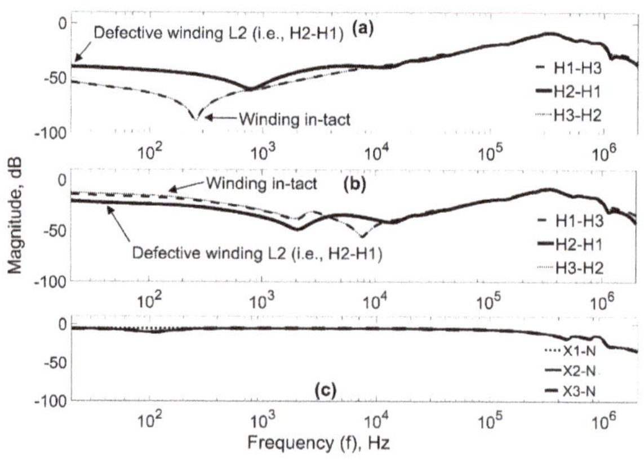

Figure 10. Phase-to-phase comparison of open and short responses of set-3/1998 defective distribution transformer. (**a**) HVOC; (**b**) HVSC; (**c**) LVOC.

Hence, it appears from this onsite study that it is essential to employ diagnostic test methods on in-service distribution transformers to monitor their operating condition and to ensure their lifetime and reliability. The major problems faced by the distribution transformers are caused by moisture ingress, intermittent loads, short-circuit, etc., which significantly affects either the insulation and/or the winding-core arrangement. In all, the current maintenance strategies exercised on the distribution transformers does not reveal the diagnostic status in timely manner, causing the failure events to be catastrophic. Naturally, the modern diagnostic methods such as DGA, DRA, and FRA tests adequately describe the integrity of the distribution transformers and help in exercising preventive measures to evade occurrence of an incipient fault condition. Such attempts are desirable as they improve the asset management.

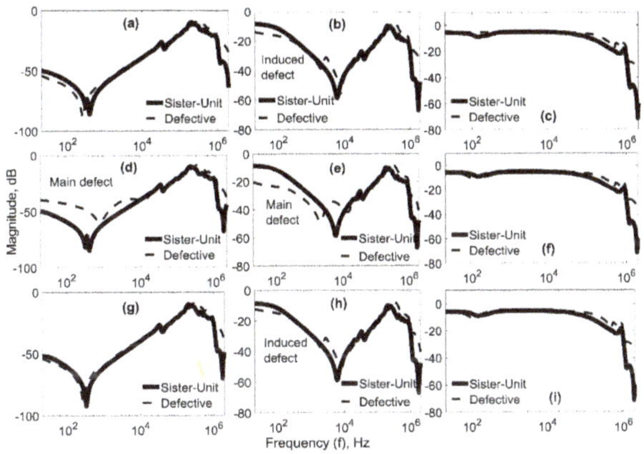

Figure 11. Open and short-circuit responses of set-3/1998 defective distribution transformer. HVOC: (**a**) A-C (**b**) B-A (**c**) C-B, HVSC: (**d**) A-C (**e**) B-A (**f**) C-B, LVOC: (**g**) a-n (**h**) b-n (**i**) c-n.

Thus, the experiments on in-service and defective transformers revealed their diagnostic status clearly and therefore their corresponding data can be further accumulated with respect to time and can be used for resolving their lifetime and reliability.

5. Discussion of Failure Root-Cause Analysis

Literature indicates that the industries and utilities have gained a vast experience on meeting load-demand conditions and to minimize operational failures (such as in arc furnace transformers, etc.,) from a consumer point of view [30,32–34]. In addition, several attempts are made to understand failure modes through maintenance test program. Most of the maintenance tests correlate the effect of loading (or overloading) to the diagnostic changes with primary focus on oil temperature, hotspot and moisture in distribution transformers. Pertinent procedures collect information such as power requirements, utilization, demand, peak and dormant hours for determining the apparent load. This information along with the operational abnormalities such as overloading, switching, unbalanced loads, etc., are collectively used as indices for prioritizing the replacement of in-service transformer units. On contrary, exercising modern diagnostic methods might provide much more clear insight into the diagnostic condition of the transformer, thereby providing an excellent opportunity to evade any premature failures. Such a protection not only enhances the reliability of the equipment but also enhances its life and helps in better asset management. Surprisingly, the modern maintenance and failure prediction and proactive reliability and asset management programs have overseen the possibilities and collect a wide-sense diagnostic data, which might help in gathering failure root-cause analysis. In this context, the present study provides an overview and significance of diagnostic measurements and the points out the quantum of information obtained from the failure root-cause analysis of the defective transformers. Two transformers suffering thermal and winding defect conditions were chosen and subjected to investigations. The defects were chosen in such a way that they reflected a common fault condition experienced by a distribution that is in-service. It is clear from the measured data that the subjected diagnostic test methods are sensitive in revealing even a minute fault in a distribution transformer. Subsequently, to understand the root-cause of the failure, a deeper investigation including a visual inspection is necessary.

5.1. Defect Condition—1: Thermal Failure (Set-3/1994)

The transformer listed in set-3/1994© was in-service in an industrial area and had been handling several intermittent and continuous overloaded situations. In addition, it was noted during the periodic maintenance tests that the oil temperature of this particular unit was always higher and at the time of failure, the temperature recorded was close to 300 °C and had raised alarm on several occasions. Prior to its thermal failure, the service provider was informed that this transformer tripped the circuit breaker several times, forcing it to the go offline on several occasions. So, in due course of this time, this transformer was subjected to a severe maintenance procedure to mitigate the possibilities of premature failure. Pertinent data was stored as reference and used for further comparative and investigative analysis. Figure 12 shows the internal view of this transformer suffering thermal failure. It is experimentally identified that the transformer grouped as 'set-3/1994©' was defective and suffered a thermal failure. The same was ascertained by observing the gases that dissolved in the transformer. Pertinent values and the possible failure modes are shown in Table 3 and Figure 8, respectively. Following this, the reason for the failure and the consequences on material degradation were further investigated. It emerges from Figure 8 and Table 3 that the thermal failure did not majorly alter the dielectric and winding-core arrangement. At the same time, it appeared from Figure 8 that the insulation geometry and cellulose (or paper insulation) may have suffered material degradation. The same was visually ensured during the refurbishing procedure of the set-3/1994 transformer. The degradation in the oil-paper insulation and insulation geometry might be due to higher oil temperature and arcing or internal discharges. In order to investigate this, the defect gas contents dissolved in the oil were studied and the same were conspicuous and indicate the following pattern. The thermal fault raised the oil temperature up to 300 °C with a gas saturation level of 107%. In addition, the oil was saturated with nitrogen gas of 107% in value. As per the previous data, the data measured during the thermal failure indicated a dramatic raise in the defect gas. This confirms active internal arcing in the transformer, which caused material degradation in paper insulation.

Figure 12. Pictorial description of the location of failure in the set-3 distribution transformer suffering thermal fault: Insulation damage, winding burnt.

Thus, it appears from the failure root-cause analysis that the intermittent and constant overloading of this transformer forced a higher operating temperature. In addition, the arcing and intermittent internal discharges caused localized weak spots in the insulation, which triggered spurious tripping of the circuit breakers. Eventually, both the adverse conditions caused the transformer to suffer a thermal failure. Nevertheless, the winding-core arrangement remained intact and henceforth the complete unit can be refurbished and re-commissioned after ensuring its diagnostic status. The same was recommended to the service provider.

16. Jaiswal, G.C.; Ballal, M.S.; Tutakne, D.R.; Vishnu, P. Intelligent condition monitoring system for distribution transformer and health status diagnosis. In Proceedings of the International Conference on Power, Instrumentation, Control and Computing Conference (PICC), Kerala, India, 18–20 January 2018; pp. 1–6.
17. Hamzeh, M.; Vahidi, B.; Askarian-Abyaneh, H. Reliability evaluation of distribution transformers with high penetration of distributed generation. *J. Electr. Power Energy Syst.* **2015**, *73*, 163–169. [CrossRef]
18. Agah, S.M.; Abyaneh, H.A. Distribution Transformer Loss-of-Life Reduction by Increasing Penetration of Distributed Generation. *IEEE Trans. Power Deliv.* **2011**, *26*, 1128–1136. [CrossRef]
19. Agah, S.M.M.; Abyaneh, H.A. Quantification of the Distribution Transformer Life Extension Value of Distributed Generation. *IEEE Trans. Power Deliv.* **2011**, *26*, 1820–1828. [CrossRef]
20. Awadallah, S.K.E.; Milanović, J.V.; Jarman, P.N. The Influence of Modeling Transformer Age Related Failures on System Reliability. *IEEE Trans. Power Syst.* **2015**, *30*, 970–979. [CrossRef]
21. Cho, N.; Yun, S.; Jung, J. Determining the reverse fault current by the type of transformer and Distributed Generation in distribution system during the single-line to ground fault. *Renew. Sustain. Energy Rev.* **2019**, *109*, 102–115. [CrossRef]
22. Shahzad, U.; Kahrobaee, S.; Asgarpoor, S. Protection of Distributed Generation: Challenges and Solutions. *Energy Power Eng.* **2017**, *9*, 614–653. [CrossRef]
23. Bakshi, A.; Kulkarni, S.V. Analysis of Buckling Strength of Inner Windings in Transformers Under Radial Short-Circuit Forces. *IEEE Trans. Power Deliv.* **2014**, *29*, 241–245. [CrossRef]
24. Bagheri, M.; Phung, B.T.; Blackburn, T. Influence of temperature and moisture content on frequency response analysis of transformer winding. *IEEE Trans. Dielectr. Electr. Insul.* **2014**, *21*, 1393–1404. [CrossRef]
25. Satish, L.; Saravanakumar, A. Identification of Terminal Connection and System function Pair for Making Sensitive SFR Measurements. *IEEE Trans. Power Deliv.* **2008**, *23*, 742–750. [CrossRef]
26. Arumugam, S. Experimental Investigation on Terminal Connection and System Function Pair during SFRA Testing on three phase transformers. *J. Electr. Power Energy Syst.* **2014**, *58*, 101–110. [CrossRef]
27. Arumugam, S. Theoretical Considerations While Applying FRA Method in Determining the Inter-Winding Capacitance of Power Transformers at Higher Frequencies. *Eng. Rep.* **2019**, *1*, e12036.
28. Arumugam, S. Experimental validation of using frequency response analysis method in measuring interwinding capacitance of power transformers. *Eng. Rep.* **2019**, *1*, e12079. [CrossRef]
29. Koch, M. Reliable Moisture Determination in Power Transformer. Ph.D. Thesis, University of Stuttgart, Stuttgart, Germany, 2008; pp. 77–79.
30. Koch, M.; Krueger, M. A Fast and Reliable Dielectric Diagnostic Method to Determine Moisture in Power Transformers. In Proceedings of the International Conference on Condition Monitoring and Diagnosis (CMD 2008), Peking, China, 21–24 April 2008.
31. Pandey, S.B.; Lin, C. Estimation of life model of transformer insulation under combined electrical and thermal stress. *IEEE Trans. Reliab.* **1992**, *41*, 466–468. [CrossRef]
32. Raetzke, S.; Koch, M.; Krueger, M. Condition assessment of instrument transformer using dielectric response analysis. *E I Elektrotechnik Und Inf.* **2013**, 1–5. [CrossRef]
33. Zhang, X.; Gockenbach, E. Assessment of the actual condition of the electrical components in medium-voltage networks. *IEEE Trans. Reliab.* **2006**, *55*, 361–368. [CrossRef]
34. Milutinovich, J.S.; Mahon, W.T. Some Thoughts on Minimizing Cost of Arc-Furnace Transformers. *IEEE Trans. Reliab.* **1975**, *24*, 251–254. [CrossRef]

energies

MDPI

Article

Temperature Distribution in the Insulation System of Condenser-Type HV Bushing—Its Effect on Dielectric Response in the Frequency Domain

Krzysztof Walczak * and Jaroslaw Gielniak

Institute of Electrical Power Engineering, Poznan University of Technology, Piotrowo 3A, 60-965 Poznan, Poland; jaroslaw.gielniak@put.poznan.pl

* Correspondence: krzysztof.walczak@put.poznan.pl; Tel.: +48-61-665-2797

Abstract: HV bushings are an important part of the equipment of large power transformers, responsible for their many serious (including catastrophic) failures. Their proper exploitation needs to apply correct and reliable diagnostics, e.g., the use of dielectric response methods, that take into account their specific construction and working conditions. In this article, based on laboratory tests carried out on a real bushing, it has been shown that the significant temperature distribution within its core significantly affects the shape of the dielectric response of its insulation; therefore, the approach to its modeling should be changed. Hence, a new method for interpreting the results, using the so-called the 2XY model, is proposed. Subsequently, based on the measurements made on the insulators in operation, a new modeling method was verified. In conclusion, it can be stated that the 2XY model significantly improves the reliability of the dielectric response analysis, which should be confirmed in the future by tests on withdrawn and revised insulators.

Keywords: HV bushing; dielectric response; paper–oil insulation; temperature distribution

Citation: Walczak, K.; Gielniak, J. Temperature Distribution in the Insulation System of Condenser-Type HV Bushing—Its Effect on Dielectric Response in the Frequency Domain. *Energies* **2021**, *14*, 4016. https://doi.org/10.3390/en14134016

Academic Editor: Zhijin Zhang

Received: 11 June 2021
Accepted: 30 June 2021
Published: 3 July 2021

1. Introduction

The high voltage bushing is a device used to take a high electric potential lead out from a grounded housing such as a transformer tank. Due to its function, its insulation system is therefore exposed to the influence of a strong electric field (especially within the vicinity of its passage through a grounded metal housing) and generally works in two environments, often with different properties. For example, in a power transformer, the top part of the bushing is in the air and the bottom is in contact with the insulating liquid. Hence, in order correctly perform its function, devices of this type are required to have quite a complex structure. In the case of solutions for voltage above 30 kV, the bushings are equipped with specially designed screens that control the electric field distribution (condenser type bushing), between which there is an appropriately selected insulation. Likewise, due to its operations in different environments, the top and bottom parts of the insulator are significantly different. The overhead part is often more than three times longer than the lower oil part to ensure adequate surface electrical strength, and is protected by an additional cover that provides resistance to changing weather conditions. In turn, the lower oil part, aside from having to meet the requirements related to electrical strength, should also be resistant to the effects of an insulating liquid and have the appropriate heat resistance class. This is because the upper layer of the tank, where the bushing is located, has a relatively high temperature during transformer operation. An example of the construction of an RIP bushing (with Resin Impregnated Paper insulation) is shown in Figure 1 [1].

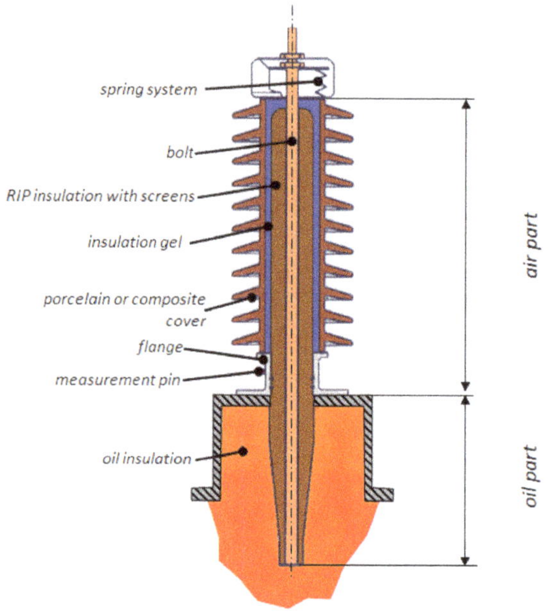

Figure 1. Cross-section of an RIP insulator.

Due to the difficult working conditions, transformer bushings for 110 kV and more are characterized by a high risk of failure. A main cause of failure is dielectric damage (due to the electric or thermal breakdown of individual layers of insulation), as well as increased moisture resulting from leakage and progressive cellulose oxidation processes [2,3]. This is confirmed by the failure rate statistics, which are periodically published by relevant scientific groups, e.g., CIGRE or IEEE. An example of such an analysis can be found in the report "Transformer Reliability Survey" of the CIGRE Working Group A2.37 published in 2015 [4]. Based on a statistically large group of power transformers (536 network transformers and 127 GSU transformers), the main causes of failures and the factors that determined them were analyzed. These statistics show that for both network and unit transformers, bushing breakdown is a frequent cause of failures. According to the report cited above, these were the third largest cause of serious transformer failures and amounted to 17.16% for network transformers, and 16.53% for unit transformers (including insulators installed on the high and low voltage sides). Therefore, based on these statistics, it can be concluded that a bushing failure may have very serious, and in many cases even catastrophic, consequences for the entire device. The analysis of numerous cases shows that insulator damage is often also the cause of the entire transformer and the surrounding infrastructure catching fire, which in turn generates huge costs and causes serious environmental contamination. Figure 2 presents selected photos from the last few years where we can see examples of catastrophic power transformer failures, which were initiated by the damage of high voltage bushing. In each of the cases, the high-power transformers (from 160 to 240 MVA) were installed at power stations belonging to the national transmission network operator. The failed HV bushings (220 and 400 kV), were made using the OIP (Oil Impregnated Paper) technology.

Figure 2. Examples of catastrophic failures of high power transformers initiated by the failure of a bushing.

Both the damage resulting from the degradation of the insulation and the increase in its moisture can be detected with the use of various diagnostic methods, including the so-called dielectric response methods [5]. Measurements of the dielectric response in the time or frequency domain have an advantage over traditional methods (e.g., measurement of tan delta and C at 50 Hz) since they provide more complete information about the tested insulation material [6–9]. Owing to this more credible type of diagnosis, it is thus possible to observe the symptoms of upcoming failure in advance [10].

Despite the fact that the dielectric response methods are becoming more and more reliable, the interpretation of the obtained results is still difficult [9,11–14]. This is mainly due to the fact that the bushing works as a parallel system of two elements located in different environments, and the measured dielectric response is their resultant. The currently used algorithms (based on the XY model) for assessing the dielectric response often give an ambiguous result, frequently due to the poor fit of the measurement curve to the model curve. In a situation where an obtained value, such as moisture, is close to the criterion value, this will be of great importance when making a decision about its further use or replacement, which in turn is a complicated and costly process.

The 2XY model proposed in this article, developed on the basis of laboratory and field measurements, is an interesting alternative to the traditional approach in assessing the condition of the bushing. It seems that in many cases it will increase the reliability of the obtained analysis results, which in turn will translate into a reduction in the number of transformer failures.

This article presents laboratory test results of a 110 kV insulator made using the OIP (Oil Impregnated Paper) technology. The insulator was subjected to uniform and non-uniform heating in order to observe the influence of temperature and its distribution on the dielectric response of the insulation system. Based on the obtained results, a new method of dielectric response modeling (2XY model) was developed in order to obtain a reliable method for assessing the degree of degradation and moisture content of the solid insulation in various types of bushings. Finally, the new method for interpreting the insulation dielectric response was verified on a few selected bushings in operation.

2. Materials and Methods

For the experimental tests, a Trench transformer bushing, model COT-250-800, was used (Figure 3a). It is an insulator made using the OIP technology, which means that the core is a paper insulation impregnated with a special liquid based on mineral oil. The bushing is equipped with screens made of aluminum foil that control the distribution of the electric field. The cover of the oil-paper insulation is made of porcelain. The head, containing the degassed liquid that fills the insulator, is equipped with an oil level indicator. The bushing was designed for 52 kV rated voltage, 800 A current, and 250 kV impulse withstand voltage.

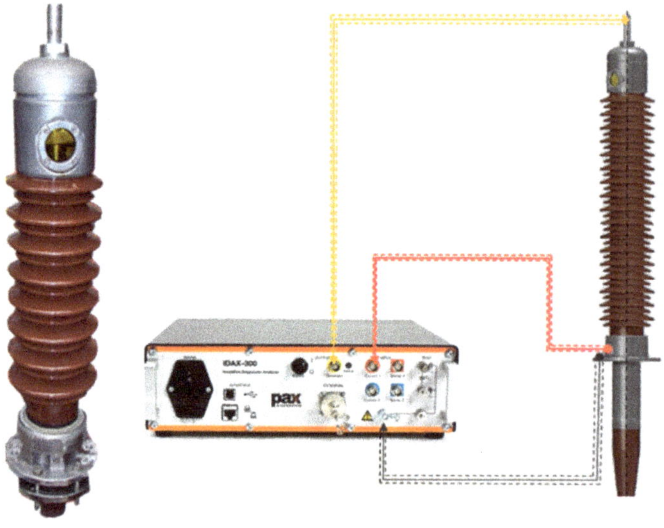

(**a**) (**b**)

Figure 3. COT-250-800 Trench transformer bushing (**a**), method for connecting the test bushing to the IDAX 300 m (**b**).

The bushing insulation dielectric response was measured using the IDAX 300 insulation diagnostic system from PAX Diagnostics. The connection system of the bushing with the IDAX 300 m is shown in Figure 3b. The measurements were performed in the frequency range of 0.001–1000 Hz, at a voltage of 200 V.

During the tests, temperature sensors were attached to the bushing, the arrangement of which is shown in Figure 4a. The sensors allowed us to control the temperature distribution and the heating of individual parts of the insulator during the experiment. The probes used to measure the temperature were thermocouples with a J-type connector (Fe-CuNi), the limit error of which, according to the manufacturer's specification, was ±1.5 °C. To be sure, the sensors' indications were previously verified through a test in a thermal chamber, performed in accordance with the standard [15]. In order to be able to control the temperature of the test object, the upper part of the insulator was placed within a glass pipe, and the lower part was immersed in a glass vessel filled with oil. The temperature of the object in the upper part was changed with the use of a fan heater, while in the lower part, the temperature was modified with the help of a heater with a magnetic stirrer. The arrangement of the temperature control systems is shown in Figure 4b.

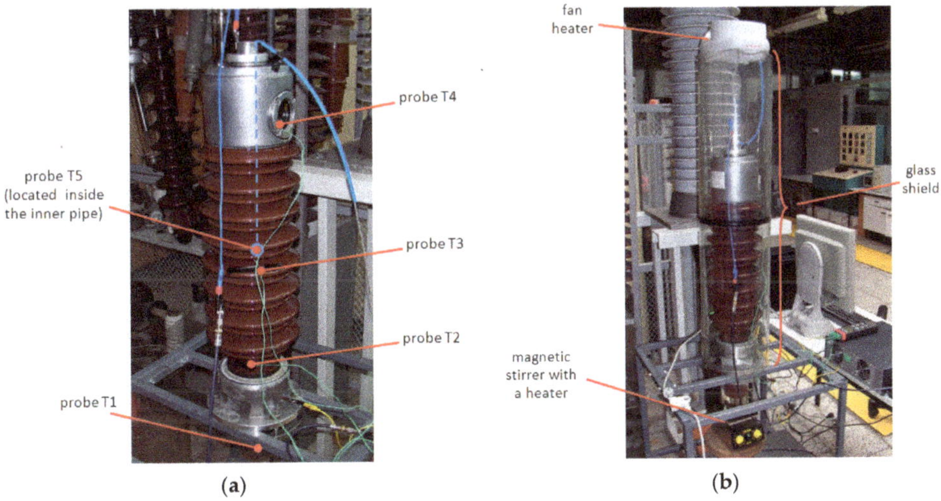

Figure 4. Arrangement of temperature sensors (**a**); insulator heating system (**b**).

The research methodology adopted in the experiment was as follows: First, the measurements of the dielectric response (tan $\delta = f(f)$) of the bushing insulation were performed under thermal equilibrium conditions, at room temperature (23 °C). This result was used as a reference in further studies. Subsequent tests included the uniform heating of the system to the temperatures 30 °C and 40 °C, and unevenly forcing the lower oil part of the bushing to 50 °C and 60 °C, while maintaining an open temperature range for the air part (the room temperature was then 23 °C). The individual test variants and the adopted temperatures of the oil and air parts of the bushing are presented in Table 1. The adopted temperature ranges correspond to the real operating conditions that occur in real power transformers in the summer season.

Table 1. Summary of cases examined during the experiment.

	Temperature (°C)	
Case N°	**Oil**	**Air**
1	23	23
2	30	30
3	40	40
4	50	23
5	60	23

3. Results

3.1. Measurements with Uniform Temperature Distribution

In the first part of the experiment, measurements were carried out at different insulation temperatures, but with a uniform temperature distribution throughout the bushing. According to the data presented in Table 2, the average object temperatures during the measurements were 23.1 °C, 30.1 °C, and 40.7 °C.

Table 2. Measurement of the bushing temperature at various points to check for uniform temperature distribution; the markings of the measurement points are as shown in Figure 4a.

	Temperature (°C)		
Sensor N°	**Case 1**	**Case 2**	**Case 3**
1	22.7	30.4	40.4
2	22.7	28.3	37.9
3	22.9	29.6	40.7
4	23.2	30.5	43.1
5	23.8	31.3	41.3
Average	23.1	30.1	40.7

After the proper preparation of the system and the connection of the measuring equipment as shown in Figure 3b, measurements of the dielectric response in the frequency domain were carried out, which is reflected in the characteristic tan $\delta = f(f)$. The individual results, for comparison purposes, are shown in one collective diagram presented in Figure 5. The obtained result is typical for a new OIP-type insulator. Comparing the individual characteristics, it can be seen that as the insulation temperature increases, they do not change their shape but rather shift to the higher frequencies, which is a typical effect described by the Arrhenius law [16]:

$$\tan\delta_1 = \tan\delta_2 \exp\left[\frac{A_e}{k}\cdot\left(\frac{1}{273-T_1}-\frac{1}{273-T_2}\right)\right], \quad (1)$$

where:

- tan δ—loss tangent;
- A_e—activation energy;
- T—temperature;
- K—Boltzman's constant.

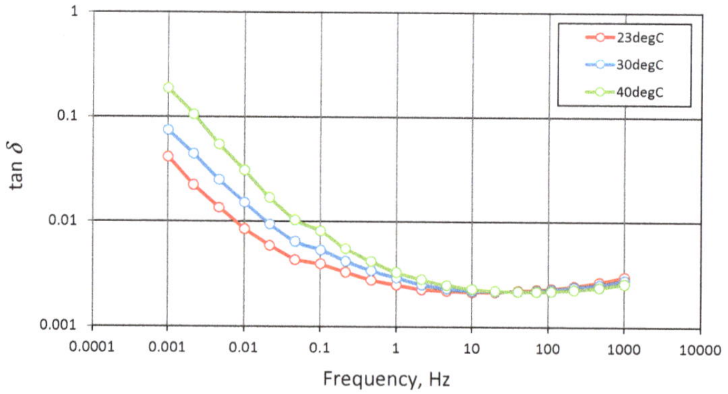

Figure 5. Dielectric response of the bushing insulation with uniform temperature distribution; measurements made for temperatures: 23 °C, 30 °C, and 40 °C.

Assuming the activation energy typical for the paper–oil insulation system, i.e., 0.9 eV [17–19], and converting the individual dependencies to the temperature of 23 °C (Figure 6), one can see that the characteristics perfectly overlap. Therefore, despite the existence of two significantly different media (air and oil), the system behaves very homogeneously; therefore, it should be assumed that with small differences in object temperature during the measurements, there will be no difficulties with modeling the system and interpreting the obtained results (e.g., determining the degree of moisture content). The method of modeling and analyzing such systems will be described in more detail in Section 4.1.

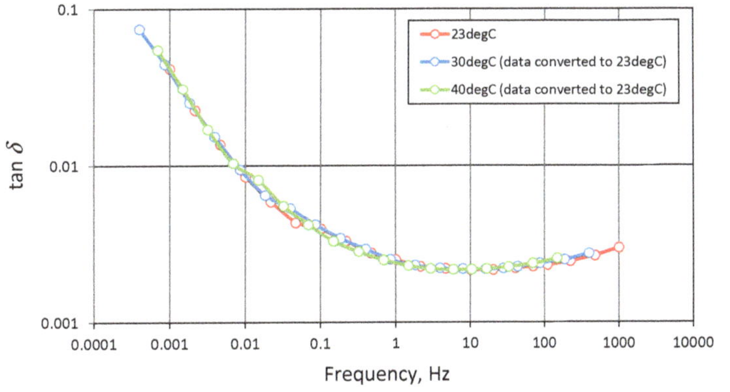

Figure 6. Dielectric response of the bushing insulation with uniform temperature distribution; measurements made for temperatures: 23 °C, 30 °C, and 40 °C, subsequently converted to 23 °C, assuming activation energy at the level of 0.9 eV.

3.2. Measurements with Nonuniform Temperature Distribution

In the second part of the experiment, measurements were carried out to simulate the situation of nonuniform temperature distribution, which was obtained by forcing different operating conditions on the air and oil part of the bushing. According to the data presented in Table 3, in both cases, the air part was operated at an ambient temperature of approximately 23 °C, and the oil part was heated to a temperature of 50 °C and 60 °C, respectively. The temperature distribution was obtained after 2 h of the heating system operation.

Table 3. Measurement of the bushing temperature at various points to check for non-uniform temperature distribution; the markings of the measurement points are as shown in Figure 4a.

	Temperature (°C)	
Case N°	**Case 4**	**Case 5**
Oil	50.0	60.1
1	25.4	29.2
2	23.6	25.1
3	23.5	25.1
4	23.7	25.2
5	25.2	30.4

For comparison purposes, the individual results are shown in one collective graph in Figure 7. Analyzing the characteristics, it can be noticed that this time the temperature change caused not only a curve shift towards higher frequencies, but also a change of its shape. This can be seen by converting the obtained results to the same temperature level (according to the Arrhenius law mentioned above). The result of conversion is presented in Figure 8. At the shift, the activation energy was assumed at the level of 0.9 eV and the object temperature was an average value calculated from temperature measurements by individual sensors (Table 3). The shifted curves do not match as well as they did in the case of the uniformly heated bushings, despite the fact that the average values of the temperatures obtained in individual cases do not differ significantly (within 9 °C).

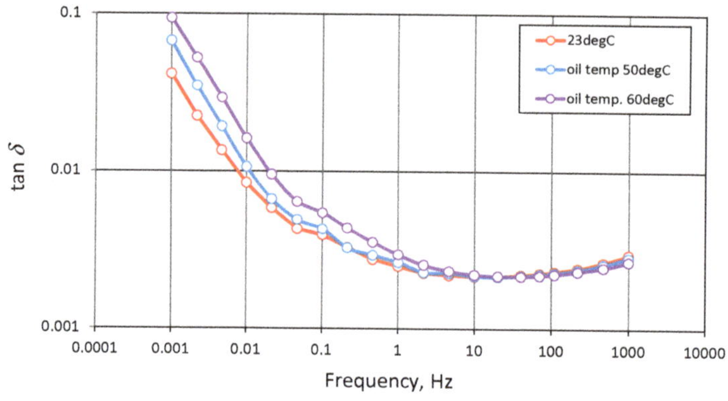

Figure 7. Dielectric response of the bushing insulation with non-uniform temperature distribution; measurements made at an ambient temperature of 23 °C; temperature of the oil part at 50 °C and 60 °C.

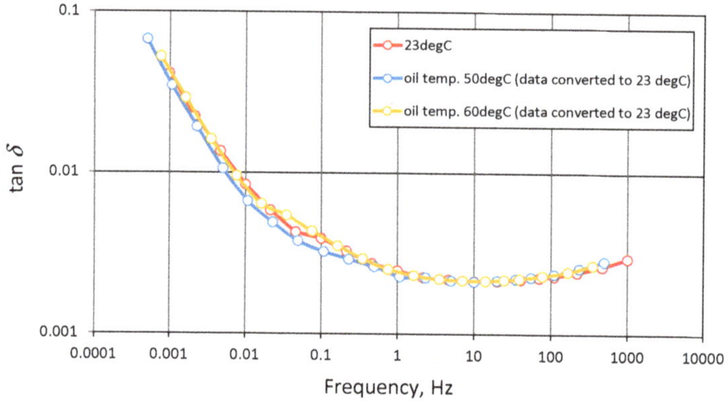

Figure 8. Dielectric response of the bushing insulation with non-uniform temperature distribution; measurements made at an ambient temperature of 23 °C; the temperature of the oil part was at 50 °C and 60 °C, subsequently converted to 23 °C, assuming an activation energy at the level of 0.9 eV.

Summarizing the obtained results, several conclusions can be drawn. Firstly, the nonuniform temperature distribution changes the shape of the tan $\delta = f(f)$ characteristic, which will certainly make the process of modeling and assessing the condition of the bushing's paper–oil insulation more difficult in the traditional way, i.e., assuming one resultant temperature. It should be emphasized that the tested insulator had a relatively large disproportion between the oil and air parts (in the ratio 1:4), which means that the temperature distribution was characterized by a sizeable gradient and was largely dominated by the ambient temperature. It should be expected that in the case of insulators for higher voltage (from 110 kV and higher), for which the proportions are different, the observed effect will be even stronger. In addition, it should also be noted that the tests were carried out on a new insulator, with dry and non-degraded insulation, which made it more homogeneous. In the case of insulators with a long service life and in which degradation processes have already occurred, the dielectric properties of the bushing will have already changed and may consequently affect the resultant dielectric response of the device. The influence of these factors, confirmed by the measurements made on insulators in service, will be presented in Section 4.2.

After analyzing the results of the studies presented above, and owing to their years of experience in the practice of high voltage bushing diagnosis, the authors were prompted to verify the standard approach in the interpretation of measurements of the dielectric response of bushing insulation; consequently, the authors were inspired to develop a new modeling method, which will improve the reliability of assessments of these important elements of transformer equipment. The developed concept and its verification will be discussed in the following sections of the article.

4. Discussion

4.1. Modeling the Dielectric Response of the Bushing Insulation in the Case of Significant Temperature Distribution along Its Axis

The basis for the interpretation of the results in the DFR (Dielectric Frequency Response) method is the comparison of the dielectric response obtained from the measurements (e.g., the relationship $\tan\delta = f(f)$) with the response of the model based on previously determined reliable moisture patterns [20]. The so-called XY model [21–24], where the parameters X and Y determine the percentage of cellulose insulation in the cellulose–oil insulation system (Figure 9), is a commonly used analysis tool, described in many publications such as in [25–27].

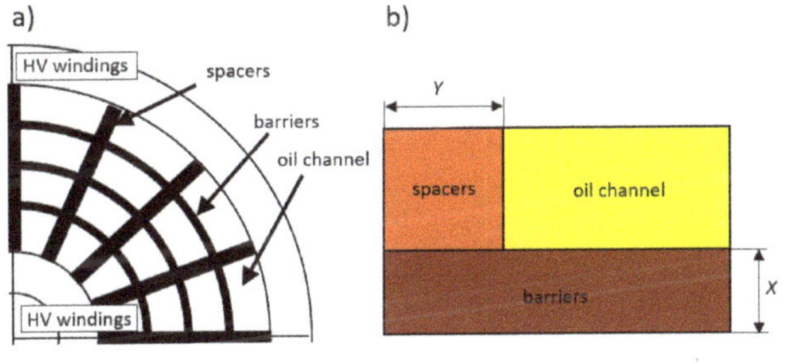

Figure 9. A portion of the transformer insulation cross-section, with a cylindrical arrangement of windings (**a**) and a simplified transformer insulation model (**b**).

In this model, information about the geometry of the insulating system as well as information on the conductivity and electric permeability of the insulating liquid and the object temperature are required. Therefore, the credibility of the obtained result will largely depend on whether the correct parameters for the model [28] are adopted. Most of the geometric and material parameters can be determined quite precisely in the process of matching the measurement curve to the model curve; however, use of the appropriate moisture standards and the adoption of the correct temperature are necessary [29].

Moisture patterns are presented as components of real and imaginary permittivity, depending on the voltage frequency for the selected level of moisture in the cellulose insulation and the selected temperature. They are usually available as a part of an analytical software, included by the manufacturer of the measuring apparatus.

However, determining the proper temperature of the object, despite its relatively small dimensions compared to the dimensions of the power transformer, may pose some difficulties. As it was proved earlier in the article, a bushing working in two environments very often has a significant temperature distribution along its axis, i.e., the temperature in the overhead section can be much lower than in the oil section. In this case, in order to model the insulation system, a resultant temperature should be assumed. Some guides suggest assuming a temperature that is two-thirds of the ambient temperature and one-third of the temperature of the upper oil layer [30]. Such an approximation produces relatively good results, provided that the temperature of the overhead and oil portions are not significantly different. However, a much better representation can be obtained if the mathematical model is modified by treating the overhead and oil parts of the bushing as two insulation systems connected in parallel (Figure 10). The 2XY model proposed by the authors therefore represents the dielectric loss factor measured between the line terminals Z_1 and measuring terminals Z_2 of the insulator. In the model thus defined, the XY1 block (overhead part) and the XY2 block (oil part) can be assigned the actual temperature, as well as separate geometric and material parameters, consistent with the construction of the bushing.

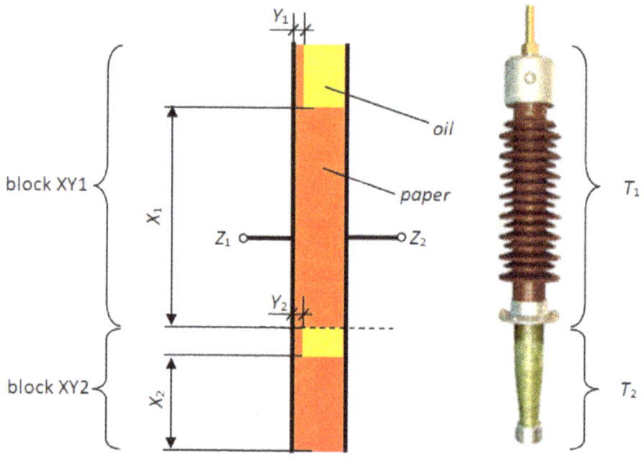

Figure 10. The 2XY model representing a bushing whose overhead part (block XY1) is at temperature T_1, and oil part (block XY2) is at temperature T_2.

For each area representing a part of the insulation of a different temperature (T_1 and T_2), complex permittivity is calculated according to the following formulae:

$$\varepsilon^*_{XY_1(\omega,T)} = X_1\varepsilon^*_{P(\omega,T)} + \frac{1-X_1}{\frac{1-Y_1}{\varepsilon^*_{U(\omega,T)}} + \frac{Y_1}{\varepsilon^*_{P(\omega,T)}}} \tag{2}$$

$$\varepsilon^*_{XY_1(\omega,T)} = X_2\varepsilon^*_{P(\omega,T)} + \frac{1-X_2}{\frac{1-Y_2}{\varepsilon^*_{U(\omega,T)}} + \frac{Y_2}{\varepsilon^*_{P(\omega,T)}}} \tag{3}$$

where:

- $\varepsilon^*{}_{P(\varpi,T)}$—complex permittivity of paper impregnated with mineral oil;
- $\varepsilon^*{}_{U(\varpi,T)}$—complex permittivity representing areas of the leakage nature (oil, shield);
- X_1, Y_1, X_2, Y_2—geometric parameters describing the participation of the paper in the whole modeled space.

Subsequently, for each area, the tangent of the dielectric loss angle is determined according to the following formulae:

$$\tan\delta_{XY1} = \frac{\varepsilon''{}_{XY_1}}{\varepsilon'{}_{XY_1}} \tag{4}$$

$$\tan\delta_{XY2} = \frac{\varepsilon''{}_{XY_2}}{\varepsilon'{}_{XY_2}} \tag{5}$$

where:

- ε'_{XY_1}, ε'_{XY_2}—the real part of permittivity calculated respectively for the areas of temperatures T_1 and T_2 according to Formulae (1) and (2);
- ε''_{XY_1}, ε''_{XY_2}—the imaginary part of permittivity calculated respectively for the areas of temperatures T_1 and T_2 according to Formulae (1) and (2).

The tangent of the dielectric loss angle is calculated for the whole insulator at the final stage, according to the formula:

$$\tan \delta_{2XY} = A \tan \delta_{XY1} + (1 - A) \tan \delta_{XY2} \quad (6)$$

where:

- A—participation of the area of temperature T_1 in the whole volume of the insulator;
- $\tan \delta_{XY1}$, $\tan \delta_{XY2}$—tangent of the dielectric loss angle of the areas of temperatures T_1 and T_2.

Figure 11 shows exemplary curves obtained from mathematical modeling for blocks XY1 and XY2 and the resultant curve for connected blocks XY1and XY2. The measurement points are plotted on the result curve. As you can see, the good fit of the XY1 + XY2 model curve as well as the measurement points prove that the methodology is correct. Using the described approach, it is possible to determine the moisture content of the bushing cellulose insulation with sufficient accuracy; this will be analyzed in the next section using the results of field measurements.

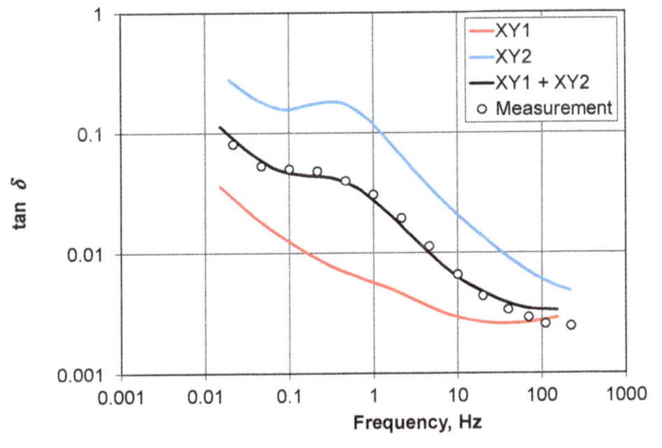

Figure 11. Loss tangent for 220 kV voltage insulator, Hitachi type OS, measured and modelled using the 2*XY* model, taking into account the axial temperature distribution. Upper part temperature, 8 °C; oil temperature in the upper layer, 29 °C.

4.2. *Verification of the 2XY Model on Insulators in Service*

In order to verify the proposed 2XY model, measurements on bushings in service were carried out. Bushing insulators, due to their high impedance, i.e., low capacitance and high resistance, are difficult elements in terms of providing information on the dielectric response and therefore require the use of an additional high-voltage amplifier, as shown in Figure 12. The equipment used for the tests are the insulation response analyzer (IDAX 300) and a VAX type amplifier supplied by PAX Diagnostic. The VAX-020 amplifier enabled measurements at voltages up to 1400 V_{RMS} in the frequency range from 10^{-4} Hz to 10^3 Hz.

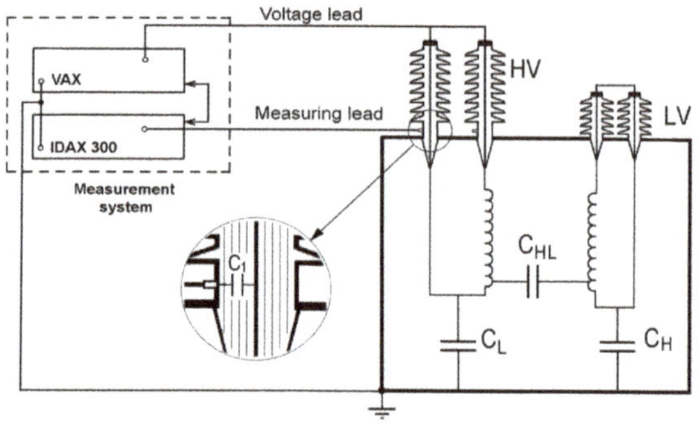

Figure 12. Test system for the measurement of the dielectric response of bushings.

The presented research concerns the measurements of the capacitance of bushing insulators equipped with a measuring tap. The UST (Ungrounded Specimen Test) system was used, in which the voltage lead is connected to the line clamps, and the measuring lead to the measuring tap. In such a configuration (Figure 12), only the capacitance C_1 of the bushing is measured (C_{HL}, C_H, and C_L capacitances are not measured and the influence of leakage current flowing on the outer surface of the insulator is eliminated).

Figures 13–15 show examples of dielectric response (tan $\delta = f(f)$) modelling using the classic XY model and the 2XY model for three different bushing types. In the lower part of each of these figures, the relative differences between the measured values and those calculated with the XY and 2XY models are shown in the form of bars. The parameters (X, Y, oil conductivity, moisture in paper insulation) resulting from the application of the XY model were obtained using the standard auto-matching procedure implemented in the MODS program, which is the measuring device used. So far, no IT tool has been developed that could similarly search for parameters for the 2XY model; hence, the adjustment process was performed manually. However, in order for the procedure to be repeatable, an algorithm was used to minimize the error. From the outset, it was assumed that the geometry of individual blocks (air and oil) may change slightly due to the specificity of the measuring system and because it is largely determined by paper insulation, thus the filling is close to 100%. After adopting the geometry of the system and selecting the appropriate temperature for individual parts of the model, conductivity and moisture were introduced to adjust the curves and minimize the value of the M_{ADM} parameter (Mean of the Absolute Difference Modules, defined below). The process was repeated until the minimum value of the criteria parameter was obtained.

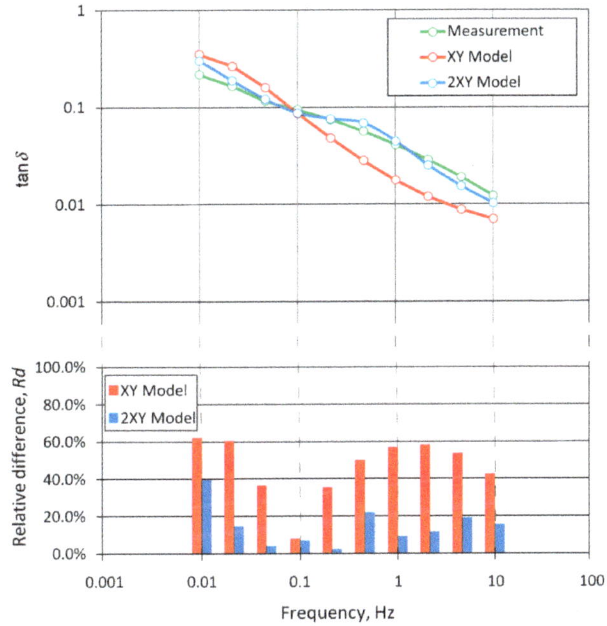

Figure 13. The dependence of the loss tangent on the frequency, measured for the COT type (U_n = 142 kV) bushing insulator and calculated based on the XY and 2XY models, with the calculated relative differences between the measured and calculated values for both models.

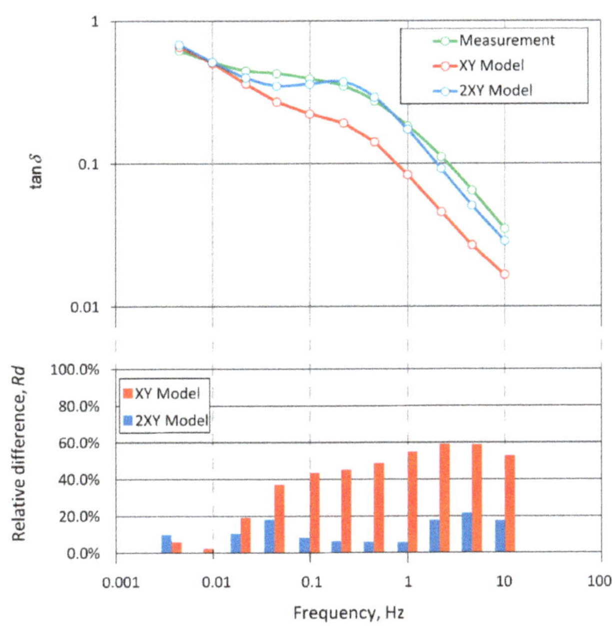

Figure 14. The dependence of the loss tangent on the frequency, measured for the OS type (U_n = 110 kV) bushing insulator and calculated based on the XY and 2XY models, with the calculated relative differences between the measured and calculated values for both models.

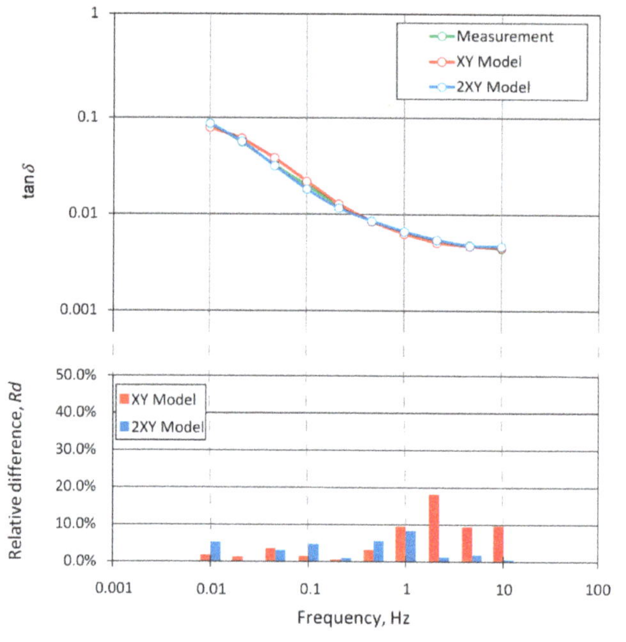

Figure 15. The dependence of the loss tangent on the frequency, measured for the COT type (U_n = 220 kV) bushing insulator and calculated based on the XY and 2XY models, with the calculated relative differences between the measured and calculated values for both models.

The mean of the absolute difference modules (M_{ADM}) was used as a first quantity to assess the quality of the curve fitting obtained from modelling and measurement. The M_{ADM} value can be calculated from the equation:

$$M_{ADM} = \frac{\sum_{i=1}^{n} \left| \frac{\tan \delta_{m_i} - \tan \delta_{f_i}}{\tan \delta_{m_i}} \cdot 100\% \right|}{n}, \quad (7)$$

where:

- $\tan \delta_m$—loss tangent measured for a specific frequency;
- $\tan \delta_f$—loss tangent calculated for a specific frequency based on XY or 2XY model;
- n—number of points for which measurements and calculations were carried out.

The second quantity we use to assess the quality of the fitting is the correlation coefficient (r). The correlation coefficient was calculated according to the formulae [31]:

$$r = \frac{\sum_{i=1}^{n} \left(\tan \delta_{m_i} - \overline{\tan \delta_m}\right)\left(\tan \delta_{f_i} - \overline{\tan \delta_f}\right)}{\sqrt{\sum_{i=1}^{n} \left(\tan \delta_{m_i} - \overline{\tan \delta_m}\right)^2} \cdot \sqrt{\sum_{i=1}^{n} \left(\tan \delta_{f_i} - \overline{\tan \delta_f}\right)^2}}, \quad (8)$$

where:

- $\tan \delta_m$—loss tangent measured for a specific frequency;
- $\tan \delta_f$—loss tangent calculated for a specific frequency based on XY or 2XY model;
- n—number of points for which measurements and calculations were carried out.

In our case, the correlation coefficient describes the relationship between the measured curve and the one calculated on the basis of the model. It can take values from −1 to 1, where 1 means full correlation, zero means no correlation, and −1 means inverse correlation. The correlation coefficient r is a good illustration of the similarity of the shapes of the analyzed curves, but it does not illustrate the differences in values. The differences in values are correctly described by the mean of the absolute difference modules (M_{ADM}).

Tables 4 and 5 present the parameters used in the calculation of the dielectric response (X/Y, temperature T, conductivity of oil δ, and moisture in paper Mc) as well as the correlation coefficients r and the mean of the absolute difference modules M_{ADM} for the three analyzed bushings, respectively, for the XY (Table 4) and the 2XY models (Table 5).

Table 4. *XY* model parameters used in modelling the dielectric response of the tested bushings.

№	XY Model					
	X/Y, %	T, °C	σ, S/m	Mc, %	r	M_{ADM}, %
Bushing 1	85/45	30	7.79×10^{-13}	2.3	0.980	46.5
Bushing 2	90/0	22	2.72×10^{-11}	2.3	0.942	38.9
Bushing 3	74/86	25	1.69×10^{-12}	1.0	0.991	5.9

Table 5. 2*XY* model parameters used in modelling the dielectric response of the tested bushings.

№	2XY Model								
	X_1/Y_1, %	X_2/Y_2, %	T_1, °C	T_2, °C	σ_2, S/m	Mc_1, %	Mc_2, %	r	M_{ADM}, %
Bushing 1	96/2	100/100	24	43	1.87×10^{-11}	2.5	2.5	0.985	14.6
Bushing 2	31/69	100/100	15	36	1.60×10^{-9}	3.7	3.6	0.983	11.3
Bushing 3	94/0	98/33	18	38	9.37×10^{-12}	1.0	1.0	0.999	3.2

The parameters with the index 1 presented in Table 5 refer to the air part of the insulator, while parameters with the index 2 refer to the oil part. Table 5 does not provide the oil conductivity for the air part because it is calculated automatically in the program based on the oil conductivity in the oil part, and the temperature difference between the oil part and the air part of the insulator, according to the formulae [16]:

$$\sigma_1 = \sigma_2 \exp\left[\frac{A_e}{k}\cdot\left(\frac{1}{273 - T_1} - \frac{1}{273 - T_2}\right)\right], \tag{9}$$

where:

- σ_1—oil conductivity in the air part (at T_1 temperature);
- σ_2—oil conductivity in the oil part (at T_2 temperature);
- A_e—activation energy equal to 0.4 eV;
- k—Boltzmann constant.

Analyzing the data presented in Figures 13–15, one can see that in each case, the fit of the model curves to the measurement curves is better when the 2XY model is used. It is especially visible for insulators with higher humidity, but also noticeable for dry insulators. The relative differences (Rd), correlation coefficients (r), and mean of the absolute difference modules (M_{ADM}) clearly indicate that the 2XY model better fits the dielectric response of insulators than the classical model.

5. Conclusions

The most important conclusion from the research results presented in this article is the need to analyze the dielectric response of bushings, taking into account the axial temperature distribution. During the work life of a bushing, the temperature distribution occurs both in the axial and radial directions. However, measurement can only be performed a few hours after the transformer has been turned off (in particular due to the need to comply with formal procedures, safety procedures, preparation of the workplace, removal of connectors from insulators, etc.). After this time, the temperature distribution in the bushing changes and is close to the axial distribution, which can be roughly described. We can distinguish the temperature of the lower part (usually equal to the temperature of the transformer oil in the upper layer) and the temperature of the upper part of bushing in the air.

The laboratory tests performed on a specially prepared measuring stand and the field tests on insulators installed on transformers indicate the influence of the axial temperature distribution on the dielectric response of the bushings. This influence is especially visible for insulators in poor technical condition (those whose insulation has a relatively high water content).

The authors proposed to extend the XY model—which assumes a uniform temperature distribution and is usually used in the analysis of the dielectric response—to the 2XY model. The 2XY model makes it possible to consider different values of the bushing temperature in its oil and air parts, as well as different water content in its paper insulation and different oil conductivity in these parts.

Comparisons of the dielectric responses calculated with the use of the XY and 2XY models show that considering the axial temperature distribution results in a much better fit of the model curves to the measured ones. This fit was assessed using the absolute difference modules (M_{ADM}) and the correlation coefficient (r). Both of these parameters clearly indicate that the 2XY model better reflects the measured dielectric response than the XY model. This better fit is also confirmed by the values of relative differences between the measured and calculated values for both models.

The examples presented in the article show that the use of the XY model to analyze the dielectric response of insulators with a significant temperature distribution results in an underestimation of the water content in the insulation, which may incorrectly indicate that the insulator is in better condition than it actually is. This effect will especially be noticeable in the case of insulators in poor technical condition, which may lead to leaving such a bushing in operation and, consequently, to a catastrophic failure.

Including the axial temperature distribution in the modeling significantly improved the quality of fitting the dielectric response. However, the authors are aware that in practice, there may be situations in which the proposed 2XY model will be less effective, e.g., when a radial temperature distribution occurs, or when the surface of the ceramic shield is heavily contaminated.

In the future, the patterns for dielectric response of insulation that can absorb water (for example RIP and RBP), should be created; it will then be possible to conduct tests for those types of insulators.

Author Contributions: Conceptualization, K.W. and J.G.; methodology, K.W. and J.G.; validation, K.W.; formal analysis, K.W. and J.G.; investigation, K.W. and J.G.; resources, K.W. and J.G.; data curation, K.W. and J.G.; writing—original draft preparation, K.W. and J.G.; writing—review and editing, K.W. and J.G.; visualization, K.W. and J.G.; supervision, K.W. All authors have read and agreed to the published version of the manuscript.

Funding: This research was financing from Ministry of Science and Higher Education, Poland, grant number 0711/SBAD/4455.

Institutional Review Board Statement: Not applicable.

Informed Consent Statement: Not applicable.

Data Availability Statement: Not applicable.

Conflicts of Interest: The authors declare no conflict of interest.

Nomenclature

Variable	Symbol	Units
Complex permitivity	ε^*	F/m
Real part of complex permitivity	ε'	F/m
Imaginary part of complex permitivity	ε''	F/m
Relative content of barriers	X	%
Relative content of spacers	Y	%
Loss tangent	$\tan\delta$	-
Mean of the absolute difference modules	M_{ADM}	%
Correlation coefficient	r	-
Relative differences	Rd	%
Temperature	T	°C
Frequency	f	Hz
Voltage	U	V
Activation energy	A_e	eV
Participation of the air part in the whole volume of the insulator	A	%
Conductivity	σ	S/m
Moisture content in paper	Mc	%

References

1. Micafil Transformer Bushings, AirRIP, 2021. Available online: https://www.hitachiabb-powergrids.com/pl/pl/offering/product-and-system/transformer-insulation-and-components/bushings/resin-impregnated-paper-rip/ac-bushing-type-airrip-flex (accessed on 30 June 2021).
2. Septyani, H.I.; Arifianto, I.; Purnomoadi, A.P. High voltage transformer bushing problems. In Proceedings of the 2011 International Conference on Electrical Engineering and Informatics, Bandung, Indonesia, 7–19 July 2011; pp. 1–4. [CrossRef]
3. Wankowicz, J.; Bielecki, J.; Szrot, M.; Subocz, J.; Malewski, R. HV Bushing Failure in Service, Diagnostics and Modeling of Oil-type Bushings. In Proceedings of the CIGRE Session, Paris, France, 22–27 August 2010; pp. 1–10.
4. CIGRÉ WG A2.37. *Transformer Reliability Survey*; CIGRE Brochure: Paris, France, 2015.
5. Chen, M.; Liu, X.; Xu, P.; Wen, T. Local Inter-foil Insulation Deterioration Diagnosis and Simulation of RIP Bushing Based on FDS Method. *Cond. Monit. Diagn. (CMD)* **2018**, 1–5. [CrossRef]
6. Huang, Z.; Fu, M.; Zhuo, R.; Jing, Y.; Luo, Y.; Wang, G.; La, Y.; Dong, X. Research on the Wetting Mechanism of Oil-Paper Insulated Bushings and the Application of FDS Method. In Proceedings of the IEEE Sustainable Power and Energy Conference (iSPEC), Beijing, China, 21–23 November 2019; pp. 1903–1908. [CrossRef]
7. Zhang, L.; Han, Y.; Zhao, H.; Zhang, H.; Zhang, D.; Mu, H. Simulation Study on the FDS Characteristics of Non-uniform Dampened Oil-paper Bushing. In Proceedings of the IEEE International Conference on High Voltage Engineering and Application (ICHVE), Beijing, China, 6–10 September 2020; pp. 1–4. [CrossRef]
8. Long, G.; Mu, H.; Zhang, D.; Li, Y.; Zhang, G. Moisture Location of Oil-Paper Insulated Bushing Based on Polarization and Depolarization Current Analysis. In Proceedings of the IEEE 3rd International Conference on Dielectrics (ICD), Virtual, Valencia, Spain, 5–31 July 2020; pp. 692–695. [CrossRef]
9. Qi, B.; Dai, Q.; Li, C.; Zeng, Z.; Fu, M.; Zhuo, R. The Mechanism and Diagnosis of Insulation Deterioration Caused by Moisture Ingress into Oil-Impregnated Paper Bushing. *Energies* **2018**, *11*, 1496. [CrossRef]
10. Yang, F.; Du, L.; Yang, L.; Wei, C.; Wang, Y.; Ran, L.; He, P. A Parameterization Approach for the Dielectric Response Model of Oil Paper Insulation Using FDS Measurements. *Energies* **2018**, *11*, 622. [CrossRef]
11. Wei, C.; Liao, C.; Du, L.; Chen, X. Influence of Moisture on Dielectric Responses of Oil-Impregnated Paper Condenser Bushings. In Proceedings of the 5th International Conference on Electric Power Equipment—Switching Technology (ICEPE-ST), Kitakyushu, Japan, 13–16 October 2019; pp. 709–712. [CrossRef]
12. Enciso, L.; Martínez, J.L.; Arce, P. Life expectancy of high voltage bushings based on incipient failure detections: A practical approach. In Proceedings of the 2020 IEEE Electrical Insulation Conference (EIC), Knoxville, TN, USA, 22 June–3 July 2020; pp. 325–328. [CrossRef]
13. Buchacz, J.; Cichon, A.; Skubis, J. Detection of conductive layers short circuit in HV condenser bushings using Frequency Domain Spectroscopy. *IEEE Trans. Dielectr. Electr. Insul.* **2017**, *24*, 552–558. [CrossRef]
14. Subocz, J.; Mrozik, A.; Bohatyrewicz, P.; Zenker, M. Condition Assessment of HV Bushings with Solid Insulation based on the SVM and the FDS Methods. *Energies* **2020**, *13*, 853. [CrossRef]
15. Standard No: PN-EN 60584-1:2014-04. In *Thermocouples—Part 1: EMF Specifications and Tolerances*; PKN: Warsaw, Poland, 2014.
16. Jonscher, A.K. Dielectric relaxation in solids. *J. Phys. D Appl. Phys.* **1999**, *32*, 57–70. Available online: https://iopscience.iop.org/article/10.1088/0022-3727/32/14/201 (accessed on 30 June 2021). [CrossRef]

17. Dielectric Frequency Response (DFR) Task Force—Final Report. IEEE Transformers Committee, 07 June 2012. Available online: https://grouper.ieee.org/groups//transformers/subcommittees/dielectric_test/F13-C57.161-DFR-TF-Final%20Report.pdf (accessed on 30 June 2021).
18. Liu, J.; Fan, X.; Zhang, Y.; Zheng, H.; Yao, H.; Zhang, C.; Zhang, Y.; Li, D. A Novel Universal Approach for Temperature Correction on Frequency Domain Spectroscopy Curve of Transformer Polymer Insulation. *Polymers* **2019**, *11*, 1126. [CrossRef] [PubMed]
19. Du, Y.; Yang, L.; Liu, X.; Wang, Y.; Gao, J.; Yao, R. Effect of moisture and thermal degradation on the activation energy of oil-paper insulation in frequency domain spectroscopy measurement. *IET Gener. Transm. Distrib.* **2016**, *10*, 2042–2049. Available online: https://ietresearch.onlinelibrary.wiley.com/doi/10.1049/iet-gtd.2015.0946 (accessed on 30 June 2021).
20. Ekanayake, C.; Gubanski, S.M.; Graczkowski, A.; Walczak, K. Frequency response of oil impregnated pressboard and paper samples for estimating moisture in transformer insulation. *IEEE Trans. Power Deliv.* **2006**, *21*, 1309–1317. [CrossRef]
21. Liu, J.; Fan, X.; Zhang, Y.; Zheng, H.; Jiao, J. Temperature correction to dielectric modulus and activation energy prediction of oil-immersed cellulose insulation. *IEEE Trans. Dielectr. Electr. Insul.* **2020**, *27*, 956–963. [CrossRef]
22. Gielniak, J.; Graczkowski, A.; Gubanski, S.; Moranda, H.; Mościcka-Grzesiak, H.; Walczak, K. Influence of thermal ageing on dielectric response of oil-paper insulation. *Mater. Sci.-Poland* **2009**, *27*, 1199–1205.
23. Ekanayake, C.; Graczkowski, A.; Walczak, K.; Gubanski, S.M. Dielectric spectroscopy measurements on oil impregnated paper/pressboard samples. In Proceedings of the International Conference APTADM, Wroclaw, Poland, 15–17 September 2004; pp. 113–120.
24. Gielniak, J.; Ekanayake, C.; Walczak, K.; Graczkowski, A.; Gubanski, S.M. Dielectric response of new and aged transformer pressboard in dry and wet states. In Proceedings of the International Symposium on Electrical Insulating Materials ISEIM 2005, Kitakyushu, Japan, 5–9 June 2005; Volume 2, pp. 386–389. [CrossRef]
25. Xie, J.; Dong, M.; Yu, B.; Hu, Y.; Yang, K.; Xia, C. Physical Model for Frequency Domain Spectroscopy of Oil–Paper Insulation in a Wide Temperature Range by a Novel Analysis Approach. *Energies* **2020**, *13*, 4530. [CrossRef]
26. Gafvert, U.; Frimpong, G.; Fuhr, J. Modelling of Dielectric Measurements on Power Transformers. In Proceedings of the CIGRE Session, Paris, France, 1998; pp. 15–103. Available online: https://e-cigre.org/publication/SESSION1998-1998-cigre-session (accessed on 3 July 2021).
27. Gafvert, U.; Adeen, L.; Tapper, M.; Ghasemi, P.; Jonsson, B. Dielectric Spectroscopy in Time and Frequency Domain Applied to Diagnostics of Power Transformers. In Proceedings of the International Conference on Properties and Applications of Dielectric Materials, Xi'an, China, 21–26 June 2000; pp. 825–830. [CrossRef]
28. Gielniak, J.; Graczkowski, A.; Moranda, H.; Przybylek, P.; Walczak, K.; Nadolny, Z.; Moscicka-Grzesiak, H.; Feser, K.; Gubanski, S.M. Moisture in cellulose insulation of power transformers—statistics. *IEEE Trans. Dielectr. Electr. Insul.* **2003**, *20*, 982. [CrossRef]
29. Gielniak, J. Influence of geometrical parameters and temperature of oil-paper insulation on analysis results by means dielectric response method. *Przeglad Elektrotechniczny* **2010**, *86*, 220–222.
30. *Framework Instruction for Operation of Transformers*; Energopomiar—Elektryka: Gliwice, Poland, 2012.
31. Johnson, R.; Kuby, P.J. *Elementary Statistics*, 11th ed.; Cengage Learning: Boston, MA, USA, 2011.

MDPI
St. Alban-Anlage 66
4052 Basel
Switzerland
Tel. +41 61 683 77 34
Fax +41 61 302 89 18
www.mdpi.com

Energies Editorial Office
E-mail: energies@mdpi.com
www.mdpi.com/journal/energies

www.ingramcontent.com/pod-product-compliance
Lightning Source LLC
Chambersburg PA
CBHW040827250926
51850CB00007BA/1280
9783036572741